Content Delivery Networks Using Artificial Intelligence and Internet of Things

Scrivener Publishing
100 Cummings Center, Suite 541J
Beverly, MA 01915-6106

Publishers at Scrivener
Martin Scrivener (martin@scrivenerpublishing.com)
Phillip Carmical (pcarmical@scrivenerpublishing.com)

Content Delivery Networks Using Artificial Intelligence and Internet of Things

Revolutionizing Content Delivery

Edited by

Tanusree Chatterjee

Rajib Banerjee

and

Souvik Ganguli

Scrivener Publishing

WILEY

This edition first published 2026 by John Wiley & Sons, Inc., 111 River Street, Hoboken, NJ 07030, USA and Scrivener Publishing LLC, 100 Cummings Center, Suite 541J, Beverly, MA 01915, USA

For more information about Scrivener publications please visit www.scrivenerpublishing.com.

Wiley Global Headquarters
111 River Street, Hoboken, NJ 07030, USA

For details of our global editorial offices, customer services, and more information about Wiley products visit us at www.wiley.com.

The manufacturer's authorized representative according to the EU General Product Safety Regulation is Wiley-VCH GmbH, Boschstr. 12, 69469 Weinheim, Germany, e-mail: Product_Safety@wiley.com.

Library of Congress Cataloging-in-Publication Data

ISBN 9781394347155

Cover image: Generated with AI using Adobe Firefly
Cover design by Russell Richardson

Set in size of 11pt and Minion Pro by Manila Typesetting Company, Makati, Philippines

Printed in the USA

10 9 8 7 6 5 4 3 2 1

Contents

Preface

In an era where the internet serves as the primary conduit for communication, entertainment, commerce, and information exchange, the efficient delivery of digital content has become more critical than ever. Content Delivery Networks (CDNs), the invisible yet indispensable backbone of modern web infrastructure, have matured significantly since their inception. However, the explosion of connected devices, real-time data demands, and evolving user expectations has revealed the limitations of traditional CDNs, which were never designed to cope with the scale, complexity, and dynamism of today's digital ecosystem.

This book, *Content Delivery Networks Using Artificial Intelligence and Internet of Things: Revolutionizing Content Delivery*, emerges from a confluence of three technological frontiers—Artificial Intelligence (AI), Internet of Things (IoT), and CDNs—collectively referred to as AIIoT. Our core motivation stems from a need to reimagine content delivery frameworks, transitioning from static and reactive models to intelligent, adaptive, and proactive systems. By embedding AI and IoT capabilities into CDNs, we unlock unprecedented potential: predictive caching strategies, real-time routing optimization, contextual content personalization, and intelligent threat detection, all underpinned by dynamic data from millions of edge-connected devices.

As veteran researchers and educators from diverse yet complementary domains of computer science, electrical engineering, and networked systems, we have witnessed firsthand the rapidly shifting paradigms of internet infrastructure and user behavior. This book represents our attempt to bridge the academic and industrial gaps by offering a comprehensive exploration of how AI and IoT can transform the way digital content is sensed, processed, and delivered. Through a synergistic lens, we discuss the architecture, design principles, algorithms, protocols, and real-world applications of AIIoT-enhanced CDNs, backed by practical case studies and forward-looking research directions.

Our target audience spans a wide spectrum—from undergraduate and postgraduate students aspiring to understand cutting-edge CDN technologies to seasoned researchers and industry professionals aiming to develop scalable, secure, and intelligent content delivery solutions. We believe this book will also be of interest to cloud architects, network administrators, application developers, and digital strategists who are looking to stay ahead in a rapidly evolving digital landscape.

We express our sincere gratitude to our institutions and academic collaborators for their constant encouragement and to Scrivener Publishing and Wiley for their belief in the relevance and timeliness of this work. We hope this book not only serves as a technical guide but also inspires new innovations in the interdisciplinary domains of networking, AI, and IoT.

We welcome readers to engage critically, apply creatively, and build purposefully on the foundation this book aspires to provide.

The Editors

1

Introduction to AIIoT and CDNs

Priyanka Pramod Pawar[1], R. Kannan[2], R. Anand[2], Deepak Kumar[1], K. Arunprasath[3*], A. Bhuvanesh[4] and Tanusree Chatterjee[5]

[1]Department of Information Technology, University of the Cumberlands, Williamsburg, Kentuchy, United States
[2]Department of Electrical and Electronics Engineering, Nehru Institute of Engineering and Technology, Coimbatore, Tamil Nadu, India
[3]Department of Mechanical Engineering, PSN College of Engineering and Technology, Tirunelveli, Tamilnadu, India
[4]Department of Electrical and Electronics Engineering, PSN College of Engineering and Technology, Tirunelveli, Tamilnadu, India
[5]Department of Computer Science and Engineering, Future Institute of Technology, Kolkata, West Bengal, India

Abstract

This chapter seeks to explore the influence of emerging technologies, specifically artificial intelligence (AI) and the Internet of Things (IoT), on content delivery networks (CDNs) in the context of today's digital landscape. Next, it outlines the fundamentals of Artificial Intelligence of Things, which is simply a means of powering the IoT with AI to build smart systems that are interconnected and can independently interpret data and make decisions. We will then look at CDNs, which are a concisely distributed collection of servers that aid with the dissemination of media and material on the World Wide Web. Considering the hassling issues like scalability, latency, and efficiency, an AIoT-based topology with CDN leads to improved data processing, real-time analytics, and content delivery. This chapter sets out the motivation and the high-level concepts behind this handbook as a guide for using the AIoT-CDN convergence to turn digital infrastructures into smart, efficient, and adaptive systems.

***Keywords*:** AIoT, CDN technology, Internet of Things, data parsing, network optimization, reducing latency, smart systems

**Corresponding author*: aruncmr12@gmail.com

Tanusree Chatterjee, Rajib Banerjee and Souvik Ganguli (eds.) Content Delivery Networks Using Artificial Intelligence and Internet of Things: Revolutionizing Content Delivery, (1–30) © 2026 Scrivener Publishing LLC

1.1 AIoT Explained: The Convergence of AI and IoT

However, a new technological paradigm is emerging by combining the greatest potential of AI and the IoT and is referred to as the Artificial Intelligence of Things (AIoT), empowering things to be intelligent, responsive, and adaptive. Artificial intelligence (AI) refers to the capability of machines to learn from patterns over time and make independent choices, while the Internet of Things (IoT) aims at interconnectedness among devices to gather, share, and analyze data. From smart homes to factory automation, from health to environmental monitoring, the union of AI and the Internet of Things (IoT) has the massive potential to transform how we live and work forever. With this mental synergy, systems will emerge that can converse, receive, and respond to requests from users more naturally. Connecting to networks and sharing data is a fundamental capability of every Internet of Things device. And with sensors, these machines collect huge quantities of data, which can be processed and analyzed to generate valuable insights. However, the Internet of Things (IoT) cannot be the sole solution to all of the challenges of complex data, unstructured data interpretation, and environmental change [1]. The concept of merging of AI with IoT is shown in Figure 1.1.

This is where AI comes into its own, making the IoT devices more "intelligent" by improving their ability to receive, interpret, analyze, and respond. AI (artificial intelligence) algorithms go through the heaps of data generated by IoT devices to present the state-erased patterns. As they receive more data, these systems could learn how to improve themselves, a field called machine learning within artificial intelligence. That enables them to make more accurate predictions, adjust their settings as needed, and even alert users on the fly. AIoT can be observed in action in industries with heavy data automation tasks. AI models are constantly analyzing sensor data to predict when maintenance is required in order to avoid potential downtimes and keep the equipment running smoothly and efficiently. This is also used in healthcare, just like the AIoT apps are used for analyzing health data, monitoring patients' vital statistics, and warning the doctor about any abnormalities. In critical care, where timing is everything, this accuracy to respond in real-time can be invaluable [2].

Another big influence is AIoT's potential to improve energy efficiency. More smart city systems scan for energy consumption data as well as inefficiencies in resource application and traffic flow with AIoT. With the help of data-driven algorithms, cities can now regulate streetlights, optimize garbage man routes, and distribute energy efficiently to build resilient and

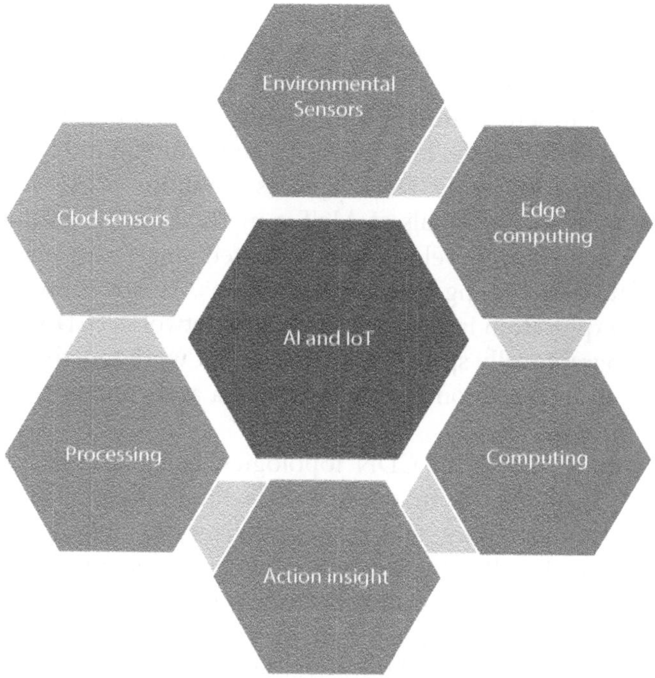

Figure 1.1 AI and IoT.

sustainable cities. Another area where AI-powered IoT applications bring benefits is environmental monitoring, where sensors track pollution and other environmental factors. People can take rapid measures in a way to save public health and the environment based on insight brought by using AI models that help quantify variables such as these [3].

In consumer goods and smart home technologies, AIoT is also essential to provide an individual experience for every single user. Smart speakers, smart thermostats, and smart lighting powered by artificial intelligence can now "learn" the tastes and habits of their customers. AIoT systems involve data-driven usage, which relies heavily on large datasets for effective functioning and adaptation to user behavior. Data breaches are a risk from this reliance on technology because appliances often log private and sensitive data. Furthermore, strong security protocols are equally important to prevent cyberattacks, unauthorized access, or tampering with an increasingly critical infrastructure [4] of AIoT systems. It also needs advanced computing resources for the integration of AI algorithms with the Internet of Things (IoT). To process data close to the source and reduce latency and dependence on cloud computing resources, edge computing solutions are generally needed. While AIoT's growth still faces significant challenges, it has been increasingly supported by the rapid development of pivotal technologies such as 5G and edge computing that work to support the establishment of intelligent systems that respond to the input of human beings. In the context of the digital environment today, the chapter "Introduction to AIoT and CDN" explores the union of AIoT and CDN. First, it describes the essentials of AIoT, a method for integrating AI with the IoT that is helping develop interconnected smart systems capable of analyzing data and making their own decisions. This is subsequently followed by an exploration into content delivery networks (CDN), which are groups of geographically spread-out servers that work together to provide fast delivery of Internet content such as media and content on the World Wide Web [5].

By leveraging AIoT with CDN topologies, challenges like scalability, latency, and efficiency can be overcome, facilitating enhanced data processing, real-time analytics, and content delivery as ubiquitous computing becomes a reality. It is the introduction chapter, summarizing the key aspects, challenges, and the transformational potential of this integration of AIoT-CDN, which can guide their utilization for creating intelligent, efficient, and responsive digital infrastructures. AI is also a big leap toward the virtualization of the self-sufficient, intelligent, and networked systems dream. Working in tandem, the Internet of Things (IoT) and artificial intelligence (AI) advance productivity, sustainability, and user experience by bringing added intelligence to otherwise stand-alone items, creating self-governing, smart-functioning systems. This surge in AIoT is likely to also provide an intrinsic factor in driving new intelligent and automated applications, driving the edge of technology forward in an ever-advancing digital age [6].

A CDN (content delivery network) is a group of geographically distributed servers that can provide web images, videos, HTML pages, and scripts to users more quickly and efficiently. Material delivery networks decrease latency by delivering material from the server that is the closest to the user as opposed to the origin server, which typically sits in a distant land. It improves loading times and user experience, and the main server is not hit as hard. Content delivery networks, which cache material on numerous edge servers, feature better scalability, load balancing, and protection against high traffic and possible Distributed Denial of Service (DDoS) attacks, among other advantages. Examples of widely recognized content delivery networks include Amazon CloudFront, Akamai, and Cloudflare. The flowchart on the application of CDN is detailed in Figure 1.2.

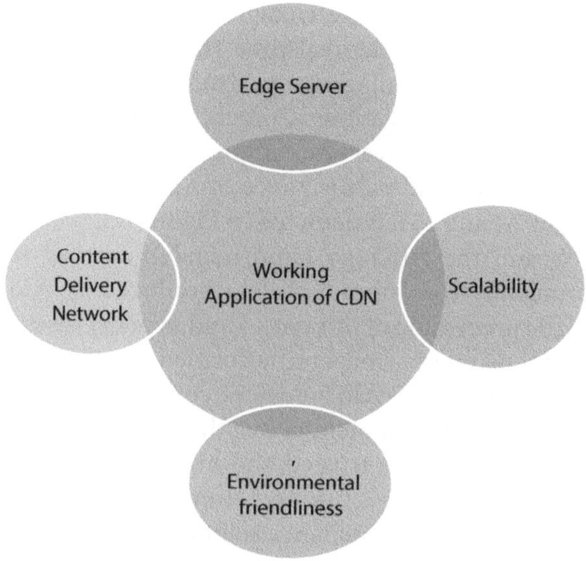

Figure 1.2 Flowchart of the working application of CDN.

1.2 The Interrelationship between AIoT and CDN

With intelligent connections and optimized content delivery, the integration between CDN and AIoT covers the fundamental arms of content and data needed for the technology of today. Combining the data-driven automation of the AIoT with the scalability of content delivery networks also enables faster, more reliable, and more flexible digital experiences across numerous applications through this partnership. An AIoT system is where AI empowers and extends the functions of the IoT, which ultimately leads to a network of devices that autonomously collect and process data and take action based on the data. Internet-of-Things devices can use the power of artificial intelligence to analyze data instantly through sensors, cameras, and integrated chips to make intelligent decisions without any human intervention [7].

Examples of dynamic environments that heavily take advantage of the combination of intelligence and connection afforded by AIoT include industrial automation, smart cities, and even autonomous cars. In these cases, a fast response is very important. This is where content delivery networks come into play; without them, the data that has been processed cannot be adequately delivered to users and other systems, slowing the potential of AIoT. Helps with real-time communication and content delivery in environments that are changing. This combination is crucial for smart city applications, like real-time traffic updates, environmental monitoring, and energy management, requiring seamless connectivity among equipment in such locations. Content delivery networks effectively reduce latency in content delivery and improve bandwidth, which, in turn, improves the scalability and performance of AIoT systems [8]. The synergic integration of AI and IoT is depicted in Figure 1.3.

Thus, CDNs and the AIoT go hand in hand as they solve different but related problems. The volume of data produced by AIoT systems is immense, and the processing of this data—and in many cases, its transfer to other devices or consumers—must be performed in real time. For instance, in healthcare environments, AIoT devices continuously monitor patient vitals and alert caregivers to potential health hazards. One example is the case of CDN, which provides trusted and high-speed access to this data, enabling skilled healthcare practitioners from afar to take timely actions. Manufacturing predictive maintenance solutions are also IoT-enabled and use such equipment data to anticipate and prevent issues. Cloud data networks enable off-site engineers to quickly access data from the manufacturing floor, which increases decision time and reduces downtimes. The synergy of AIoT and CDN will improve both data processing efficiency and network resilience. In addition, content delivery networks can enable edge computing for the Internet of Things (IoT), which moves decisions about data closer to the moment they were collected, reducing latency and

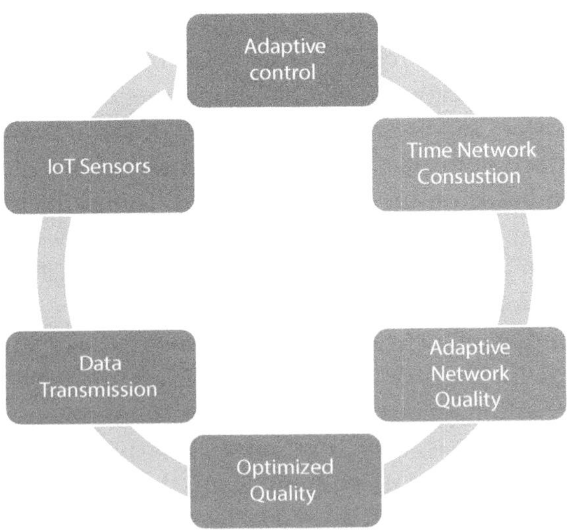

Figure 1.3 Integration of AI and IoT.

relieving the workload of main servers or the cloud. Edge computing inside CDN frameworks is enabling faster AI-based on-the-fly decision-making for applications such as autonomous driving, where rapid action is paramount, since the model can evaluate and react to incoming data in real time [9]. AIoT and CDNs can share a lot in terms of security. As the number of connected devices grows, the security of sensitive information is even more crucial than ever. AI-powered security protocols in Internet-of-Things (IoT) devices can recognize and respond to threats instantaneously, and content delivery networks add another layer of protection by distributing data in a way that makes it more difficult for attackers to focus on congested sites. The future of the Geo5G AIoT-CDN Union partnership seems promising with the advent of 5G technology! The improved data transfer speeds brought in by 5G networks will allow CDNs and AIoT systems to be implemented even more effectively. Cloud delivery networks (CDNs) offer seamless data transfer between users across the world for data collected and processed by AIoT devices. The fusion of these two technologies may transform fields requiring high-speed processing and low latency, such as remote-controlled robots, AR and VR [10]. As modern digital ecosystems become ever more complex, the integration of AIoT with CDN serves as an innovative solution to tackle the challenges that arise. By combining the data-gathering, processing, and responding capabilities of AIoT with the efficient, low-latency content delivery of CDN, this alliance opens the gate to smarter, faster, and more robust infrastructure for the digital world. We also discuss how the integrated CDN + AIoT stack will form the bedrock in supporting the future data-driven applications to enable real-time use from service providers to industries.

1.3 Advantages of AIoT in Content Distribution

AIoT content distribution systems have revolutionized the management and distribution of digital content by integrating the power of artificial intelligence with the pervasiveness of the Internet of Things. The combination of AI's data interpretation capabilities and IoT's data distribution will change how information is spread, for the betterment of suppliers and consumers. These benefits, which change the allocation of physical worldwide connect space, include better interactivity, velocity, security, and operational economy, to name a few. So, one of the biggest benefits that AIoT can give content here is enhancing the speed and efficiency of distribution. As AIoT is added behind content, the distribution of the content will become increasingly responsive, going beyond traditional content distribution networks, which focus on reducing latency of contentions and caching data on the CDNs in proximity to consumers. Based on the algorithms in artificial intelligence, Internet-of-Things devices might learn from the actions of users and require them to evaluate application usage components to predict traffic spikes and to adapt server loads dynamically [11]. Dynamic resource allocation also means that users benefit from reduced load times and uninterrupted content access. You may benefit from enhanced responsiveness and smoother performance when accessing digital experiences like video streaming, online gaming, and cloud-based apps, even at peak times. AIoT also enhances content distribution customization, which is crucial as user expectations for personalized content intensify. AI-based analytics might analyze records collected from IoT devices to maximize user experience with personalized content. For instance, in video streaming, AIoT can analyze a user's watching history, device, location, and activity level and recommend personalized content accordingly. AI does much more than just drive recommendations, as it can also make the viewing experience more personalized (adapting the quality of the content or advertising based on user characteristics). This makes for a far more immersive experience, as it is personalized to their own preferences and habits, so users are more engaged and satisfied with the experience as a whole. In addition, the introduction of AIoT brings the added benefit of making content distribution networks more scalable. Due to the growing number of connected devices, traditional networks are susceptible to the risk of congestion, particularly during peak usage times. AIoT addresses this problem by regularly monitoring the network and adjusting the consumption of bandwidth [12].

In the interest of maintaining high service quality, real-time analytics can flag potential bottlenecks to AI-driven systems that can reallocate resources or adjust traffic to address that issue. Its scalability enables content distribution systems to handle demand surges, as well as scale to accommodate the needs of further users without compromising efficiency. This element is particularly important in industries experiencing exploding user growth, because it enables the distribution of quality information as demand continues to expand. It also optimizes the management of resources in content delivery systems to gain operational and environmental benefits with AIoT. With the advent of artificial intelligence in cloud systems, these resources may be managed much better, as its ability to monitor several metrics in real time, like server load, bandwidth usage, energy utilization, etc., allows for much more accurate analysis and predictions. To reduce waste and operational costs, AI can, for example, shut down some servers during times of low traffic or put them to use for other purposes. Additionally, AI could adjust the resolution of the content on demand based on the bandwidth available at the time, thus saving data with no compromise on user enjoyment. By reducing power consumption, it can also reduce overhead costs for organizations [13], which may minimize their environmental footprint as a sustainable practice. As the volume of data grows and more devices are connected to the Internet of Things, cyberattacks pose a serious threat, and AIoT can play a key role in stopping them before they reach content distribution networks. Artificial intelligence systems monitor data that travels across networks, watching for unusual patterns that can signal a security breach. By using IoT, with the data that can be collected from a variety of sources, AI offers a comprehensive view of potential threats along with real-time responses. Such a configuration enables the prevention of attacks, such as detecting intrusion attempts, thwarting malicious actors, and isolating damaged areas of the network so that these areas do not spread damage to other parts of the network itself [14].

Cost saving is another big perk that AIoT brings to content distribution. AI as a way to optimize these operations is leading to more efficient operations, which leads to significant cost savings. As AIoT can anticipate fluctuations in content demand, the technology enables providers to optimize their resources accordingly to avoid overspending bandwidth or server capacity that will lie unused. AIoT's potential to dynamically manage infrastructure is only possible by lowering an organization's operating expenses and enabling its budget to cover the cost of being an AIoT content provider. AIoT can gather insights about user preferences and behavior, which can be useful when it comes to prioritizing content, investing in infrastructure, improving ad targeting, and more. Ultimately, it all contributes to a more efficient, lower-cost, more profitable content-delivery system. Eventually, the interoperability of AIoT with upcoming technologies such as 5G and edge computing escalates its place in content delivery. These improvements enable faster processing of data, thus resulting in lower latencies, which significantly improve the efficiency of content distribution using AIoT [15].

Real-time data is a critical component of applications like virtual reality (VR) and live streaming, and AIoT can enable such applications by harnessing the low-latency and high-speed characteristics of 5G. Shifting processing closer to the consumer *via* edge computing can help reduce latency even further, as it eliminates the need for data to flow over long distances. These technologies, when optimized together, complement AIoT by facilitating low latency and high-speed content transfer rates, both critical for high-performance applications. Figure 1.4 explains the advantages of AIoT in content distribution.

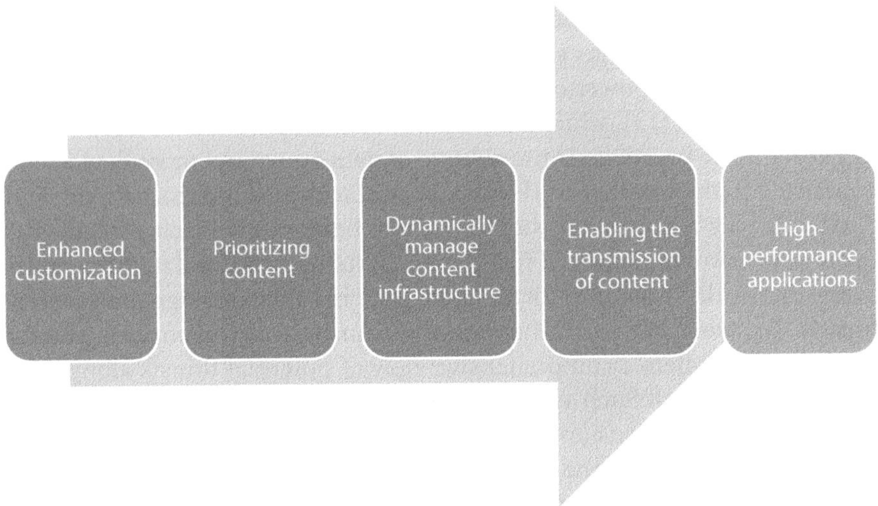

Figure 1.4 Benefits of AIoT in content distribution.

1.4 Major Challenges in the Integration of AIoT and CDN

With this enhanced efficiency, security, customization, and speed, content delivery networks (CDNs) augmented with Artificial Intelligence of Things may redefine content delivery. While the added connectivity and AI-powered data analytics of IoT are advantageous, there are several challenges that must be addressed to make the most of CDNs. These problems are all the features like data security, scalability, interoperability, latency, infrastructure cost, and regulatory compliance. To incorporate AIoT into the CDN well, these companies should therefore have a strong understanding of these problems to develop solid strategies, use advanced technology solutions, and invest in holistic visions to overcome the challenges. Data security and privacy are great challenges in AIoT and CDN integration. AIoT is built on the collection, processing, and communication of large amounts of data from disparate IoT devices. As a result, there are more potential attack surfaces for cybercriminal activity. An additional challenge is that the processing power and storage capacity of IoT devices are so low in many cases that they cannot support strong security standards [16].

As a result, the entire CDN infrastructure is vulnerable to attacks against it. In addition, an AI algorithm needs to have access to and analyze sensitive data of the user so it can provide a personalized experience, which further increases the potential of data breaches, unauthorized access, or misuse of data. Adding new layers of security increases the sophistication of the AIoT-CDN architecture required to protect data over such a broad network. Some of these mechanisms are end-to-end encryption, real-time monitoring, and anomaly detection. Scalability is another significant challenge standing in the way of AIoT-CDN integration. As the number of IoT devices ever in existence increases and the amount of content ever-increasing, the delivery networks require effective resource scalability. Even if AI algorithms can improve traffic pattern prediction and the optimization of server load, scaling an AIoT-enabled content delivery network requires a significant amount of processing power, bandwidth, and storage space. To deliver programming without latency or repeated downtimes, video streaming networks with millions of daily users might need to grow quickly during peak usage times. This scalability is managed dynamically, as the demand for content can become unpredictable [17].

Poor scaling may lead to network congestion, slower delivery speeds, and a bad user experience, while overestimating demand can result in high operating expenses. AIoT-CDN is deployed, and a compromise might be hard to achieve in terms of scalability and efficiency. However, high-quality content delivery is prioritized for AIoT-CDN integration, which is more dependent on low-latency data processing. Besides latency, IoT devices send data to a central server for analysis and decision-making, a core component of the architecture of AIoT systems. This process may have a delay if the servers are located far away from the devices. Even small delays spoil the experience for users of real-time content, such as streaming video, playing games, or using augmented reality. Deploying and monitoring edge equipment is challenging and costly, but edge computing, or processing closer to the point of consumption, can reduce latency. Setting up and maintaining dispersed edge nodes is a burden for enterprises where low latency is a must-have. This brings additional complexity and operational cost to the system of AIoT-CDN [18].

The AIoT-CDN environment involves complexity due to expensive infrastructure. The implications of this in terms of high-performance infrastructure to manage massive amounts of data and analytics are clearly apparent, as AI and the Internet of Things, where respondents plan to distribute content, require new frameworks. Building such an infrastructure requires a massive outlay of capital for computers, storage, network capacity, and AI programs. Because these resources require specialist people and ongoing maintenance to drive high levels of performance with minimal downtime, the costs are even higher. Small and medium-sized enterprises may face an additional challenge to adoption, which is clearly heavily

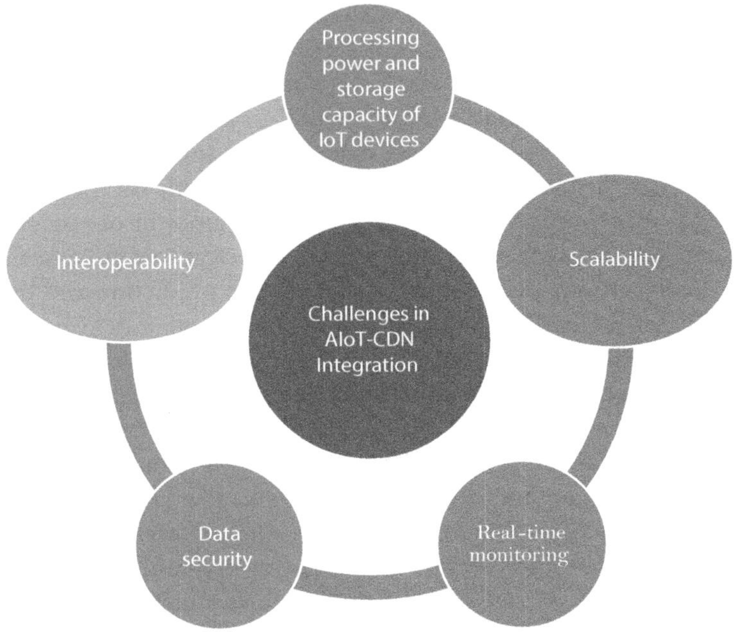

Figure 1.5 Challenges in AIoT-CDN integration.

related to the cost burden of deploying an AIoT-enabled content delivery network. For organizations looking to implement such integration, it is advised to consider the significant upfront cost of implementation versus the long-term gain of AIoT [19].

The integration of AIoT and CDN is further complicated by the need to comply with regulations. The growing emphasis on data privacy and protection, combined with progressive rules in key regulations such as those adopted by the EU (GDPR), has led organizations into a complex web of compliance obligations. With the considerable amounts of user data that AIoT systems accumulate, a fundamental challenge to AIoT systems is safeguarding user information while delivering personalized content. In order to be compliant, data protection is only part of the equation; transparency and giving control to consumers over how organizations are using their data are key. A violation can result in a blow to a person's reputation, finances, and legal standing. For socio-technical AIoT-enabled CDN, adopting data management frameworks according to both the local and international legislation can be a costly and time-consuming process [20].

In consideration of data security, scalability of the system, interoperability, reducing latency, management of infrastructure costs, regulatory adherence, and protocol standardization, many of these areas need to be resolved. Supply of AIoT in content delivery requires a careful step—leveraging advanced technology and following industry best practices is essential for organizations looking to adopt AIoT in content delivery. Addressing these issues properly is essential to getting the full benefits of AIoT-CDN integration and building smarter, faster, and more resilient digital content delivery systems. The challenges in AIoT-CDN integration are shown in Figure 1.5.

1.5 Optimizing Latency and Network Efficiency

As consumers now expect constant, frictionless access to information and services online, optimizing network performance and reducing latency is an absolute must in today's digital age. Latency, which is the time between when a command is issued and when data transport begins, is a major driver of experience in real-time communication, online gaming, and video streaming. Some of the strategic and technological advancements that could enhance network efficiency and latency include leveraging edge computing, content caching, data traffic prioritization, and AI-driven analytics. These solutions will complement each other accommodate the growing data demands, as networks are looking to enable more complicated, data-intensive applications [21]. One of the best methods to reduce latency is to deploy edge computing, which brings data processing close to the consumer. Delays, especially over long distances, can occur in conventional architectures because data must first be sent to a central data center and then returned to the consumer. Edge technology is a necessary part of modern-day network design with the least latency and better user experiences, but it requires a heavy amount of money and maintenance during deployment. Content caching is another way to improve efficiency and reduce latency on the network. Caching enables page loads to speed up the process [22] through distributing copies of commonly used material over many of the nodes available in the network.

Networks might store data in local caches instead of repeatedly reaching out to distant servers for user requests. This saves on latency and lightens the load on the main server, alleviating potential bottlenecks. This method is commonly used by content delivery networks, which cache data to servers that are strategically positioned to ensure that users all over the world receive digital assets such as videos, images, and music in a timely fashion. Content delivery networks are important in increasing network efficiency and maintaining speeds in high usage periods by localizing content. Using

Quality of Service (QoS) settings, for example, network managers may prioritize bandwidth for low-latency services such as video conferencing and online gaming. This allows essential services to be provided without additional hops falling within the same network while also optimizing resource usage.

Traffic prioritization won't eliminate latency on the networks entirely, but it will help to channel resources where they are genuinely required, resulting in more reliable and more efficient service delivery overall. Data traffic that needs to be prioritized is important; however, it needs to be monitored effectively so that policies are adapted flexibly without any disruption to services for other users, ensuring equal services for every user [23]. Another advancement geared toward optimizing latency and network efficiency is utilizing adaptive streaming technologies. Adaptive streaming dynamically alters the quality of the content depending on the user's internet connection and the resources present on the network. Artificial intelligence can quickly sift through mountains of network data, spot trends, and predict impending traffic jams—all in real time. Using machine learning methods, networks can serve to reallocate resources proactively at peak demand times to minimize the chances of delays. Another AI-enabled feature is automated traffic routing, which uses real-time network conditions and availability to calculate the best paths for data transfer [24].

This level of intelligence allows for proactive management of network resources, making it easier to reduce latency and maintain efficiency even during periods of peak traffic. Network Function Virtualization (SDN) enables operators to have greater control over data flows, which aids in balancing out latency-sensitive applications and mitigating traffic congestion. In addition to enhanced performance, SDN facilitates rapid fault detection and recovery, helping networks to minimize downtime and ensure steady performance. Software-defined networking (SDN) is an effective way to allocate resources and reduce latency in modern networks, using real-time analytics and the capability to construct network pathways "on the fly" [25].

Optimizing routing protocols is yet another method for reducing latency in long-distance data transfer. Even though the internet does have complex routing algorithms to determine how data can actually be sent from point A to point B, these methods aren't always the fastest or most effective way to do it. Routing protocols can be optimized for improved network performance and reduced latency, in addition to using Multi-Path TCP (MPTCP) techniques. These techniques permit the data to travel in many directions at once. MPTCP enables significantly faster data delivery, where traffic can be shifted to alternative paths in the case of congestion, which helps in providing more robustness as well. Just a few real-world high-performance use cases around which the 5G networks promise to cluster the data phenomenon with extremely low latency are autonomous vehicles, smart cities, and industrial automation. The faster transmission rate and more bandwidth of the 5G network make it possible for 5G networks to transmit more data with less latency as compared to previous generations of wireless networks [26].

Better connectivity with reduced latency and increased efficiency helps facilitate the expanded Internet-of-Things (IoT) ecosystem, where countless connected objects rely on real-time communications, thanks to 5G's enhanced capabilities. These content delivery networks will be enhanced with the rollout of 5G networks, optimizing and reducing latencies around the network for more reliable and faster content delivery.

1.5.1 Key Applications in Which AIIoT Can Enhance CDN Significantly

- Smart surveillance is streaming from multiple cameras and processing footage upfront to deliver security insights. An AIIoT-enabled CDN dynamically adjusts streaming quality according to available bandwidth to ensure low latency and uninterrupted streaming.
- Using information from Internet-of-Things sensors, the content delivery network can adjust the way it stores and distributes material in real time based on conditions like temperature and human movement analyzed by AI algorithms that evaluate video feeds for anomalies. Content delivery networks that are AIIoT-enabled can boost immersive environments in AR/VR by tailoring content to varying states of device performance and network conditions.
- Artificial intelligence can predict user interests, such as interaction modes and preload materials in advance, and Internet-of-Things devices ensure adaptive quality control. Self-driving cars depend on collecting and sharing data continuously.
- A CDN enhanced with AIIoT might be inundated with data from the many sensors (LiDAR, cameras, and so forth) in it and the infrastructure surrounding it. Employing AI for navigation and harnessing inputs from IoT, CDNs are capable of

quickly routing vital information and ensuring that updates come through at low latencies.

- Telemedicine uses the transmission of patient data and video consultations in real time. Smart IoT CDNs can give priority to urgent medical information requests and adjust video quality in real time based on the state of the network. AI processes data consistently transmitted by IoT health-monitoring devices so that healthcare professionals have useful information and can identify anomalies.
- Content delivery networks use advanced Internet-of-Things features to improve broadcasting of international events by controlling the quality of streamed video based on real-time monitoring of sorted-out network issues. AI models are used to ensure uninterrupted delivery, preallocate resources, and predict peak streaming times.

1.6 Real-Time Data Processing and Analytics

Real-time analytics makes the world we live in today possible, where data is used in real time at every step in order to make faster and more informed choices, improve workflows, or tailor experiences for customers. Real-time analytics helps to sift through these huge amounts of data and get meaningful insights almost instantly due to the explosion of devices generating data and advances in technology. This capability has potential advantages across dozens of industries, from autonomous cars' ability to navigate safely to financial organizations' ability to track trading activity to e-commerce users' experience with the platform. In the medical sector, for example, real-time data analysis could continuously observe patients' vital signs to signal doctors of inconsistencies in those vital signs before they become harmful [27].

It is difficult to process them in real time because of the rapidity, diversity, and volume of this data. Organizations now use modern data processing architectures such as Apache Spark Streaming, Apache Kafka, and Flink to address such complexities. The ability of these systems to perform high-throughput, low-latency computing makes it possible to manage massive datasets and rapidly execute complicated processes. Real-time processing frameworks are designed such that they can combine both short-term data storage for quick access and analysis as well as long-term data storage for revisiting and integrating data into more complex studies [28].

Real-time analytics allow businesses to operate better and continuously improve performance, resulting in operational efficiency. In manufacturing, real-time analytics are used to sense deviations from expected outcomes and avoid costly equipment failures. By utilizing this information in real time, firms can provide instant responses to customer queries, customized offerings, and personalized recommendations [29].

An example of this would be displaying products that are related or discount codes based on what is being viewed by a customer in e-commerce. This kind of real-time suggestion based on watching behavior can also increase user happiness and loyalty on streaming services. These interactions help businesses establish a stronger relationship with their clients and increase conversion rates. Traditional systems may take hours or even days to detect fraud, but with real-time capabilities, suspicious transactions can be detected and stopped before they get out of hand. Likewise, cybersecurity systems scan all incoming and outgoing data in real time to detect malware attacks, network vulnerabilities, and unauthorized access [30].

Empowered with real-time analytics, organizations can respond quickly to security problems, enabling them to prevent breaches and minimize damage. Real-time data processing can only succeed if it has a robust infrastructure like distributed systems, cloud computing, and edge computing. Edge computing approaches help bring the processing power closer to the data-creation point, which is crucial in situations that require low latency, while cloud platforms provide the scale and flexibility necessary to process enormous quantities of information. For example, self-driving cars will need to analyze sensor data on the fly to make safe driving decisions. Real-time data can be used to train AI systems, making them more accurate and useful over time. This is beneficial for use cases such as demand prediction, dynamic pricing, and targeted advertising [31].

Machine learning models can analyze data patterns and provide predictions almost in real time, enabling businesses to be proactive rather than reactive. In energy management, for example, AI enables the optimization of energy consumption and the reduction of power outages by adjusting real-time energy distribution according to consumption trends. Additionally, ensuring consistency and high quality in the data is crucial since errors in real-time data streams can lead individuals to reach incorrect conclusions, which in turn leads to poor decisions being made [32]. Real-time data processing and analytics are shown in Figure 1.6.

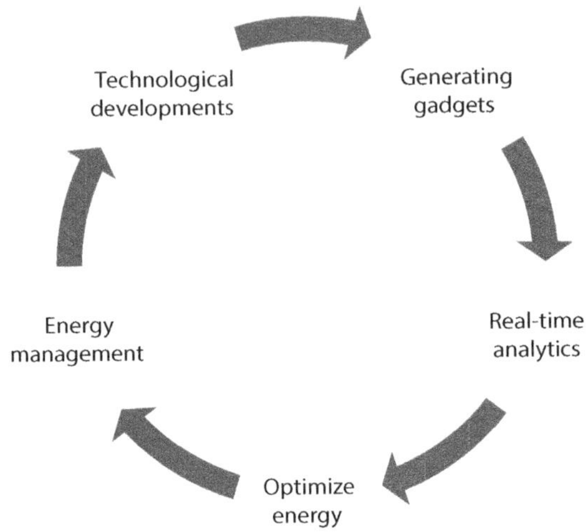

Figure 1.6 Structured streaming for data processing and analytics.

1.7 Scalability and Adaptability of AIoT-CDN Systems

With the emergence of data-driven applications, devices, and user expectations, the performance, scalability, and agility of AIoT-CDN systems are critical to achieving the utmost performance. AIoT-CDN systems are an advanced technology framework that synergizes AI with IoT and content delivery network technologies. The AI-powered IoT devices generate large amounts of data every day, which the network architecture and data processing capabilities need to support. Scalability denotes how well a content delivery network, which serves as the backbone of data dissemination, is able to handle an extensive volume of content queries from distinct geographic regions with minimal latency [33].

Its data is spread across a network of edge servers, which cache and serve information from closer to the end consumers, improving load times and reducing latency. The scalability is critical in AIoT-CDN systems when it comes to retrieving and processing data from millions of devices simultaneously. Secondly, due to their flexibility, AIoT-CDN systems can respond immediately to dynamic network conditions, user activity, and sudden traffic spikes. An adaptive AIoT-CDN system utilizes adaptive AI algorithms to modify data processing and content distribution dynamically based on current conditions. Adaptability to new technological breakthroughs in areas such as resource facilities includes data types, protocol integration, and device integration without having to upgrade or replace the existing systems to adopt this kind of technology [34].

The integration of cloud and edge computing into AIoT-CDN systems has significantly enhanced their scalability and flexibility. AIoT applications that require real-time reaction, such as autonomous cars, industrial automation, or healthcare monitoring systems, significantly benefit from this hybrid approach [35]. AI-based CDNs can quickly respond to demands for processing and add devices by balancing cloud and edge resources. AI really excels in the scaling and flexibility of these systems. AI-powered content delivery networks are also providing predictability in application performance and stability, foreseeing potential security breaches, device failures, or traffic spikes, and allocating resources [36].

To relieve pressure and expedite response times, machine learning models might predict which CDN edge servers would be experiencing significant stress and aggressively cache assets on other proximate servers. This certainly is a boon for streaming services, online stores, and digital marketers, as providing relevant and timely content makes customer happiness and engagement soar. Taking user experience to the next level, AIoT-CDN systems can predict user behavior patterns by analyzing user interactions, enabling them to optimize their delivery strategies to proactively meet user needs [37].

Scalable and adaptive AIoT-CDN systems provide significant leverage, but the path is not facile. One of the challenges facing Internet-of-Things devices is how to handle the volume and variety of data these devices generate. Every device generates data with different structures, formats, and data processing requirements; thus, real-time standardization and processing capability of data are essential for AIoT-CDN systems. Such systems must also be capable of making split-second decisions for processing and distributing data due to application latency constraints, such as in the cases of smart grids and autonomous vehicles. Moreover, to cope with these challenges, we require frameworks for data management that are reliable, edge devices that have excellent performance, and algorithms capable of processing large volumes of data with high performance and accuracy [38].

To support the integration of AIoT and CDN systems into industries such as smart cities, as well as healthcare and finance, modern networks need to be scalable and adaptable. The Internet of Things (IoT), combined with content delivery networks, empowers healthcare organizations to not only capture and evaluate patient data in real time but also to utilize the uniqueness of each patient and scale to potentially millions of devices. The result is enhanced performance, user satisfaction, and operating effectiveness. Next-generation digital infrastructure will rely heavily on AIoT-CDN systems that

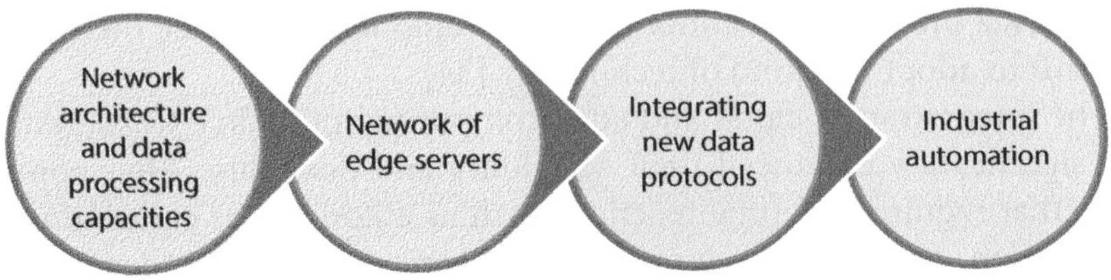

Figure 1.7 Scalability and adaptability of AIoT-CDN systems.

will be scalable and adaptive, solving challenges in areas such as data diversity, latency, or energy efficiency. And all this has been made possible due to the ongoing enhancement in AIoT and CDN technology. Because of the ability of these technologies to stand at the forefront of opening new avenues of use across sectors, the method of processing, delivering, and experiencing data will go through radical transformational shifts [39]. Figure 1.7 implicitly shows the scalability and adaptability of AIoT-CDN systems.

1.8 AIoT and CDN Networks Security and Privacy

The growing application of AI, IoT, and CDN has taken us to the next phase of technical innovation, where systems have become smarter and more efficient. But these innovations also come with major concerns, particularly around privacy and security. Providing scalable and adaptive security and privacy mechanisms is paramount in protecting AIoT and CDN networks from evolving threats and preserving sensitive data. An effective security architecture must be developed with AIoT and CDN systems by keeping up with the ever-evolving threat landscape. In the context of network security, scalability refers to the ability of security measures to accommodate and adjust to changes in the network. Alone, organizations cannot scale without security solutions that consume more public utilities and bandwidth without causing a bottleneck. To that end, automated security processes may be established to keep an eye on numerous endpoints and devices in real time for indications of dubious behavior [40].

Machine learning methods are essential for offering scalable security. Since they observe the traffic in real time, these algorithms are able to identify trends in network traffic and detect potential threats such as unauthorized access and strange data flows. Modern artificial intelligence systems can learn from past data, thereby improving their ability to predict and respond to security threats in real time. Adaptability in privacy and security focuses on the system's ability to respond to changing risks and compliance requirements. Attacks against AIoT and CDN networks are continuously developing and can take several forms, such as including data breaches, ransomware, and advanced persistent attacks [41].

A flexible security architecture is needed to identify these new risks and adapt security measures accordingly. By adopting techniques for continuous learning, AI models can adapt to real-time intelligence, enhancing their ability to identify emerging threats. By following privacy-by-design principles, privacy protections can be built directly into the system from day one, enabling rapid alterations to comply with new rulings. AIoT and CDN networks involve a wide variety of devices and user scenarios, making it critical for security and privacy in their use to be scalable and flexible. To ensure the safety of all the Internet-of-Things (IoT) devices, we need to have a security model that can dynamically adapt to their individual requirements. More robust devices can apply advanced authentication and encryption processes when edge devices have low computation capability that must be deployable in a lightweight secure protocol without affecting performance [42].

Based on the specific capabilities and risk profiles of the different elements in the AIoT ecosystem, companies can use a tiered security approach to protect individual devices. Building this flexible and extensible framework, however, does not mean that user privacy and data security are a lesser priority. In an AIoT and CDN environment, data reporting and processing can take place in many different locations; this raises several issues regarding where this data is stored, processed, and whether it can be shared and how. To mitigate these concerns and to ensure that confidential data is adequately protected, strong access restrictions and data anonymization strategies should be employed. Moreover, companies should simplify the way they authorize user permissions so that users can control their data utilization in CDN and AIoT networks. The transparent and immutable properties of blockchain technology build a trustworthy atmosphere for managing sensitive data, as all the access and modifications of data are verifiable [43].

This approach can enhance privacy and security, allowing businesses to scale while upholding their dedication to user data protection. While powerful, scalable, and adaptive security and privacy frameworks have many benefits, implementing them can prove challenging for enterprises. Another challenge is ensuring that the various platforms and devices within the AIoT ecosystem can interact with one another. Since different devices may not employ standardized protocols, adopting a unified security strategy can be tricky. It is essential for organizations to cooperate and standardize so as to provide common frameworks that can safely accommodate a range of devices [44]. Figure 1.8 shows the security and privacy in AIoT and CDN networks.

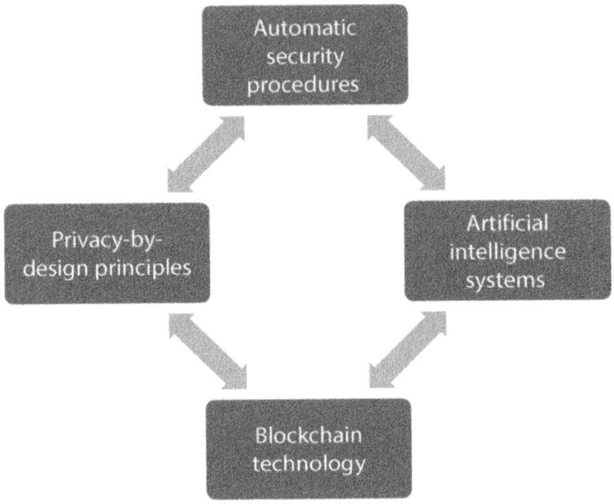

Figure 1.8 Security and privacy in AIoT and CDN networks.

1.9 Future Trends in AIoT and CDN Technologies

Edge computing is gaining traction within the fields of artificial intelligence (AI) and content delivery networks. As the number of IoT devices continues to grow, edge computing, which resides physically closer to the data source, is taking on increasing importance. Latency is reduced, consumption of bandwidth is cut to a minimum, and data transmission efficiency is improved with edge computing. By discovering wide-scale processing and analysis of data at edge devices, organizations can improve their responsiveness and speed of decision-making. It allows data to be analyzed well and thus helps the companies to gain knowledge and analytic strategies of data [45].

For instance, machine learning could optimize content delivery in a content delivery network (CDN) scenario by analyzing demand patterns and predicting user behavior. Beyond reducing costs, it also improves the user experience by utilizing existing resources more efficiently. The convergence of 5G, AIoT, and CDNs is set to transform the connected device landscape. With blazing speeds and near-zero latency, data can be exchanged seamlessly between devices on 5G networks, facilitating real-time applications never before possible. This better connection will allow for more beautiful experiences to be delivered through smart apps, such as AR and VR. As 5G can deliver data to consumers with near-zero latency, even under high-demand scenarios, content delivery networks (CDNs) will see their performance enhanced, with content being more easily propagated across the network. An increased focus on sustainability and energy efficiency will also see the development of greener AIoT and CDN solutions. More and more companies are trying to reduce their negative environmental impacts from the amount of energy and CO_2 consumed for the processing and communication of data [46].

The future path of CDN and AI is also being shaped by the concerns of data privacy and security. The number of connected devices is indeed rising; hence, attack surfaces for cyber threats are widening, and strong security measures are mandatory. Future enhancements will focus on implementing advanced security protocols, including those based on blockchain technology and decentralized identity management, to enhance data protection and ensure user privacy. AI-powered security solutions will be critical for real-time detection and response to threats so that trust can be built between enterprises and their consumers. Another interesting trend occurring with CDN and AIoT technology is the spread of decentralized architectures. This change improves fault tolerance [47] and allows for more openness and accountability in data processing.

AIoT and CDN will be increasingly user-centric, as both enterprises and consumers alike demand more personalized experiences. Companies can study customer habits and preferences through AI algorithms and personalize content distribution and Internet-of-Things (IoT) services. Further, it will help in improving the user experience by making better content suggestions and targeted advertising when interacting with smart devices. Cross-device data analysis makes service integration possible and will ensure a unified and natural user experience. The collaborative effort among enterprises is also expected to drive innovation in CDN and AIoT technologies. These converged activities and programs will promote innovation and accelerate implementation, enabling seamless integration of CDN and AIoT solutions across a range of sectors [48]. Future trends in AIoT and CDN technologies are depicted in Figure 1.9, which shows and describes the future trends and their integration with AIIoT and CDNs. Table 1.1 displays the comparison of future trends and their integration.

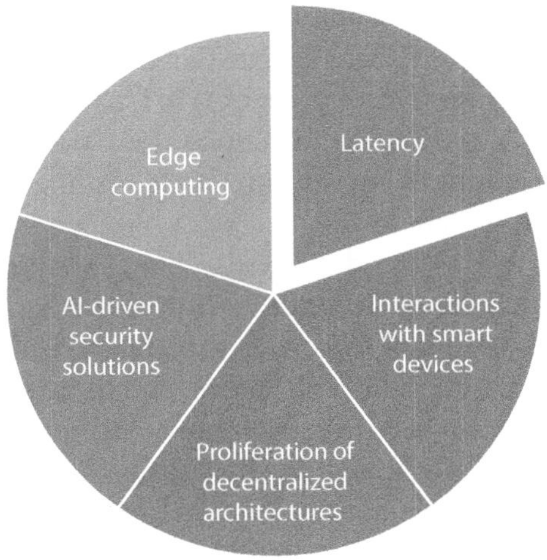

Figure 1.9 Trends for the future of AIoT and CDN technology.

Table 1.1 A comparison table of future trends and their integration.

Future trend	Description	Integration with AIIoT	Integration with CDNs
Edge computing	Low latency is achieved by utilizing decentralized data processing at the edge of the network	Empowers swift data decisions by processing AI in real-time on IoT devices	Lowers latency by storing content nearer to end-users
5G connectivity	Networking Technology with High-bandwidth and Low-latency	Enables smooth communication and data transfer between the IoT devices	Improves the efficiency of CDN and allows for faster delivery of quality content
AI-driven predictive Maintenance	Intelligent algorithms to predict and mitigate system downtimes	IoT devices use AI to monitor and predict hardware problems	Facilitates preemptive management of content delivery infrastructure

(*Continued*)

Table 1.1 A comparison table of future trends and their integration. (*Continued*)

Future trend	Description	Integration with AIIoT	Integration with CDNs
Blockchain for data security	Data integrity and security Decentralized ledger technology	Ensures secured communication of IoT data and enhances trust in as well as between interconnected systems	Bridges trust gaps in connected products and IoT systems
Smart cities and autonomous systems	Autonomous applications and IoT-enabled urban infrastructure	Bringing together IoT data from sensors	Improves the provision of real-time data streams for urban uses
Sustainable and Green Technologies	Focus on energy-efficient and sustainable tech consumer solutions	Artificial intelligence (AI) in IoT devices optimizes power usage and emissions	CDNs use energy-efficient data centers to minimize carbon
Personalized content delivery	Content recommendations based on User personalization using AI	Machine learning in interconnected devices creates tailored experiences for the users	CDNs deliver content based on user preferences
Quantum computing	Groundbreaking computing technology with significant processing capacity	Artificial Intelligence Lab in Internet of Things (IoT): Improves complex solutions	Cuts the path of data in CDNs and encrypts it
Interoperability standards	Standard protocols: interoperability of varied technologies	Allows different types of IoT to interact seamlessly	Encourages cross CDN service compatibility
Cybersecurity enhancements	Machine learning based algorithms to detect and combat threats	Defensive against IoT weaknesses and breaches	Protects CDN infrastructure from malicious attacks

1.10 Conclusion: Toward Smarter, Faster Digital Infrastructures

This merger continues to build a more AIoT-powered and faster digital infrastructure, the next chapter of the digital transformation story. From health care and education through entertainment and logistics, this convergence not only enhances operational efficiency but also yields dramatic improvements in customer experiences. The emergence of 5G connectivity, computing at the edge, and powerful machine learning is turning the processing, transmission, and consumption of data on its head, delivering real-time insights on anything, anywhere in seamless convergence. From energy-efficient gadgets to smart homes, this upward trend toward sustainable technology innovations signals a changing tide toward a more ethical form of tech development.

Improved digital infrastructures are closer than you think, though challenges await on the way there. With so many devices being so interconnected, security and privacy are big concerns. To win the trust of users & confidence of the organization, strong security measures & privacy policies should be the first priority. Decentralized architectures and advanced encryption technology are critical to protecting sensitive information from cybercriminals. The collaboration and interoperability of CDN and AIoT tech would be the key characteristics of future tech as we have envisioned. Building collaborations and setting standardized standards will promote innovation as organizations will allow seamless integration of diverse systems and devices. Finally, the combination of CDN and AIoT technologies heralds a new age in digital infrastructure development. Leveraging these technologies to the fullest, we would be able to build more intelligent, efficient systems that will help us respond to our dynamic world and anticipate what our customers will need next. Through continuing innovation and adaptation, we can make our way to a future that has the highest levels of intelligence and connection possible, providing efficiency and opportunity in our increasingly digital world.

We have worked collaboratively in writing the chapter where Introduction to AIIoT and CDNs is covered. My contributions mainly came down to exploring recent advancements in AI-driven IoT designs that make for such a symbiotic relationship, pushing on data optimization for IoT-dominated environments, and pushing toward CDNs, as they are a pivotal function toward faster data retrieval at scale. I helped write some segments of the chapter and edited the contents to fit with the trends and academic flow of the coming years. As such, we hope that it will be the AI specialist's and the AIIoT experts' condensed way of introducing all aspects when dealing with the fast-dynamic AIIoT world in a content distribution fashion.

AI disclosure statement: The author acknowledges the use of AI-based tools for minor language editing or grammatical correction. However, all intellectual contributions, analyses, and conclusions presented in this chapter are entirely the author's own.

References

1. Kumar, N.A. and Daniel, R., Artificial intelligence-based energy efficiency models in green communications towards 6G, in: *Towards Wireless Heterogeneity in 6G Networks*, pp. 158–179, CRC Press, Florida, United States, 2024.
2. Govind, M. and Pinheiro, A., Multifunctional Urban Agriculture and Sustainability: A Global Overview, in: *Sustainable Urban Agriculture*, pp. 20–39, 2024.
3. Sharma, N., Thota, V.P., Yuvaraj, T., Tripathi, S., Pandey, O.J., OptRISQL: Towards Performance Improvement of Time-Varying IoT Networks Using Q-Learning. *IEEE Trans. Netw. Serv. Manage.*, 3008–3020, 2024.
4. Rus, A.C., El-Hajj, M., Sarmah, D.K., NAISS: A reverse proxy approach to mitigate MageCart's e-skimmers in e-commerce. *Comput. Secur.*, *140*, 103797, 2024.
5. Zong, H., Wu, R., Cha, J., Wu, E., Li, J., Tao, L., Shen, B., Advancing Biomedical Text Mining with Community Challenges, arXiv preprint arXiv:2403.04261, 2024.
6. Paliwal, S., Jain, A., Sharma, M., Jamwal, V., Vig, L., Custom Text: Customized Textual Image Generation using Diffusion Models, *arXiv preprint arXiv:2405.12531*, 2024.
7. Zong, H., Wu, R., Cha, J., Feng, W., Wu, E., Li, J., Shen, B., Advancing Chinese biomedical text mining with community challenges. *J. Biomed. Inf.*, 104716, 2024.
8. Poddar, R., Naik, T., Punnam, M., Chourasia, K., Pandurangan, R., Paali, R.S., Swaminathan, M., Experiences from Running a Voice-Based Education Platform for Children and Teachers with Visual Impairments. *ACM J. Comput. Sustainable Soc.*, *2*, 3, 1–35, 2024.
9. Vedachalam, S.K., Siromany, V.A., VanderEnde, D., Malpiedi, P., Vasquez, A., Dikid, T., Mathur, P., Implementing a Healthcare-Associated Bloodstream Infection Surveillance Network in India: a Mixed-Methods Study on the Best Practices, Challenges and Opportunities. *Antimicrob. Resist. Infect. Control*, 13, 1, 1–14, 2022.
10. Shamil, M.M., Gooneratne, D.W., Gunathilaka, D., Shaikh, J.M., The effect of board characteristics on tax aggressiveness: the case of listed entities in Sri Lanka. *J. Account. Emerg. Econ.*, 14, 4, 747–770, 2024.

11. Vert Martin, N. and Pineau, P.O., Overallocation in the California-Québec carbon market: a non-constraining cap until 2030. *Environ. Econ. Policy Stud.*, 1–26, 2024.
12. Malekijou, H., Hakami, V., Javan, N.T., Malekijoo, A., Q-learning-based algorithms for dynamic transmission control in IoT equipment. *J. Supercomput.*, 79, 1, 75–108, 2023.
13. Hajiakhondi-Meybodi, Z., Mohammadi, A., Abouei, J., Deep reinforcement learning for trustworthy and time-varying connection scheduling in a coupled UAV-based femtocaching architecture. *IEEE Access*, 9, 32263–32281, 2021.
14. Mondejar, C.L.C., Bello, G.E., Canto, K.V., Galvan, C.K.L., Paclibar, J.N.F., Assessment of Variants of Organic-Based Rice Production Systems in Negros Occidental, Philippines. *Philipp. Agric. Sci.*, *107*, 3, 4, 2024.
15. Lai, X., Hu, Q., Wang, W., Fei, L., Huang, Y., Adaptive resource allocation method based on deep Q network for industrial Internet of Things. *IEEE Access*, 8, 27426–27434, 2020.
16. Jala, C., Financial Literacy and Retirement Preparedness of Public-School Teachers: Basis of Developmental Plan for Retirement. *Psychol. Educ.: Multi. J.*, *26*, 10, 1122–1137, 2024.
17. Huang, H.C. and Chen, Y.X., Evolutionary optimization of fuzzy reinforcement learning and its application to time-varying tracking control of industrial parallel robotic manipulators. *IEEE Trans. Ind. Inf.*, *19*, 12, 11712–11720, 2020.
18. Liang, F., Yu, W., Liu, X., Griffith, D., Golmie, N., Toward deep Q-network-based resource allocation in industrial Internet of Things. *IEEE Internet Things J.*, 9, 12, 9138–9150, 2021.
19. Ibričević, A., Decided Return and Reintegration in a Post-Conflict Society, in: *Decided Return Migration: Emotions, Citizenship, Home and Belonging in Bosnia and Herzegovina*, pp. 75–119, Springer International Publishing, Cham, 2024.
20. Eyeleko, A.H. and Feng, T., A critical overview of industrial internet of things security and privacy issues using a layer-based hacking scenario. *IEEE Internet Things J.*, 10, 24, 21917–21941, 2023.
21. Ramya, N., Ajay, H., Balaji, J., Book Cave–Online Book Streaming Platform Based On Web Application, in: *2023 Intelligent Computing and Control for Engineering and Business Systems (ICCEBS)*, IEEE, New Jersey, United States, pp. 1–6, 2023.
22. Chilamkurthy, N.S., Karna, N., Vuddagiri, V., Tiwari, S.K., Ghosh, A., Cenkeramaddi, L.R., Pandey, O.J., Energy-efficient and QOS-aware data transfer in q-learning-based small-world lpwans. *IEEE Internet Things J.*, 10, 24, 22636–22649, 2023.
23. Xu, M., Chen, X., She, Y., Jin, Y., Wang, J., Time-varying weights in multi-reward architecture for deep reinforcement learning. *IEEE Trans. Emerging Top. Comput. Intell.*, *8*, 2, 1865–1881, 2024.
24. Tariq, U., Jaafar, F., Malik, Y., Detecting and Mitigating Adversarial Perturbations to Improve E-Commerce Security, in: *2024 IEEE/ACM*

International Conference on Big Data Computing, Applications and Technologies (BDCAT), IEEE, 334–341, 2024.
25. Racehorse, V. and Hohag, A., Achieving climate justice through land back: An overview of tribal dispossession, land return efforts, and practical mechanisms for# landBack. *Colo. Nat. Resour. Energy Env't. L. Rev.*, 34, 175, 2024.
26. Naveen Kumar, R., Janani, M., John William, A., Pavithra, R., Assessing performance metrics for e-commerce reverse logistics and circular economy initiatives in supply chains: the five R's model approach. *IIMBG J. Sustain. Bus. Innov.*, 3, 188, 2025.
27. Llaulli-Romero, N.R., Quispe-Perez, J., Nuñez-García, A.I., Perez-Campomanes, G., Blocks with the addition of Mucilago de Nopal: Compressive Strength, Waterproofing and Automation in Factory IO, in: *2023 IEEE World AI IoT Congress (AIIoT)*, IEEE, New Jersey, United States, pp. 0098–0104, 2023.
28. Karnani, S. and Shakya, H.K., Mitigation strategies for distributed denial of service (DDoS) in SDN: A survey and taxonomy. *Inf. Secur. J.: Global Perspect.*, *32*, 6, 444–468, 2023.
29. Al-Heety, O.S., Zakaria, Z., Abu-Khadrah, A., Ismail, M., Mahmood, S.N., Shakir, M.M., Alsariera, H., Traffic Control Based on Integrated Kalman Filtering and Adaptive Quantized Q-Learning Framework for Internet of Vehicles. *CMES-Comput. Model. Eng. Sci.*, *138*, 3, 2024.
30. McCully, G.A., Hastings, J.D., Xu, S., Fortier, A., Comparing Unidirectional, Bidirectional, and Word2vec Models for Discovering Vulnerabilities in Compiled Lifted Code, *arXiv preprint arXiv:2409.17513*, 2024.
31. Krishnamoorthy, Y., Subbiah, P., Rajaa, S., Krishnan, M., Kanth, K., Samuel, G., Sinha, I., Barriers and Facilitators to Implementing the National Patient Safety Implementation Framework in Public Health Facilities in Tamil Nadu: A Qualitative Study. *Global Health Sci. Pract.*, *11*, 6, 2024.
32. Khandkar, V.S. and Hanawal, M.K., Fairnet: A measurement framework for traffic discrimination detection on the internet. *IEEE Trans. Netw. Serv. Manage.*, *20*, 4, 4097–4109, 2024.
33. Hussain, S.M., Brunetti, A., Lucarelli, G., Memeo, R., Bevilacqua, V., Buongiorno, D., Deep learning based image processing for robot assisted surgery: a systematic literature survey. *IEEE Access*, *10*, 122627–122657, 2022.
34. Samuel, G. and Sinhaa, I., Barriers and Facilitators to Implementing the National Patient Safety Implementation Framework in Public Health Facilities in Tamil Nadu: A Qualitative Study. *Global Health*, *11*, 6, 1, 2022.
35. Prayogo, Dimas, H., Roman, D., Paulina, G.D., The Key Factors for Improving Returns Management in E-Commerce in Indonesia from Customers' Perspectives—An Analytic Hierarchy Process Approach. *Sustainability*, 16, 17, 7303, 2024.
36. Ramzan, M., Shoaib, M., Altaf, A., Arshad, S., Iqbal, F., Castilla, A.K., Ashraf, I., Distributed denial of service attack detection in network traffic using deep learning algorithm. *Sensors*, *23*, 20, 8642, 2023.

37. Mehrad, A., Brian Paul, M., Meriem, B., Neema George, R., A Qualitative Approach to Reverse Logistics in the Retail E-commerce Sector in Barcelona, Spain. *Eur. J. Manag.*, 32, 2, 67–74, 2024.
38. Cogburn, D.L., Ochieng, T.A., Wong, H.M., Towards an understanding of global 'private ordering' in ICANN: text mining 23 years of Uniform Domain-Name Dispute-Resolution Policy (UDRP) Decisions. *J. Cyber Policy*, 8, 2, 186–217, 2023.
39. Ojobor, R.C., 1 Blockchain Technology for Library Services: Challenges and Opportunities for Libraries from a Nigerian Perspective, in: *Information Services for a Sustainable Society: Current Developments in an Era of Information Disorder*, vol. 183, p. 7, 2023.
40. Jujjavarapu, C., Suri, P., Pejaver, V., Friedly, J., Gold, L.S., Meier, E., Jarvik, J.G., Predicting decompression surgery by applying multimodal deep learning to patients' structured and unstructured health data. BMC Medical. *Inf. Decis. Making*, 23, 1, 2, 2023.
41. Rey, W.P., Face Recognition (FR) Integration on MABIS: A Mobile Automated Biometric Identification System for Law Enforcement in the Philippines, in: *Proceedings of the 2023 6th International Conference on Electronics, Communications and Control Engineering*, pp. 8–16, 2023.
42. Tougui, I., Jilbab, A., El Mhamdi, J., Impact of the choice of cross-validation techniques on the results of machine learning-based diagnostic applications. *Healthcare Inf. Res.*, 27, 3, 189–199, 2023.
43. Kumar, Y. and Mahajan, M., Recent advancement of machine learning and deep learning in the field of healthcare system, in: *Computational intelligence for machine learning and healthcare informatics*, vol. *1*, p. 77, 2020.
44. Tian, X., Xu, X., Wang, C., Ruan, T., Wu, B., Que, M., Towards Explainable Search Results in E-commerce, in: *Companion Proceedings of the ACM on Web Conference*, 76–484, 2025.
45. Choi, S.W., Ko, T., Hong, K.J., Kim, K.H., Machine learning-based prediction of Korean triage and acuity scale level in emergency department patients. *Healthcare Inf. Res.*, 25, 4, 305–312, 2019.
46. Fyffe, I., Sorensen, J., Carroll, S., MacPhee, M., Andrews-Paul, A., Crooks, V.A., Mithani, A., Long COVID in long-term care: a rapid realist review. *BMJ Open*, *13*, 12, e076186, 2023.
47. Denis, A., Volkova, T., Grima, S., Noja, G.G., Cristea, M., Adapting Information Security Policy Messaging Styles to the Targeted Audience, in: *Digital Transformation, Strategic Resilience, Cyber Security and Risk Management*, vol. 111, pp. 115–131, Emerald Publishing Limited, Bingley, United Kingdom, 2023.
48. Wang, S., Liu, Z., Yang, W., Cao, Y., Zhao, L., Xie, L., Learning-Based Multimodal Information Fusion and Behavior Recognition of Vascular Interventionists' Operating Skills. *IEEE J. Biomed. Health. Inf.*, 27, 9, 4536–4547, 2023.

2

Predictive Caching Strategies Using Machine Learning for Seamless Content Access

Jagendra Singh[1]*, Monika Dandotiya[2], Shivani Agarwal[3], Kannan Srinivasan[4], Rahul Jadon[5] and Guman Singh Chauhan[6]

[1]School of Computer Science Engineering and Technology, Bennett University, Greater Noida, India
[2]Department of Computer Science & Engineering, Poornima University, Jaipur, India
[3]Department of Information Technology, Ajay Kumar Garg Engineering College, Ghaziabad, India
[4]Saiana Technologies Inc., Piscataway, New Jersey, United States of America (USA)
[5]CarGurus Inc., Boston, Massachusetts, United States of America (USA)
[6]John Tesla Inc., Dallas, Texas, United States of America (USA)

Abstract

The recent demand for seamless access to content within educational institutions, especially with the environment involving servers in colleges, has emerged as an urgent call for methods that could be more efficient in delivering content. Classical caching schemes lack the feature of predicting user behavior, which induces latency or causes server overload when accessing content. This chapter presents how prediction-based caching strategies can be implemented using various techniques in machine learning to enhance efficiency in accessing content. This research uses three machine learning models: artificial neural networks, support vector machines, and decision trees, using the kind of usage behavior and access patterns, content types, and frequency by the users. The dataset for this work was collected over a period of an academic semester from a college server system comprising one main server, which hosts most of the educational content available, such as videos, PDFs, and articles, on five local servers. Data preparation involved

**Corresponding author*: jagendrasngh@gmail.com

Tanusree Chatterjee, Rajib Banerjee and Souvik Ganguli (eds.) Content Delivery Networks Using Artificial Intelligence and Internet of Things: Revolutionizing Content Delivery, (31–52) © 2026 Scrivener Publishing LLC

data cleaning, feature selection, and transformation to improve model accuracy. ANN reached the highest prediction accuracy of 97.67%, followed by SVM at 92.6% and DT at 87.45%. Again, the reduction in terms of percentage was maximum with ANN, at around 64%, followed by SVM at 55% and DT at 48%. Improvements in cache hit rates and latency reduction were also observed.

***Keywords*:** Predictive caching, machine learning, content access, server optimization, artificial neural networks

2.1 Introduction

It has been for a long time a crucial solution in improving the speed of server systems, especially in those environments where users frequently request data, such as educational institutions. Traditional techniques of caching are widely adopted, but with different degrees of efficiency depending on the accessibility frequency of the content [1, 2]. The conventional approaches, which include the use of LRU and FIFO methods, work on the principle of temporary storage, based on the understanding that data accessed repeatedly would be in demand again. Although useful in static settings, these approaches, however, are not found to be efficient in dynamic settings where the behavior of users keeps changing rapidly, thus making the prediction of content demand uncertain. The exponential growth of digital content and the increasingly complex user access patterns give rise to the need for dynamic caching strategies that can respond instantly to real-time user behavior. Predictive caching, which relies on using machine learning techniques, may be considered one potential solution to this problem. Therefore, predictive caching improves the performance of servers, eliminates latency, and helps provide for smoother user experiences; it takes into account patterns of user access as well as their content consumption behaviors [3, 4].

In contrast, traditional caching techniques cannot optimize server performance because they lack a predictive system to ascertain prospective demands for content. The primary basis of such techniques is historical information, and they lack the capability to forecast future content requests [5, 6]. For instance, LRU is based on the assumption that what was accessed recently will be requested again soon, whereas FIFO stores information in the cache purely based on the order of the request without any consideration for future use. Although such algorithms sound useful when there is an environment for either a simple or low-content server, they do not work effectively in cases where access to the content is rather unpredictable, increasing the load on the server, latency, and more cache misses.

The main limitation in the traditional caching approach lies in its failure to predict dynamically what the user could be interested in, thus it is much less desirable in environments that have very high variability in the demand for content, which includes cases such as educational institutions where students can access different resources at different times based on their study needs [7, 8]. This deficiency points out the areas where more advanced caching techniques are needed in order to respond to changes in the real behavior of users and enhance the efficiency of content delivery.

Machine learning makes predictive caching possible. In this way, it optimizes server systems by making proactive predictions about the type of content that will likely be served due to the patterns detected in the huge datasets. Some research has been done on machine learning for caching, which proves that these techniques may significantly decrease server loads and improve response times. These approaches can broadly be classified into three major categories: time-series analysis, behavior-based prediction, and content-based prediction—all based on machine learning [8, 9]. The idea of time-series analysis relies on historical data. It will indicate future access patterns based on time intervals. Behavior-based prediction focuses on the habits of individual users, comparing how these users interact with the content throughout time. Lastly, content-based prediction categorizes and differentiates various types of content and predicts future demand according to the nature of the content. All the approaches have their own advantages. The selection of one over others depends on the type of server environment and many other constraints. However, hybrid solutions that combine facets of all three approaches prove to be most effective in achieving the best accuracy [10, 11].

2.2 Related Work

Machine learning has transformed various fields by enabling computers to learn from data and make predictions or decisions without explicit programming. This transformation is especially evident in the realm of data analysis and predictive modeling, where traditional statistical methods often fall short in handling complex, nonlinear relationships inherent in real-world datasets. Machine learning encompasses a variety of algorithms and techniques, each with unique strengths and weaknesses, making it a versatile tool for tackling diverse problems across numerous domains [12, 13].

For instance, clustering algorithms like k-means can group similar data points together, enabling insights into the underlying structure of the dataset [14]. Unsupervised learning is particularly valuable in exploratory data analysis, where the goal is to uncover hidden patterns and relationships without predefined labels. This approach can reveal significant insights, especially in domains such as marketing and customer segmentation, where understanding customer behavior is crucial [15, 16].

In this paradigm, an agent learns to make decisions by interacting with an environment, receiving feedback based on its actions. Over time, the agent optimizes its decision-making strategy to maximize cumulative rewards [17, 18]. For example, deep reinforcement learning algorithms have achieved remarkable success in mastering complex games like chess and Go, showcasing the potential of this approach to tackle challenging decision-making problems.

The flexibility of machine learning algorithms allows them to be applied across a broad spectrum of industries. In healthcare, predictive models can analyze patient data to assist in diagnosis, treatment recommendations, and resource allocation. In finance, machine learning is used for credit scoring, fraud detection, and algorithmic trading. Retailers leverage machine learning to optimize inventory management, enhance customer experiences, and personalize marketing strategies [19, 20].

A lot of comparative studies have been performed over different machine learning models in predictive caching, emphasizing the trade-offs related to accuracy, computational complexity, and interpretability. ANN always outperforms the other models with respect to accuracy, especially in complex and non-linear environments in terms of user behavior. In contrast, it's less feasible in those server environments because of its high computational demands [21]. SVM is a good balance between accuracy and efficiency; it excels in environments with more structured access patterns to content, which can better be classified. DT is less accurate, but simplicity and low computation make it valuable in environments where interpretability and quicker deployment are vital concerns. Each model has its strengths and weaknesses, so the choice of which to implement depends on the specific needs of the server environment, particularly the volume of data, complexity of user behavior, and available computational resources [22, 23].

Although promising machine learning has shown promise for developing enhanced caching strategies, the present literature still contains several gaps that must be addressed. One such gap is the scarcity of studies focusing on educational institutes, where content access patterns may highly vary due to heterogeneity in student needs. Although many works studied the use of machine learning in areas like e-commerce or media streaming, only a few of them have focused on how such techniques may be applied to optimize content delivery within educational settings. Moreover, most of the currently existing studies mainly focus on a single model and do not provide comprehensive comparisons between different models in light of their performance in predictive caching. This limits our knowledge of which models are suitable for which server environment. In relation to those gaps, this research has made contributions to the existing literature by testing three different types of models of machine learning, SVM, and DT, on the problem of predictive caching in a college server environment. It compares the effectiveness of the models in predicting content access patterns by measuring their effectiveness regarding accuracy, server load reduction, cache hit rates, and latency reduction [24].

2.3 Methodology

The research first determines the current configuration of the server architecture and also the type of content most students access on a frequent basis. Such fundamental information is essential in the design of a forecasting mechanism of caching that is responsive to users' expectations and within the feasible and reasonable capabilities of operations. In this research, the configuration of the server systems has one primary server where the videos, learning materials, and PDF files are uploaded, and five local servers monitor the mode of accessing the aforementioned content by students.

Data is collected first in the form of access logs from both the main server and the local server. It captures information such as user IDs, timestamps of access, content types including videos, PDFs, and articles, and whether the content served came from the local cache or was fetched from the main server. This dataset covers the entire semester, meaning that it has an opportunity to capture a wide range of user behavior. The local servers handle multiple users daily, and they accompany a comprehensive dataset ready for analysis.

After the collection of data, preprocessing comes in, which allows cleaning up the dataset and removing irrelevant or duplicate entries, as well as handling missing values within it and transforming data to an efficient format for use. Features such as "time of day," "day of the week," and previous access patterns are appended to make the models more predictive. One-hot encoding is a must for categorical variables, and normalization of data is also a must for machine learning models that learn properly from the dataset. The working of the proposed research is shown in Figure 2.1.

Feature selection is the next step in the critical optimization of machine learning models. It tries to find the most important predictors of content access with techniques like correlation analysis or recursive feature elimination.

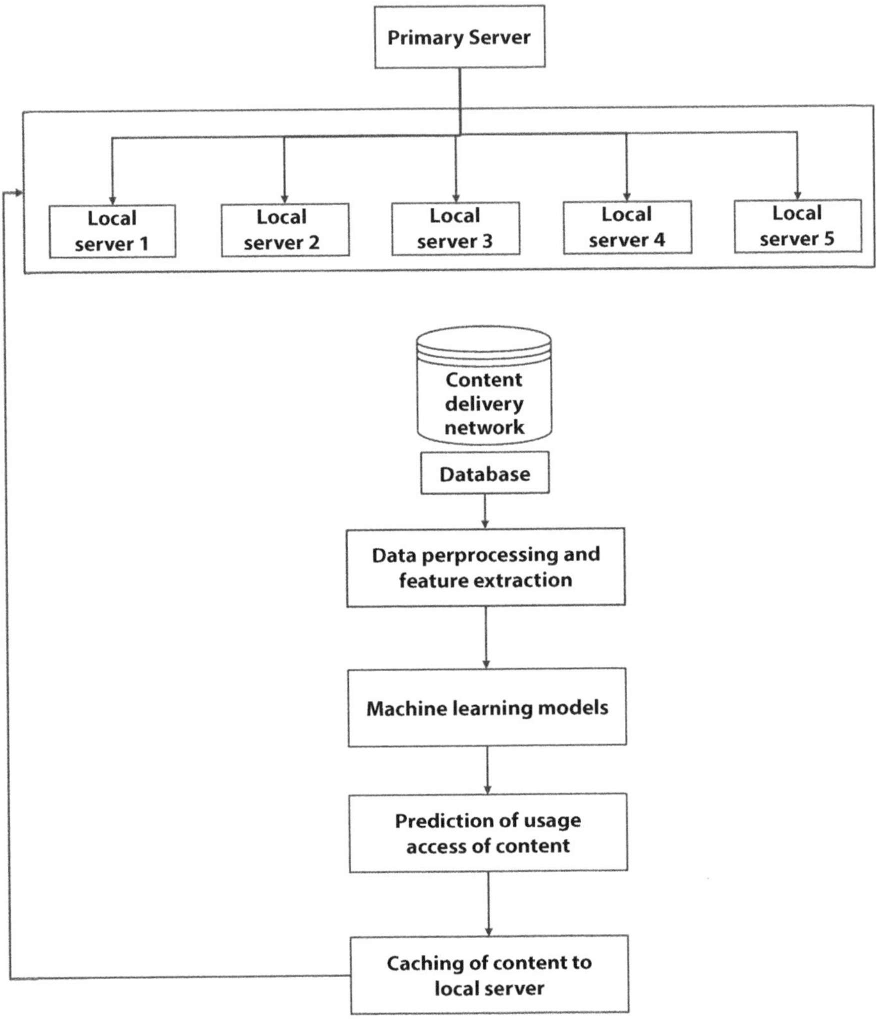

Figure 2.1 Working of the proposed system.

The reduced dimension of the dataset makes way for better model performance and interpretability, thereby minimizing the risk of overfitting.

The core of the research is training three different machine learning models—ANN, SVM, and DT. This was done on a dataset split into training, validation, and testing sets, where models were trained on 70% of the data, with better performance achieved through hyperparameter tuning techniques like grid search and random search. Finally, the models are tested with multiple performance metrics such as predictive accuracy, precision, recall, F1 score, and confusion matrix to provide a fair assessment of their predictive capabilities on content access patterns.

2.3.1 Collection of Datasets

For this research, a systematic approach was adopted in capturing user interactions within the college server environment over the course of an academic semester through the use of Google Analytics. The core data source was by integrating Google Analytics with the college's web portal, where educational content—videos, PDFs, and articles—is hosted. In the case of Google Analytics tracking codes included in the website, logs of activity were accumulated automatically, detailing unique user IDs, timestamp content access, type of content requested, and whether it had been served from the local cache or retrieved from the main server.

The collection of data was done over a period of months with the expectation of collecting approximately 20,000 data points. This phase allowed the acquisition of a large and rich dataset by the students who interacted with the materials at their disposal, Google Analytics, which recorded in real time each content access and substantially provided much information on user behavior. The system not only recorded the access frequency to content but, through this process, also captured other contextual information, like device type—mobile versus desktop—used for accessing content by students, ensuring that usage scenarios across the local servers were diverse.

In addition to the regular access logs, features in Google Analytics gathered supplemental data. This included metadata about the content itself, such as video lengths, file sizes for PDFs, and categories of educational materials. Furthermore, the dataset also included insights from Google Analytics about geographical location and engagement, thereby making it richer and possible to understand and interpret interactions better. The breadth of information proved vital in developing predictive models, as it provided much finer context around user behavior and engagement with content.

The data collection exercise adhered strictly to data privacy and ethics. At the beginning of the semester, the students were informed that the accessing patterns would be recorded for research purposes through Google Analytics, and appropriate consent was sought based on that. For the dataset, anonymity was ensured by substituting real user IDs with unique identifiers that were used for analyzing patterns without revealing personal identities. By being very careful, the research study was able to be in line with institutional guidance and utilize Google Analytics' powerful capabilities in collecting data that would be used in the study.

2.3.2 Data Processing

Machine learning's quality and usability in building robust models depend on the preprocessing of the dataset. The phase is essential for the following: removal of noise, reduction of redundancy, treatment of missing values, and transformation of raw data for analysis. The dataset, containing almost 20,000 data points acquired from Google Analytics, comprised various features: user IDs, timestamps, content types, and metadata on content access. Systematic preprocessing was performed with the aim of improving predictive caching models that ensure optimally structured data for machine learning through this data. The various steps in preprocessing are shown in Figure 2.2.

Data cleaning is the first step of preprocessing, which focuses mainly on the elimination or correction of missing or incomplete entries in the dataset. For any large dataset, missing data are likely to occur due to network issues, incomplete user sessions, or errors while tracking. This research is mainly affected by missing values in timestamps or in content type fields. Missing entries are dealt with both by deletion and imputation methods. In scenarios involving missing information, such as time or type of content, the whole record is excluded from the dataset because that would not give the models enough context to function. Other fields with non-critical missing data, such as user type, have imputation applied. The gap is filled using the mean, median, or mode of the available points in those categories in order to lose the minimum information and keep the integrity of the dataset.

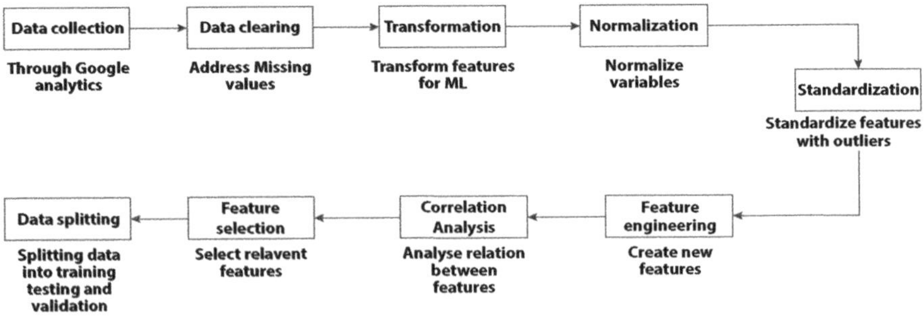

Figure 2.2 Various steps in preprocessing.

Once all these are cleaned, the data is then transformed so that features can be in a proper format for machine learning algorithms. Since the input of the machine learning model shall be numerical, categorical features such as content type that may be video, PDF, or article are encoded into numerical representations *via* techniques such as one-hot encoding. This step elaborates on inputs that create binary variables for any category, in this way allowing easy understanding of the type of content of features by such models as DT and SVM. Other fields are also encoded according to access mode, like mobile and desktop, regarding device type for uniform data representation.

Normalization and standardization of the data is another task within the preprocessing phase to make sure that all the features contribute equally towards making predictions for the model. The feature should be normalized for any continuous variable, like timestamps and file sizes, that could otherwise have great differences in scale. For instance, access times of contents are stored in raw timestamp format, which are translated into more meaningful representations such as hours, minutes, or relative time differences. Such a transformation decomplexifies the temporal patterns, thus making the data easily interpretable. Mathematically, the normalized timestamp data is symbolically presented as follows:

$$X_{nrm} = \frac{X - X_{mn}}{X_{mx} - X_{mn}} \tag{2.1}$$

where X is the original value, X_{mn} and X_{mx} are the minimum and maximum values of the feature, and X_{nrm} is the normalized value. This process scales the data between 0 and 1, ensuring that timestamps do not dominate the model's learning process due to their large numerical range.

Standardizing applies to variables such as file sizes. Such variables may have their distributions contain outliers or be skewed. Standardizing scales the data so all of them have a mean equal to zero and a standard deviation equal to one, as expressed in the following equation:

$$X_{std} = \frac{X - \mu}{\sigma} \tag{2.2}$$

where X is the original value, μ is the mean, σ is the standard deviation, and X_{std} is the standardized value. By standardizing features like file sizes, the models are able to focus on relative variations rather than absolute magnitudes, which is important for features with large value ranges.

A very important task of the preprocessing stage is feature engineering, by which, based on existing data, new features are designed in order to enhance the predictability capability of the machine learning models. In this research, several extra features are created in order to obtain temporal and behavioral patterns in access made by the users. For example, time-of-day, day-of-week, and session-duration features can be extracted from the timestamps to provide richer context to models on user behavior. These further derived features will enable the model to understand patterns like peak time accesses or user preferences in a particular type of content during particular times better than before. Feature engineering is undertaken with the guidance of domain knowledge because only through understanding the nature of temporal content access can accurate predictive caching strategies be developed.

The relations between different features and their relevance in predicting content-access patterns are uncovered by performing correlation analysis. During this process, the extremely correlated features introduced into the models could ultimately decrease their performance because of multicollinearity. The Pearson correlation coefficient is applied as a measure of linear relationships between continuous variables, like file size and access frequency. Features that depict high correlation values above 0.8 will be candidates for removal or transformation since they may contain redundancy. Just like categorical features have to be evaluated by the chi-squared tests to determine the independence of the provided variable with respect to the target variable, making sure that only significant predictors remain in the model.

Feature selection also forms part of the preprocessing step with regard to picking the most relevant features that contribute towards an accurate prediction without any redundancy. Recursive feature elimination (RFE) and random forest feature importance are techniques applied in ranking the features based on their contribution towards the accuracy of the model. On the other hand, reducing the number of features reduces the possibility of overfitting and makes the models computationally efficient. The retained features will be those that will have the highest predictive powers, thus making the models both efficient and precise.

2.3.3 Machine Learning Models

The first machine learning model applied in this research is an Artificial Neural Network (ANN). ANN is a robust model inspired by the structure and operation of the human brain. It is composed of several interconnected layers of nodes called neurons. The general architecture of an ANN consists of an input layer that receives the inputs, one or more hidden layers in which computation actually happens, and the output layer that actually produces the output or prediction. The architecture of the ANN is shown in Figure 2.3. In a network, each neuron has its weight, and the weights times the inputs by their corresponding weights are summed. This is then propagated through an activation function, which introduces a non-linearity to the model that allows it to learn complex patterns. In this research, the ANN is useful specifically because it lends itself to modeling nonlinear relationships between user access patterns and the content they are likely to request [25]. The update of the weights is obtained through backpropagation, which is a method that minimizes the difference between the predicted and actual values. In this process, a series of weight updates is calculated iteratively in such a way that the model produces predictions for content access that are more accurate over time. In general, an ANN with learnability from large datasets can be applied for complex interactions in server environments as well as for variability in access behavior.

Another model used in this study is SVM. SVM is a supervised learning model that can perform the tasks of classification or regression, which are classified, and is also very effective in high-dimensional spaces. The core concept of SVM is finding out the most optimal hyperplane to separate data into discrete classes. In the context of this research, SVM is implemented

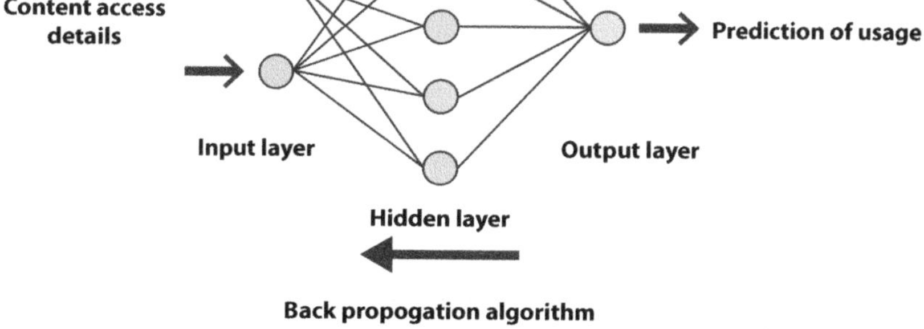

Figure 2.3 ANN architecture.

in order to classify the access type: whether accessed from cache or fetched from the main server, assuming that the input features determine this type. Here once more, it also maps the input data into a higher-dimensional space using the kernel function in which it treats the features entirely in a non-linearly separable manner. The most widely used kernel is the Radial Basis Function (RBF), and this function maps the data into some kind of feature space where the classes are linearly separable. SVM then searches for the hyperplane with maximum separation between the two classes. The critical points closest to the hyperplane are called support vectors; hence, they define the decision boundary. SVM has good generalization in high-dimensional data for the purpose of predicting the access patterns of content, since it would possibly be able to handle a wide variety of features and complex relationships within the dataset.

Another important algorithm used in this research is the decision tree model due to its simplicity and interpretability. A decision tree is a tree-like structure in which every internal node is attached with a decision about one of the features, and every leaf node is labeled with an output class or value. The architecture of the decision tree is shown in Figure 2.4. The process starts with splitting the dataset at the root node based on the feature resulting in the highest information gain or Gini index. The process repeats indefinitely until the tree grows up to its maximum depth or all of the data points in a node are of the same class. In this research, based on both user behavior and characteristics of the content, DT is applied to determine whether content should be cached or not. An essential merit of decision trees is that they can represent the decision process clearly through visualizations. For instance, given the case that a user frequently accesses video at certain hours of the day, the tree will be learned from such cases, and a

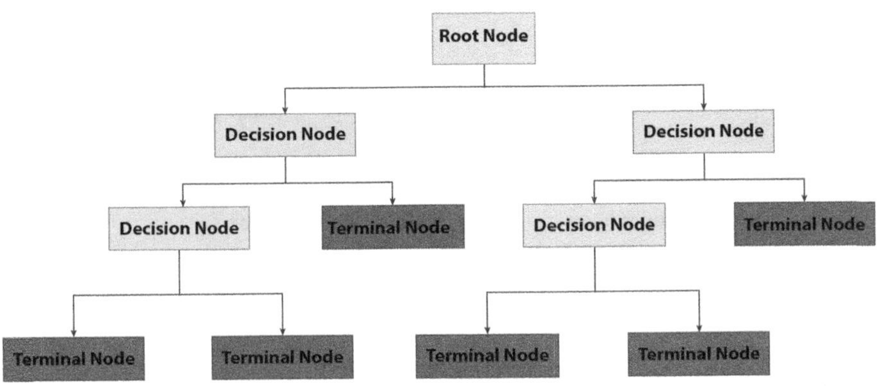

Figure 2.4 Architecture of decision tree.

prediction for the next time will be made. However, a particular drawback of decision trees is overfitting, whereby, with large datasets and large quantities of features, this tendency becomes predominant. In addition to the above technique, pruning is applied so that any of these branches that do not contribute immensely toward prediction power are removed from the model, so that the generalization capability improves.

2.4 Result and Discussion

The performance in predicting the pattern of access of user content within the college server environment was thoroughly analyzed after training each model of machine learning. The accuracy of each model is shown in Figure 2.5. Results indicated that ANN provided the highest accuracy at 97.67% regarding correct predictions regarding different accesses of user content. The main reason why the ANN excels in performance is its capability to model such intricate patterns among the characteristics, not strictly linear, between the input features, such as the behavior of users, frequency of accesses, and type of content. Its architecture with multiple layers of hidden units enables an ANN to capture complex patterns and subtle interdependencies in the data, thereby increasing the accuracy of its predictions.

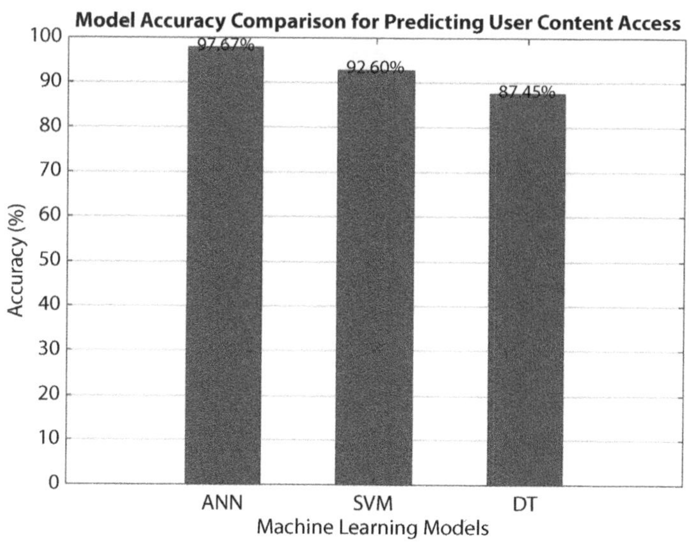

Figure 2.5 Accuracy of each ML model.

The model, through the backpropagation mechanism, minimizes the error during training, thereby generalizing very well on unseen data.

The SVM model also performed quite outstandingly. It was able to attain a 92.6% accuracy. The strength of SVM in the classification of high-dimensional data arises from the ability to find a separable hyperplane that is optimal to classify those classes. Although not as high as ANN, the accuracy found with SVM means that its differentiation between cached and non-cached content may rely on the right input features with which it is provided. The use of the RBF kernel probably helped out in its performance, as it enabled the model to handle non-linear data distributions.

The decision tree (DT) model is simpler and more interpretable; it performed better with ab accuracy of 87.45%. Though not as good as other models, DT shows how a decision is made. It can easily interpret the rules generated based on its ability to recognize basic patterns among user behavior, yet it tends to overfit, and even techniques that prune may cut off its ability to be a predictive tool in comparison with ANN and SVM.

The performance of the three machine learning models, namely, ANN, SVM, and DT, has been measured through key metrics such as accuracy, precision, recall, and F1 score, as depicted in Figure 2.6. Among them, ANN has shown the excellent ability to reach up to 97.67% accuracy in predicting user content access patterns within college server environments.

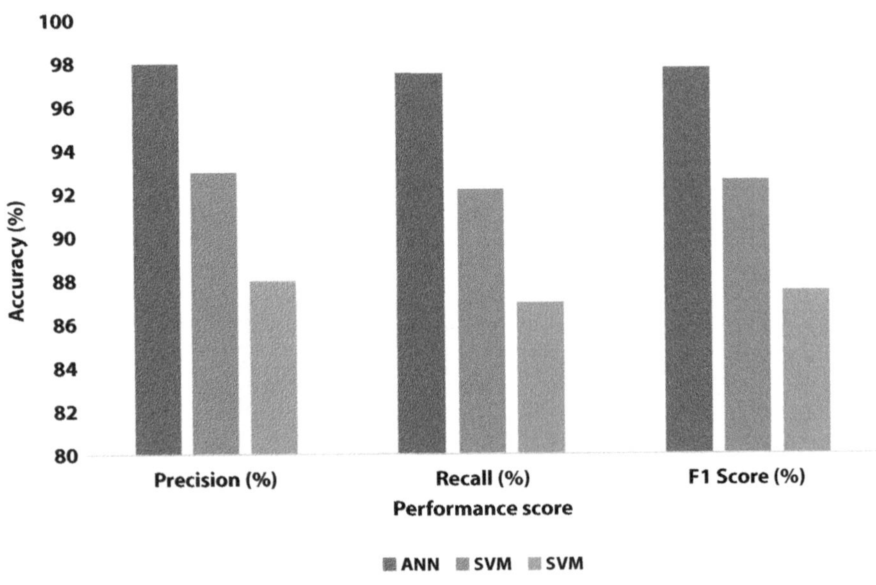

Figure 2.6 Performance score of each ML model.

High accuracy can be seen in this, with the results showing 98.00% precision and 97.50% recall. This proves that ANN accurately predicts almost every time, both when identifying cached content and when not missing the relevant instances. The F1 score was 97.75%, which confirms that ANN attained a balance between precision and recall and that it, therefore, became the most robust model for this application.

Even though it lacks the precision of ANN, SVM still achieved accuracy of 92.60%. Precisely, its precision at 93.00% and recall at 92.20% show that SVM is reliable in making a distinction between cached and noncached content. Its F1 score indicates that SVM is best suited for situations where it is imperative to classify the right content, although it is slightly less complex than what an ANN can handle. The precision of DT at 87.45% was less accurate than 88.00%, and the recall was 87.00%. Though very effective, its F1 score at 87.50% shows that DT easily overfits and will lose all the competitions with this accuracy in prediction compared to those by ANN and SVM models.

The confusion matrices can be used to serve as the actual information regarding the prediction performances of each model for the case of correctly or incorrectly classified content access patterns. The results of the confusion matrices are shown in Figure 2.7. For the artificial neural network, the model achieved 960 true positives, whereby it happened to predict the cache of the content rightly, while getting 980 true negatives whereby it correctly identified the content than was not cached. The model had only 20 false positives (FP) and 40 false negatives (FN), which portrayed minimal errors because it generally had an accuracy of 97.67%. These values presented high effectiveness of the ANN model at predicting content access patterns with low error rates.

For SVM, the confusion matrix reports 910 true positives and 960 true negatives, and also 40 false positives and 70 false negatives. Although SVM

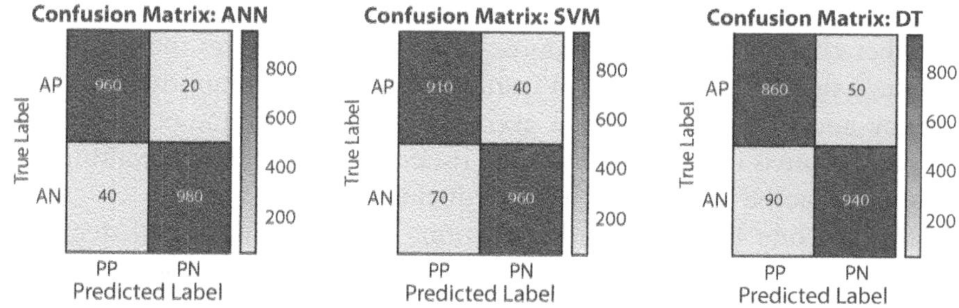

Figure 2.7 Confusion matrices of each model.

had a good performance with an accuracy of 92.6%, more false negatives and false positives occurred in SVM in comparison to ANN, which depicts that SVM was less efficient at distinguishing between cached and non-cached content. In DT, there were 860 true positives and 940 true negatives, along with 50 false positives and 90 false negatives. This indicates that the model was more often confusing both the cached and the non-cached content and had a generally lower accuracy of 87.45%. The confusion matrices reflect relative performances and error rates of each model, but, in comparison, overall, the ANN performed the best.

Results of the study indicate that all machine learning models used in this study—ANN, SVM, and DT—show an efficient prediction capability for user content access patterns. Between the results of these models, it seems that ANN is the best, as it has scored greater accuracy and reliability in prediction. Though SVM and DT show commendable capabilities, their performance metrics showed lesser effectiveness compared to the case with ANN. This conclusion points out the fact that, though all models are certainly positive in adding value to the prediction process, the performance of the ANN is highly remarkable. We further perform analysis on the caches' performance in several important metrics, such as latency reduction, cache hit rate, cache miss rate, response time, and server load reduction.

Latency reduction is measured by decreases in time to content retrieval from the server after predictive caching is implemented. This is measured based on the average latency before and after the implementation of caching strategies. Low latency means users can get to content faster, which in today's high-tech environment is deemed important in keeping the engagement and satisfaction of the users. The cache hit rate measures the ratio of requested content that is served from the local cache instead of the main server.

A low cache miss rate is essential because it reflects the effectiveness of the caching strategy in predicting future need fulfillment or user requirements and ensuring that content frequently accessed is readily accessible. Another crucial measure is response time, which measures the total time it takes a user to receive content from a request—an amount that includes latency and processing time. It averages over many requests. Improved response time is central to enhancing user experience because waiting for long can become irritating to users, reducing usage.

The last one is server load reduction, showing diminished demand on the central server after applying caching strategies. The metric can be found in determining the number of requests managed by the local servers in comparison with those forwarded to the main server. The formula used in the computation for server load reduction is as follows: A significant reduction in server load is beneficial because it alleviates pressure on the main server, allowing it to perform more efficiently and effectively manage resource allocation.

Table 2.1 expresses the average latency for the users before and after caching with the aid of the models. For the ANN, average latency measures about 250 ms before caching and is reduced down to 80 ms after caching. Thus, it reduced by an amount of 170 ms, which simply means that the latency could be reduced by a staggering 68%, thereby further reasserting ANN as a machine learning model that gives a great speed boost to the retrieval of contents. Similarly, the SVM model also presented a latency reduction from an average latency of 240 ms down to 100 ms, thus displaying a latency reduction of 140 ms as well as a percentage reduction of 58.33%. That being said, this result depicts that SVM is less potent than ANN for latency reduction. Contributions from the decision tree (DT) model were also positive—an average latency drop from 260 ms to 120 ms, indicating a 140 ms reduction and a percentage reduction of 53.85%. While DT provided the smallest percentage drop of all three models, its latency-reduction impact is still significant.

Figure 2.8 shows the cache hit rate and cache miss rate for each type of machine learning model, ANN, SVM, and DT, over 24 hours, captured on an hourly basis. The cache hit rate is the percent of content requests satisfied locally from its cache, while the cache miss rate refers to the percent of requests that needed to be fetched from the main server.

The ANN model cache hit rate was at 85% in the first hour and gradually improved until it could attain a high of 94% by the 24th hour. This, therefore, really means that the ANN model was really effective, as it could substantially predict content access patterns in terms of holding a consistently

Table 2.1 Average latency of each model.

Model	Average latency before caching (ms)	Average latency after caching (ms)	Latency reduction (ms)	Latency reduction (%)
ANN	250	80	170	68%
SVM	240	100	140	58.33%
DT	260	120	140	53.85%

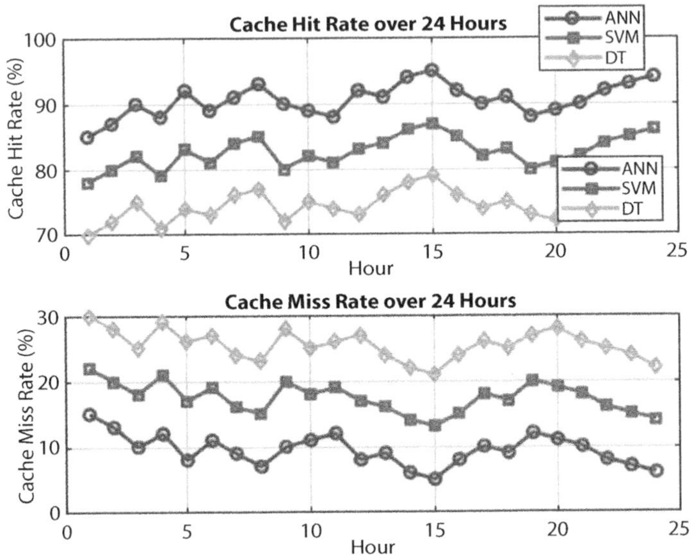

Figure 2.8 Cache hit and cache miss rate of each model.

high hit rate. Overall, the miss rate of ANN decreased from 15% within the first hour to just 6% by the end of the period, thus reflecting its capability to decrease dependency on the primary server. The SVM model also did well, having a cache hit rate of 78% initially, moving to 86% after the 24th hour. Cache miss rate fell from 22% to 14%, indicating how SVM really improved the performance of the cache, though it was not as successful as ANN. The weakest model was the decision tree (DT) model, whose hit rate started at 70% and ended at 78%, while the miss rate declined from 30% to 22%.

Figure 2.9 shows the percentage reduction in server load achieved by each machine learning model over 24 hours, with the time shown in ANN, SVM, and DT. Here, server load reduction is defined as the pressure relieved from the main server considering the prediction and, consequently, the caching of very frequently accessed contents on the local servers. Among the ANN models, reduction was highest, as it could reduce from 55 percent during the first hour to 64 percent after the 24th hour of the experimentation process.

The SVM model is also performing quite well, having started from a 47% decline and reaching up to 55% at hour 24, even though its performance was always lower than that of ANN, indicating a less efficient but still worthwhile approach to predictive caching. This shows that the DT model had the least amount of reduction, initially at 40%, then peaking

Figure 2.9 Server load reduction of each model.

by the 24th hour to 48%. This was a relatively less accurate predictor and reduced the server load with low efficiency.

2.5 Conclusion

The research shows clear evidence that predictive caching strategies do enhance the efficiency of content access within a college server environment. Thus, the developed ANN model is quite impressive in accuracy and has been shown to predict user patterns of content accesses with up to 97.67% accuracy. These skills are reflected in the reduction of server load by up to 64% and the high cache hit rate that resulted in optimized content delivery with minimal latency. The SVM model, although slightly less accurate at 92.6%, still held great server load reduction, peaking at 55% and was therefore effective in improving the overall caching strategy. The least accurate was the DT, which achieved only an 87.45% prediction rate but still managed to reduce the server load to 48% and thus demonstrates that a DT model may also be useful in less complex caching scenarios. Overall, this research concludes that machine learning-driven caching strategies can significantly reduce server strain and response times while enhancing the user experience through the concept of pre-caching, or prefetching, content that is frequently accessed. Among these models, the ANN is particularly noteworthy in terms of its efficiency and hence would thus be a strong contender for immediate use in real-time applications to optimize server performance.

AI disclosure statement: The author confirms that no artificial intelligence (AI)-generated content or AI-assisted writing tools were used in the preparation of this chapter (including summaries and the abstract). All text, analysis, and conclusions are the result of the author's original work.

References

1. Tan, K., Bremner, D., Le Kernec, J., Zhang, L., Imran, M., Machine learning in vehicular networking: An overview. *Digital Commun. Networks*, 8, 1, 18–24, 2022.
2. Rafique, W. and Qadir, J., Internet of everything meets the metaverse: Bridging physical and virtual worlds with blockchain. *Comput. Sci. Rev.*, 54, 100678, 2024.
3. Meybodi, Z.H., Mohammadi, A., Hou, M., Rahimian, E., Heidarian, S., Abouei, J., Plataniotis, K.N., Multi-content time-series popularity prediction with multiple-model transformers in MEC networks. *Ad Hoc Networks*, 157, 103436, 2024.
4. Zyrianoff, I., Gigli, L., Montori, F., Sciullo, L., Kamienski, C., Di Felice, M., Cache-it: A distributed architecture for proactive edge caching in heterogeneous iot scenarios. *Ad Hoc Networks*, 156, 103413, 2024.
5. Singhal, P., Gupta, S., Singh, J., An Integrated Approach for Analysis of Electronic Health Records Using Blockchain and Deep Learning. *Recent Adv. Comput. Sci. Commun. (Formerly: Recent Pat. Comput. Sci.)*, 16, 9, 1–10, 2023.
6. Lin, C.T. and Li, D.L., Ranks aggregation and semantic genetic approach based hybrid model for query expansion. *Int. J. Comput. Intell. Syst.*, 10, 1, 34–55, 2017, https://doi.org/10.2991/ijcis.2017.10.1.4.
7. Guezouli, L., Guezouli, L., Benaggoune, S., Bahri, M.M., Adaptive Caching Strategies for IoV Based on LTE Signal Quality Analysis. *Int. J. Networked Distrib. Comput.*, 1–23, 2024.
8. Prasad, M., Prasad, O.K., Meng Joo, E., Saxena, A.K., Lin, C.T., A novel fuzzy logic model for pseudo-relevance feedback-based query expansion. *Int. J. Fuzzy Syst.*, 18, 980–989, 2016.
9. Bilal, K. and Sajid, M., Blockchain technology: Opportunities & challenges, in: *2022 International Conference on Data Analytics for Business and Industry (ICDABI)*, IEEE, pp. 519–524, 2022.
10. Priya, P.I., Muthurajkumar, S., Daisy, S.S., Data fault detection in wireless sensor networks using machine learning techniques. *Wireless Pers. Commun.*, 122, 3, 2441–2462, 2022.
11. Sharan, A., Co-occurrence and semantic similarity based hybrid approach for improving automatic query expansion in information retrieval,

in: *Lecture Notes in Computer Science*, vol. 8956, 2015, https://doi.org/10.1007/978-3-319-14977-6_45.

12. Gupta, T., Singhal, D., Lal, B., Turukmane, A.V., Fake News Detection Using Transfer Learning, in: *Recent Trends in Image Processing and Pattern Recognition, RTIP2R 2023*, Communications in Computer and Information Science, vol. 2026, Springer, Cham, 2024, https://doi.org/10.1007/978-3-031-53082-1_2.
13. Sahu, D.P., Murkute, S., Yadav, U., Agarwal, M., Kumar, P., Deep Learning Based Bug Detection in Solidity Smart Contracts, in: *Recent Trends in Image Processing and Pattern Recognition, RTIP2R 2023*, Communications in Computer and Information Science, vol. 2026, Springer, Cham, 2024, https://doi.org/10.1007/978-3-031-53082-1_9.
14. Garg, N., Kukreja, S., Saini, M., Singh, I., Mall, S., IoT-Enhanced Agricultural Water Management System Utilizing Conventional Machine Learning Methods, in: *2023 International Conference on Modeling, Simulation & Intelligent Computing (MoSICom)*, IEEE, pp. 554–559, 2023, doi: https://doi.org/10.1109/MoSICom59118.2023.10458818.
15. Petch, J., Nelson, W., Di, S., Balasubramanian, K., Yusuf, S., Devereaux, P.J., Bangdiwala, S.I., Machine learning for detecting centre-level irregularities in randomized controlled trials: A pilot study. *Contemp. Clin. Trials*, 122, 106963, 2022.
16. Mobarak, M.H., Mimona, M.A., Islam, M.A., Hossain, N., Zohura, F.T., Imtiaz, I., Rimon, M.I.H., Scope of machine learning in materials research—A review. *Appl. Surf. Sci. Adv.*, 18, 100523, 2023.
17. Rey, V., Sánchez, P.M.S., Celdrán, A.H., Bovet, G., Federated learning for malware detection in IoT devices. *Comput. Networks*, 204, 108693, 2022.
18. Jiménez, A.F., Ortiz, B.V., Lena, B.P., Real-time mapping of crop canopy temperature using a wireless network of infrared thermometers on a central pivot. *Measurement*, 230, 114570, 2024.
19. Singh, N.P., Hasan, D.S., Shelke, N.A., Namdev, A., Girdhar, N., An Integrated Machine Learning and IoT Based Approach for Enhanced Healthcare Efficiency and Personalized Treatment, in: *Intelligent Systems Design and Applications, ISDA 2023*, Lecture Notes in Networks and Systems, vol. 1046, A. Abraham, A. Bajaj, T. Hanne, P. Siarry (Eds.), Springer, Cham, 2024, https://doi.org/10.1007/978-3-031-64813-7_16.
20. Kaliyar, R.K., Kumari, R., Sharma, N., Manjunath, S., Prasad, P.D., Advancements in Early Detection of Cervical Cancer using Machine Learning and Deep Learning Models for Cervicography Analysis, in: *2024 International Conference on Emerging Innovations and Advanced Computing (INNOCOMP)*, 2024, May, IEEE, pp. 252–256, doi: 10.1109/INNOCOMP63224.2024.00049.
21. Zhang, Q. and Zhang, K., Protecting location privacy in IoT wireless sensor networks through addresses anonymity. *Secur. Commun. Netw.*, 2022, 1, 2440313, 2022.

22. Alsamhi, S.H., Almalki, F.A., Al-Dois, H., Ben Othman, S., Hassan, J., Hawbani, A., Saleh, H., Machine learning for smart environments in B5G networks: Connectivity and QoS. *Comput. Intell. Neurosci.*, 2021, 1, 6805151, 2021.
23. Jha, A.V., Appasani, B., Ghazali, A.N., Pattanayak, P., Gurjar, D.S., Kabalci, E., Mohanta, D.K., Smart grid cyber-physical systems: Communication technologies, standards and challenges. *Wireless Netw.*, 27, 4, 2595–2613, 2021.
24. Kamel, N.E., Mohamed, E., Lmoumen, Y., A Smart Agent Design for Cyber Security Based on Honeypot and Machine Learning. *Secur. Commun. Netw.*, 8, 9, 2020.
25. Tian, Y., Liu, A., Dai, Y., Nagato, K., Nakao, M., Systematic synthesis of design prompts for large language models in conceptual design. *CIRP Ann.*, 2024.

3

Artificial Intelligence in Context Delivery

Vaishnavi[1]* and Navneet Kumar[2]

[1]*CSE Department, Birla Institute of Technology, Mesra, Jharkhand, India*
[2]*CSE Department, Birla Institute of Technology, Patna, Bihar, India*

Abstract
Artificial intelligence (AI) is one of the revolutionary components that has emerged in terms of improving how people gather information, services, and content. Context delivery is the timely and tailored delivery of content based on the needs, choices, and circumstances of a user. AI searches data from various sources, like user interactions, behavior, and environmental factors, to understand and effectively respond to specific situations. AI techniques like computer vision and others are natural language processing, and also machine learning enables systems to interpret data in real-time, adapt dynamically, and forecast human behavior. Applications included e-commerce, where various product recommendations are made, and healthcare, where patient care is customized based on real-time health data. AI context delivery enhances marketing strategies and smart home systems by boosting scalability, user satisfaction, and effectiveness. Despite its promise, problems, including algorithmic bias, data privacy, and the challenge of real-time processing, must be fixed. As AI advances and offers increasingly adaptable solutions for a variety of industries, its contribution to context delivery increases.

***Keywords*:** Artificial intelligence (AI,) content delivery network (CDN), machine learning (ML), data privacy, 6G wireless networks, smart systems, IoT integration

**Corresponding author*: vaishnavi.000111@gmail.com

Tanusree Chatterjee, Rajib Banerjee and Souvik Ganguli (eds.) Content Delivery Networks Using Artificial Intelligence and Internet of Things: Revolutionizing Content Delivery, (53–68) © 2026 Scrivener Publishing LLC

3.1 Introduction

Artificial intelligence has become a crucial element of technological innovation and is having an impact on the delivery of knowledge and services across different sectors [1]. Context delivery, which also involves providing information according to the user's situation, has gained popularity as AI has advanced. Artificial intelligence encompasses a wide range of techniques, which include computer vision, machine learning, and natural language processing [2]. In the domain of storage management, AI methods are used to create intelligent systems that can learn from previous data and access patterns, system resource constraints, and user behavior. AI enhances user experiences by understanding and responding in real-time to contextual nuances, such as tailored advertisements and dynamic healthcare recommendations. It is anticipated that the 6th generation of wireless cellular network (WSN-6G) will offer limitless wireless connections and widely used latency-sensitive AI facilities. This is an unprecedented transformation when compared to previous generations of wireless communication systems [3]. These cutting-edge apps typically come with a lot of data streams, which contribute to the rise in mobile data traffic worldwide. More specifically, there is increased competition for the communication resource as users have varying demand patterns and ask for content to be sent often and concurrently. Holding the network edge is recognized as a possible option to address these extraordinary issues. A significant amount of data is visible, pervasive, and very close to the UEs in edge-caching–based networks because the edge facilities, including access points and user equipment (UEs), may pre-fetch and store a lot of popular content [4, 5]. This chapter highlights the transformational potential of AI in context delivery by examining its principles, methods, applications, advantages, difficulties, and future directions.

3.1.1 Context Delivery

Context delivery is an intelligent method to deliver digital content, like websites and videos, that considers who is asking for it and where they are. Instead of transmitting the same thing to everyone, it customizes the experience. It uses knowledge from multiple sources, which are user interactions, location, time, and environmental factors, to understand the context in which the user operates [6]. The main goal is to provide the important and relevant information that enhances user experience. AI/ML systems evaluate vast amounts of data to detect patterns and forecast user behavior.

This enables systems to adapt to changing circumstances and personalize user experiences [7]. Suppose you are watching a live sports game. A traditional CDN holds a single broadcast tower miles away. You might get the game, but the signal could be weak. Context delivery also makes use of real-time processing to respond quickly and efficiently, which makes interactions more intuitive and natural. Machine learning models trained on a range of datasets can progressively improve the accuracy of their responses through continuous learning. Context delivery in today's CDNs is like having fewer broadcast towers strategically placed around the city.

3.1.2 Content Delivery

Content delivery is the method of efficiently delivering digital content like websites, videos, and apps to users worldwide. Assume it like this: instead of everyone downloading a film from a single server, content delivery networks (CDNs) store copies of the movie on servers closer to viewers. This resulted in faster loading times, smoother streaming, and better overall user experience. CDNs play an important role in today's digital world, ensuring that websites and online services are accessible to users everywhere. Content Delivery Networks (CDNs) are like a global street system for digital content. They are designed to provide web pages, videos, and other online resources quickly and efficiently to users worldwide.

Figure 3.1 represents the continuous delivery of machine learning, an essential technique that enables organizations to develop, deploy, and

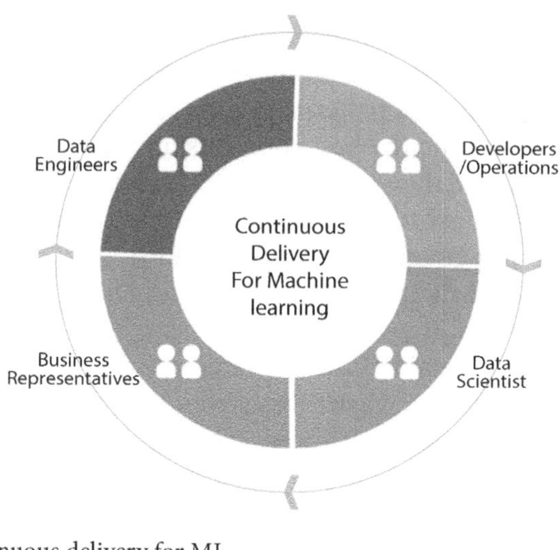

Figure 3.1 Continuous delivery for ML.

iterate on ML models rapidly and efficiently. This technique emphasizes teamwork and communication among groups to ensure a smooth and successful ML workflow.

Data engineers deliver high-quality data to data scientists, who then provide feedback on data quality and requirements. These details are then shared with developers, who offer model performance and scalability feedback. This feedback loop influences the design and development process. Business representatives provide guidance and feedback throughout the entire lifecycle, ensuring the ML models deliver actual business value.

3.1.3 Role of Artificial Intelligence

The foundation of context delivery is AI, which uses machine learning, natural language processing (NLP), and algorithms to evaluate data and forecast customer requirements. In several fields, artificial intelligence (AI) is revolutionizing the way that activities are completed, choices are made, and services are provided [8]. Fundamentally, artificial intelligence (AI) allows robots to simulate human intellect through data analysis, pattern recognition, and well-informed decision-making. Important AI techniques consist of:

- **Machine Learning (ML):** A subset of AI, machine learning is concerned with various techniques and algorithms where computers learn from previous data without the need for explicit programming. Historical datasets may be used to train machine learning algorithms to find trends, correlations, and patterns. Predictions or choices can then be made using this acquired knowledge on fresh, unseen data [9]. When it comes to storage management, machine learning algorithms may be trained on past use patterns to forecast future data requirements, determine the best compression methods for certain data kinds, and dynamically modify caching and prefetching plans in response to system conditions in real time.
- **Natural Language Processing (NLP):** NLP creates and decodes human language for individualized communication. It makes it possible for AI systems to comprehend, decipher, and react in human language. It offers automated customer support, virtual assistants, and powers chatbots. Speech recognition, chatbots, sentiment analysis, and language translation are important uses. Machines can process language structure and meaning with the aid of techniques like tokenization, syntax parsing, and semantic analysis.

- **Computer Vision:** Computer vision perceives pictures or recordings to deduce the setting. Applications incorporate facial recognition, object detection, image characterization, and autonomous vehicles. For instance, self-driving vehicles use PC vision to distinguish traffic signs, paths, and obstructions. It is likewise utilized in medical services for clinical imaging and in assembling for quality examination.
- **Reinforcement Learning:** Reinforcement learning (RL) is a sort of AI where a specialist figures out how to make choices by interacting with an environment. The specialist gets criticism as remuneration or punishment in light of their activities, planning to amplify combined compensations over the long run. Persistently further develops content conveyance procedures in light of criticism.

3.1.4 AI and Machine Learning for Storage Optimization

The limits of customary storage of board methods in asset-driven frameworks feature the requirement for additional dynamic and versatile arrangements. Artificial intelligence (AI) and machine learning (ML) offer a promising avenue for tackling these difficulties and transforming storage optimization within asset-obliged conditions [10]. Artificial intelligence envelops a wide scope of strategies, including AI, natural language processing, and PC vision. With regard to capacity, artificial intelligence can be utilized to foster smart frameworks that can gain from verifiable information access designs, framework asset limitations, and client conduct. This capacity to learn and adjust empowers artificial intelligence–powered capacity to make dynamic choices that improve capacity usage and execution within the restrictions of asset-driven frameworks.

3.1.4.1 Dynamic Decision-Making

Customary procedures frequently depend on static principles or heuristics for assignments like information pressure, reserving, and prefetching. These guidelines may not catch the subtleties of genuine information access designs, prompting less than ideal capacity use and execution. Artificial intelligence/ML, conversely, can gain from authentic information access examples and framework asset requirements. This permits them to go with dynamic choices that are custom-fitted to the particular requirements of the framework at any time. For example, an AI-powered storage management system could analyze access patterns to determine the optimal compression algorithm for different data types in real time. This powerful methodology can fundamentally further develop pressure productivity compared with static procedures that depend on a single pressure calculation for all information.

3.1.4.2 Adapting to Evolving Access Patterns

Client conduct and application jobs in asset-driven frameworks can be exceptionally powerful. New information access examples might arise over time, rendering customary prefetching methods inadequate in view of verifiable information. Artificial intelligence/ML calculations can constantly learn and adjust to advancing access designs. By dissecting late access patterns and client conduct, they can progressively change prefetching techniques to guarantee that the most important information is promptly accessible on quicker capacity levels. This flexibility guarantees effective prefetching even as access designs change over time.

3.1.4.3 Context-Aware Optimization

Context-Aware Optimization: Traditional methods often fail to consider the overall system context when making storage management decisions. For example, a reserving system could focus on getting information as often as possible, disregarding the available extra storage on the storage level. Artificial intelligence/ML algorithms can be trained to consider different context-oriented prompts, such as available storage space, handling power, and energy imperatives, while making management decisions. This setting mindful methodology permits artificial intelligence/ML to advance capacity usage while sticking to the limits of asset-obliged frameworks.

3.2 Benefits of Machine Learning for Caching

- **Dynamic Caching Decisions:** ML calculations can be prepared on verifiable access examples to anticipate future information needs [14]. This considers dynamic reserving choices that focus on information prone to be accessed soon, guaranteeing that the most important information dwells on the quicker storage level [15]. This proactive methodology essentially diminishes information access inertia and further develops recovery throughput for clients.

- **Fleeting Transformation:** ML models can be intended to catch transient patterns in information access [16]. By persistently gaining from late access patterns, they can adjust the stored information to mirror the latest access designs. This guarantees that the store stays applicable and receptive to client needs as access designs advance over the long haul.
- **Context-Aware Caching:** AI calculations can be prepared to consider different context-oriented signs, for example, available extra space, handling power, and energy limitations while making storage choices. This considers astute concessions, ensuring that the store remains viable while adhering to the framework's asset limitations.

3.3 Understanding Advanced CDN Architecture

A Content Delivery Network (CDN) is a geographically distributed server network designed to deliver web content efficiently. Its preliminary goals are the following:

- Enhance and speed up webpage loading, specifically for data-heavy applications.
- Enhance delivery reliability and security.
- Decrease costs associated with distribution.

CDNs process by hosting files around users to reduce lag time and boost speed. They distribute static pages, dynamic content, and streaming media. CDNs enhance the experience by offering faster access regardless of users' locations. Figure 3.2 illustrates the working process of CDNs.

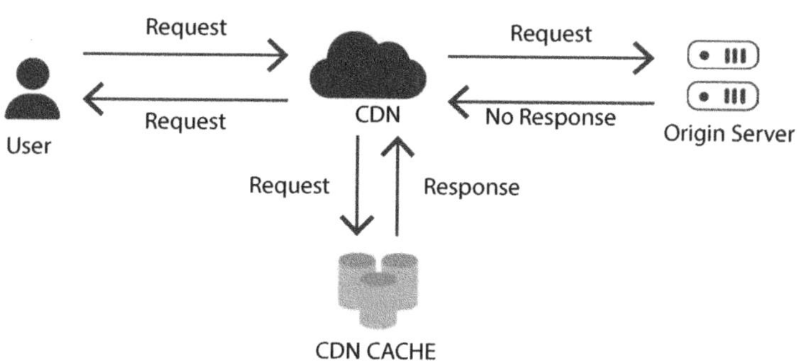

Figure 3.2 Working process of CDNs.

A user requests a resource like a picture or a webpage. Instead of moving directly to the origin server where the resource is stored, the request is intercepted by a nearby CDN server. The CDN first scans its local cache to see if it already has a duplicate of the resource. CDN delivers the resource directly to the user if it does, resulting in a much faster experience. However, if the resource is not found in the cache, the CDN requests it from the origin server. Once accepted, the CDN keeps a copy of the resource in its cache for future demands, providing faster delivery for subsequent users. Working globally, this distributed network of servers brings content closer to users, leading to reduced latency, enhanced performance, and a smoother overall user experience.

3.3.1 CDN Workflow

This workflow diagram displays how a Content Delivery Network (CDN) works to deliver content to users, as shown in Figure 3.3. A user requests content, such as a webpage or an image. A Content Delivery Network (CDN) server then intercepts this request. The CDN will notice as soon as possible if it already has a copy of the requested material in its local cache.

A quick response is achieved if the CDN delivers the thing straight to the user if it is discovered in the cache (a "cache hit"). The origin server

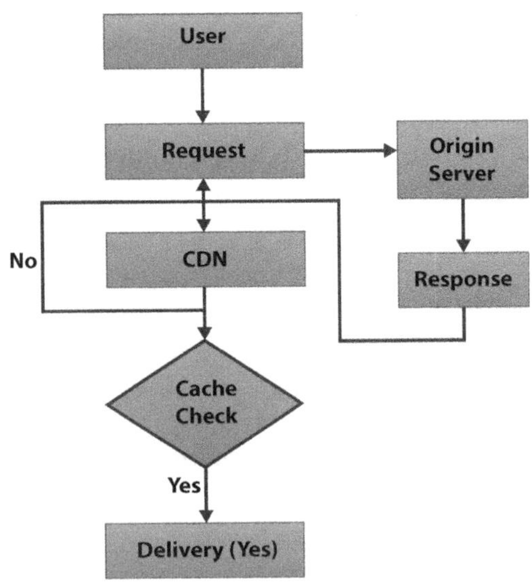

Figure 3.3 CDN workflow.

houses the original material and accepts a request from the CDN if the content cannot be found in the cache (a "cache miss"). The requested content is then returned to the CDN by the origin server. By keeping a copy of the content in its cache for later requests, the CDN improves efficiency for users who come after it.

3.4 Mechanisms Behind Context-Aware AI Systems

3.4.1 Information/Data Collection and Analysis

AI gathers information from multiple sources:

- User profiles and preferences.
- Interaction history with applications.
- Environmental data like time, location, and weather.

3.4.2 Context Recognition

AI makes use of information to determine the user's circumstances, including

- Emotional state through sentiment analysis.
- Current activity (e.g., traveling, working, or relaxing).
- Physical environment through IoT sensors.

3.4.3 Delivery and Personalization

After determining the context, AI makes the following adjustments:

- Recommendations (e.g., products, services, or learning materials).
- Dynamic content like advertisements or notifications.
- User interfaces optimized for specific devices and conditions.

3.5 Applications of AI in Context Delivery

In order to provide the appropriate information or services to the appropriate individuals at the appropriate time, artificial intelligence (AI) is essential [17]. Context delivery is the term used to describe this flexibility and responsiveness to particular circumstances. These are a few basic instances of how AI is utilized for context delivery in various sectors.

3.5.1 E-Commerce

According to customer choice, what customers want here is AI to provide better experiences. AI also makes product recommendations on the basis of user searches and previous purchase records. AI modifies discounts according to user activity, for example, by providing discounts to users who regularly peruse without making a purchase. Emails and ads are more pertinent when they are customized to each recipient's preferences.

3.5.2 Education

AI improves education in various sectors and research fields. It improves students' strengths. Here, the improvement of students is more about giving them feedback on an assignment.

3.5.3 Healthcare

By using a patient's condition and their previous medical history, AI helps to provide better healthcare solutions. AI programs also use patients' schedules to remind them to take their medications on time. AI-enabled devices, like smartwatches that function like a smartphones, include chatbots that assist patients by scheduling appointments and providing basic health advice.

3.5.4 IoT and Smart Homes

AI also improves smart home appliances to make our work easier. Depending on the context, various gadgets like Alexa or Google Assistant react to voice commands by turning off lights or playing music.

3.6 Benefits of AI in Context Delivery

AI in the context of delivery improves user-tailored services and information in various ways.

i. **Enhanced Personalization:** AI adapts goods, services, and content to each user's unique requirements. This guarantees that users get recommendations that are pertinent to them, like customized learning resources/shopping suggestions.

ii. **Improved Efficiency:** AI also improves efficiency and accurate information delivery by processing enormous volumes of data in real-time.

iii. **Better User Experience:** By using AI, users enjoy more natural interaction and increased creativity across various fields. These features enhance the overall user experience.

iv. **Scalability:** AI is used in various sectors like e-commerce, healthcare, and education, which are beneficial to all users.

v. **Real-Time Adaptation:** AI adapts services in real time to changing circumstances. For example, in smart home systems, users can remotely adjust the system according to their preferences.

3.7 Challenges in Context Delivery (CD)

During context delivery, 4 common challenges are the following:

i. **Data Security and Privacy:** Context delivery depends on user performance, location, and behavior. One of the major challenges is making sure that the data is used securely.

ii. **Real-Time Processing:** Context-aware systems need to process large volumes of data instantly, which necessitates powerful computers. Processing delays display a bad user experience, especially for time-sensitive applications such as medical alerts/navigation.

iii. **Data Quality and Availability:** Big and high-quality datasets are necessary for training machine learning models. The higher the accuracy, the better the model performs. If there is less data, then it creates overfitting problems.

iv. **Complexity of Context Understanding:** AI systems still struggle to capture and interpret complex user contexts, such as emotion detection, facial expressions, etc.

3.8 Summary and Discussion

Table 3.1 represents the comprehensive overview of AI in context delivery and indicates how AI improves system performance by utilizing context

Table 3.1 Comprehensive overview of AI in context delivery.

Category	AI application	Key benefits	Technologies used	Industries benefiting	Future potential
Personalization	Recommender systems tailoring content based on user behavior and preferences	Enhanced user experience, increased engagement	Machine learning, collaborative filtering	E-commerce, entertainment, education	Hyper-personalized experiences driven by deeper data insights
Real-time analysis	AI algorithms analyzing context (e.g., location, time, or activity) to adapt responses dynamically	Faster decision-making, relevant responses	Big data analytics, geospatial AI	Transportation, logistics, smart cities	Fully automated real-time systems for urban infrastructure
Context-aware systems	NLP for understanding user intent based on	Improved communication, seamless interactions	NLP models (GPT, BERT), sentiment analysis	Customer service, healthcare, retail	Universal, multilingual conversational agents

(Continued)

Table 3.1 Comprehensive overview of AI in context delivery. *(Continued)*

Category	AI application	Key benefits	Technologies used	Industries benefiting	Future potential
Behavior adaptation	ML models adjusting system behavior to match environmental changes	Flexibility, adaptability in operations	Reinforcement learning, IoT integration	Energy, smart homes, education	Self-optimizing systems adapting to dynamic environments
Content delivery	Automated delivery of relevant content based on real-time contextual cues	Efficient resource usage, user satisfaction	Recommendation engines, deep learning	Media, advertising, social media	Immersive and adaptive content ecosystems
Decision support	AI assisting decision-making by providing context-aware suggestions	Enhanced decision quality, reduced error rates	Decision trees, expert systems	Healthcare, finance, agriculture	AI as a co-pilot for critical decision-making

to increase functionality across industries. Important uses include context-aware systems that are driven by natural language processing (NLP) and improve communication in customer service and also in healthcare sectors; in real-time analysis, it enables the fastest decisions.

Meanwhile, behavioral adaptation enables systems like smart homes to adapt dynamically to environmental changes. In various industries like healthcare and agriculture, decision support techniques are used to improve decision quality by offering context-aware assistance. The foundation of these applications is made by technologies like machine learning, big data analytics, and IoT, which are useful in everything. AI is commonly used in automated city infrastructure and self-optimizing systems that can easily adjust to changing conditions and is also used in our daily lifestyles.

3.9 Conclusion

In this paper, we explore the concept of context delivery by using AI/ML. AI is converting how we deliver context, allowing more personalized, predictive, and contextually aware experiences. As AI continues to develop, we can expect even better innovative and impactful applications of context delivery across different industries. Context grounding is necessary for opening the full potential of generative AI. Leveraging contextual knowledge can lead to more precise, reliable, and transparent AI models. This improvement is essential for applications ranging from content generation to virtual assistants. Continuous delivery for ML is not only about technology; it's about people, processes, and teamwork. This chapter also includes explanations of the mechanisms behind context-aware AI systems, their challenges, and benefits, describing how ML/AI techniques streamline processes and save time. It is also used in the fields of e-commerce; the destiny of business-to-business connections is being shaped by artificial intelligence in setting conveyance. It has become essential to modern innovation due to its ability to gradually adapt services and content to specific needs. However, addressing issues such as security, complexity, and predisposition is essential to unlocking its full potential. As artificial intelligence advances, it will rethink the way that data and services are tailored to each individual, fostering progress and customer-focused experiences. The table also highlights how AI can improve decision-making, efficiency, and personalization. With the help of technologies like machine learning, natural language processing, the Internet of Things, and predictive analytics, its real-time data analysis, trend prediction, and context-aware adaptations help industries like healthcare, finance, logistics, and smart cities.

These developments address important issues like data privacy, algorithmic bias, and system complexity while promoting innovation and customer-centric experiences.

AI disclosure statement: The author acknowledges the use of AI-based tools for minor language editing or grammatical correction. However, all intellectual contributions, analyses, and conclusions presented in this chapter are entirely the author's own.

Bibliography

1. Giordani, M., Polese, M., Mezzavilla, M., Rangan, S., Zorzi, M., Toward 6G networks: use cases and technologies. *IEEE Commun. Mag.*, 58, 3, 55–61, 2020.
2. Letaief, K.B., Chen, W., Shi, Y., Zhang, J., Zhang, Y.-J.A., The roadmap to 6G: Ai empowered wireless networks. *IEEE Commun. Mag.*, 57, 8, 84–90, 2019.
3. Li, Y., Liu, J., Cao, B., Wang, C., Joint optimization of radio and virtual machine resources with uncertain user demands in mobile cloud computing. *IEEE Trans. Multimedia*, 20, 9, 2427-2438, 2018.
4. Peng, M., Li, Y., Jiang, J., Li, J., Wang, C., Heterogeneous cloud radio access networks: a new perspective for enhancing spectral and energy efficiencies. *IEEE Wireless Commun.*, 21, 6, 126–135, 2014.
5. Chen, M., Saad, W., Yin, C., Debbah, M., Echo state networks for proactive caching in cloud-based radio access networks with mobile users. *IEEE Trans. Wireless Commun.*, 16, 6, 3520–3535, 2017.
6. Zhao, Z., Peng, M., Ding, Z., Wang, W., Poor, H.V., Cluster content caching: an energyefficient approach to improve quality of service in cloud radio access networks. *IEEE J. Sel. Areas Commun.*, 34, 5, 1207–1221, 2016.
7. Peng, M., Yan, S., Zhang, K., Wang, C., Fog-computing-based radio access networks: issues and challenges. *IEEE Network*, 30, 4, 46–53, 2016.
8. Fu, Y., Wen, W., Zhao, Z., Quek, T.Q.S., Jin, S., Zheng, F.-C., Dynamic power control for NOMA transmissions in wireless caching networks. *IEEE Wireless Commun. Lett.*, 8, 5, 1485–1488, 2019.
9. Zhao, Z., Xu, M., Li, Y., Peng, M., A non-orthogonal multiple access-based multicast scheme in wireless content caching networks. *IEEE J. Sel. Areas Commun.*, 35, 12, 2723–2735, 2017.
10. Chatzieleftheriou, L.E., Karaliopoulos, M., Koutsopoulos, I., Jointly optimizing content caching and recommendations in small cell networks. *IEEE Trans. Mob. Comput.*, 18, 1, 125–138, 2019.
11. Liu, Q., Luo, J., Jiang, H., Xu, J., Wang, C., Chen, Y., ML-prefetch: Machine Learning Enhanced Data Prefetching for Resource-Constrained Systems.

Proceedings of the 2020 ACM SIGMOD International Conference on Management of Data, pp. 1755–1768, 2020.
12. Shye, P., Jain, S.C., Abouzeid, A.A., Cognitive radio: A statistical machine learning approach. *IEEE Trans. Wireless Commun.*, 6, 4, 1148–1159, 2007.
13. Tang, X., Zeng, S., Li, K., Data prefetching based on access pattern similarity. *J. Supercomput.*, 73, 12, 5343–5358, 2017.
14. Rissanen, J., Compression and Information Retrieval, Information Theory, Coding and Statistics, pp. 857–916, 2009.
15. Fan, Q., Xu, X., Li, R., Sun, Y., Liu, H., A survey of machine learning for network caching. *ACM Comput. Surv. (CSUR)*, 51, 5, 1–36, 2018.
16. Mao, Y., Youn, C., Zhang, J., Srinivasan, K., Netravali, R., Mao, Z.M., Rome: Replica, migration, or eviction: A machine learning approach for web content caching. *ACM SIGCOMM Comput. Commun. Rev.*, 42, 4, 70–85, 2012.
17. Youn, K., Mao, Y., Mao, Z.M., Rexford, J., Naik, V., A cost-aware approach for web content caching using machine learning. *IEEE Trans. Parallel Distrib. Syst.*, 24, 12, 2532–2541, 2013.

4

Monetizing Cyberliterature through AI-Enhanced Networks in English Education System

Muthmainnah Muthmainnah[1*], Luis Cardoso[2], Ahmad Al Yakin[1], Khristianto Khristianto[3], Widya Nirmalawati[3] and Titik Wahyuningsih[3]

[1]Teacher Training and Education Faculty, Polewali Mandar, Sulawesi, Barat, Indonesia
[2]Department of Language and Communication Sciences, Portalegre, Portugal
[3]English Dept., Faculty of Cultural Science and Communication, Universitas Muhammadiyah Purwokerto, Central Java, Indonesia

Abstract

The purpose of this study is to investigate the use of AI for the production and monetization of cyberliterature from the perspective of educational transformation 7.0 using trifecta models. Specifically in literature lessons, this study seeks to enhance student engagement, creativity, and monetization opportunities using AI. The researchers used a survey design to conduct a quantitative study, sending questionnaires to 125 university students in Indonesia. The selected sample included representatives from Universitas Muhammadiyah Purwokerto, Universitas Al Asyariah Mandar, and the State Islamic Institute of Madura. The study analyzed responses about the use of AI tools for writing, enhancing creativity, and monetizing cyberliterature using descriptive statistics. The study found that students generally understand AI tools and their potential applications in creative writing. There are still a few students who monetize their writing using sites like Webtoon or Instagram, although many have used AI to write and are aware that their cyberliterature work can be monetized. This study yields mixed results on AI's ability to foster creativity. This study shows how AI can help people from all backgrounds understand and enjoy literature in our increasingly interconnected society. The long-term impact of AI-enhanced pedagogical resources on student interest and originality should be the subject of future study.

**Corresponding author*: muthmainnahunasman@gmail.com

Tanusree Chatterjee, Rajib Banerjee and Souvik Ganguli (eds.) Content Delivery Networks Using Artificial Intelligence and Internet of Things: Revolutionizing Content Delivery, (69–100) © 2026 Scrivener Publishing LLC

Keywords: Cyberliterature, AI network, monetization, education 7.0, AI literature

4.1 Introduction

The development of teaching materials, especially in the field of literature, has undergone a rapid transformation due to the presence and integration of artificial intelligence (AI) into various fields. The field of educational literature is developing rapidly within the framework of Education 7.0, as digital resources and artificial intelligence are revolutionizing the way students learn. Cyberliterature, which is mostly found in digital and multimedia formats, greatly benefits from AI's capacity to create, evaluate, and organize content practically [53]. However, there are clear obstacles in Indonesia; in general, the development of teaching materials is still traditional, only based on existing textbooks, which causes students to generally not want to write for fun. The reality is that many students in today's fast-paced digital environment fail to recognize the value of classic literary genres, even though they are steeped in Indonesia's rich cultural heritage and storytelling traditions. Literature education seems outdated and disconnected from contemporary technological realities, which is exacerbated by the limited use of digital platforms in literature classes and the outdated teaching strategies of lecturers.

With the integration of AI for curriculum development, especially in the framework of Education 7.0, the use of AI in cyberliterature lesson plans presents a great opportunity to revolutionize literature teaching. Several studies have explored and demonstrated that students can enhance their creative abilities in literary expression by leveraging AI-powered tools to create digital-based learning materials. These materials can go beyond simple text and become interactive, active engagement and immersive experiences [16, 53] for students. By incorporating visuals, audio, and video into their writing, students can make digital literary works more engaging and accessible to a wider audience. AI robot technology can generate inspiring story concepts, characters, and plots and enhance the creativity of new types of digital storytelling.

Yakin *et al.* [4] and Kong and Yang [39] demonstrate the feasibility of personalized learning and feedback. Evaluating student writing and offering suggestions for improvement using artificial intelligence systems, such as ChatGPT-based robotics, can personalize the learning process [37]. Using AI-based analytics, students can receive feedback on the style, structure, and coherence of their stories, which can help them create more engaging and enjoyable cyberliterature. Students benefit from this type of personalized literary education because it encourages them to learn at their own pace. Based on the findings of Muthmainnah *et al.*'s [52] study on literary heritage and social media, they concluded that educational materials can undoubtedly generate revenue through these platforms. Students can optimize their cyberliterature for different platforms and audiences with the help of AI, increasing their visibility and reader engagement.

There are several digital channels where students can publish their work, such as Instagram, Pinterest, and Webtoon. In addition to helping students improve their writing skills, this strategy also paves the way for them to earn money through digital sales, sponsorships, and advertising. To succeed in the global digital economy, students need a curriculum that focuses on cyberliterature and artificial intelligence (AI). With the creative industry and content creation continuing to shift to digital platforms, students who learn to create materials that work in today's connected world have an advantage in the digital literacy and media production industry [1]. With the use of AI-enhanced lesson plans, students from diverse backgrounds can actively participate in the creative process of literature.

Cyberliterature has the ability to transcend cultural and linguistic boundaries by leveraging AI's storytelling and multimedia translation skills, thereby expanding the reach of literary works. By opening up the online literary community to students from all walks of life and with all levels of writing proficiency, this approach encourages diversity in literature. Students can gain a better understanding of the relationship between traditional literary forms and modern digital narratives through cyber literature and AI [21]. Classic Indonesian literary works can be a source of fresh narrative ideas and techniques. This blend of the old and the new not only keeps cultural heritage alive but also revitalizes literary education and adapts it to the digital world. The cyber literature lesson plans in Instruction 7.0, powered by AI, are a major step forward for literature teaching in Indonesia. By leveraging AI, we can foster creativity, spark literary inspiration, and provide students with the tools they need to disseminate their work globally, both for personal fulfillment and to contribute to the ever-growing digital economy. This approach is in line with the principles of Education 7.0, which states that at the intersection of technology, creativity, and individual learning, children will have unlimited opportunities to succeed in an ever-changing and unpredictable world.

This study suggests creating learning materials using AI assistance and adhering to the principles of cyber literature to overcome this difficulty. A literature curriculum that uses AI-powered tools and platforms can help students discover new forms of creative expression. With the help of this

AI-powered lesson plan, students can create their own animated fairy tales, complete with 3D comics, which they can then publish on platforms such as Instagram, Webtoon, and YouTube, and possibly make money. We aim to encourage our students to see literature from a new perspective in this dynamic field, where digital storytelling and multimedia narratives provide endless options. In addition, the development and design of these teaching materials can help students enhance their creativity and innovation without limits.

4.2 Literature Review

4.2.1 From Text to Hyperliterature through the AIIoT Network

The impact of GenAI has positively correlated with a more technology-oriented culture and society, increased diversity of collectivities, and substantial advances in information and communication technologies [2]. Technological advances are transforming people's daily lives around the world. They will become modern cultures that rely on online communication and constant information gathering. The impact extends to workplaces, households, and educational institutions. Internet sources are a significant part of their data acquisition efforts [33]. Users can read, write, and access information using internet-connected devices in these locations. Barroso [10] claims that the World Wide Web is the most popular and widespread use of hypermedia in terms of inter-unit relationships and relationship-based information organization.

The internet and related technologies produce cyberliterature. This new literary form emerged from the rise of electronic publishing [72]. Cybertext [6], literary machine [7], electronic literature [50], digital literary [11], and digital literature [51] are some of the terms used by researchers interested in studying cyberliterature. Furthermore, the discussion of cyberliterature can delve deeper into the topics of aesthetics and terminology. Cyberliterature paved the way for new forms of literature, such as fan fiction [24].

Babaryka [8] argues that in the realm of fan fiction, computers can function as a medium for distribution and a means of creation. Fan pen works that serve as "pseudo-sequels" to established fictional works like novels, comic strips, TV series, and films. The use of electronic space as a medium facilitates the rise of cyberliterature, enabling people to read, write, and access literary works. According to Geçgel [29], many cyber literary communities promote and publish their literary works through homepages, emails, forums, and blogs as part of the active production of cyberliterature. They are free to write whatever they want and publish their works online.

This demonstrates that the internet archives the author's work, making it accessible to everyone worldwide. The term "cyber and literature" can refer to any and all literary works published online, be it prose or poetry, anthologies of digitally preserved prose or poetry, online literary magazines, or collections of classic texts; second, it can refer to literary works published online but not by professionals; and third, it can refer to hypertext and cybertext literature, which deal with literary texts with more complex structures [67]. The proliferation of online literature is broadening our understanding of cyber literature and fostering the development of new literary concepts like cybertext and hypertext. If cyberliterature becomes more widely accepted, there may be encouraging signs for the field of literature. The increase in technological capabilities has opened many doors for writers. They are free to distribute their writings as they wish. Sharing authors' knowledge raises awareness about cultural diversity around the world. Cyberliterature has advanced significantly, and modern teaching materials tailored to students' learning styles have emerged.

Integrating cyberliterature into Indonesia's current curriculum has the potential to have pedagogical and educational implications. Therefore, transforming literature into hypertext may present numerous challenges [13]. Writing literary works using AI-based tools [34], which involves students taking on several roles in an online conversation about a literary work, is one such strategy. This method involves students writing literary works and assigning specific research tasks related to their lives. We expect them to pretend to be the assigned personalities in online conversations. Using the AI Story application, which has a lot of room to grow as a teaching tool, is another option [65]. Students can also use Pinterest to create new design units for their literary works. Building a unit to teach literature through hypertext requires collaboration with literary hypertext creators. For students to express their own ideas, criticisms, and comments about cyberliterature, AI as a medium for literary hypertext requires a wide variety of formats. Students can provide feedback on literary works that are hypertext representations of human and artificial intelligence collaboration [56].

In essence, offering a commenting option can pave the way for a new approach to appreciating literary works, namely, reader-writer engagement in cyberspace. Focusing on channeling attention that is accessible to other readers and possibly requiring digitally traceable comments are two possible outcomes. Such engagement may encourage readers to reflect and immerse themselves. Sharing and commenting on texts can also spark new ideas. Additionally, the development of online technologies exposes readers to new text-reading methods that shift from traditional linearity to a navigational approach [43]. With the use of navigation aids on the internet, linear reading has given way to hypertext, which is characterized by electronic text links. When reading hypertext literature, one thing readers do to receive more information is to click on the "show me" button. By utilizing hypertext as a medium, cyberliterature promotes literature as a means of expression. Educators face new challenges and opportunities when teaching literature in the digital age [59]. Some innovative approaches include social media networks that encourage reader-writer interaction and role-based online discussions about cyberliterature.

4.2.2 The Multi-Model of Gen AI Tool in Education 7.0

The pedagogical connections between education and digital media technology components of the twenty-first century provide credence to the idea that AI could improve certain aspects of education. However, finding a complete solution, especially in areas where AI robots necessitate digital flexibility and knowledge transfer, remains a challenge [62]. As far as artificial intelligence technologies are concerned, humans provide the ideal example to follow [28]. This theory also requires replicating one of the most advanced technological advances enabled by human language. Obaid *et al.* [54] assert that ChatGPT, utilizing generative AI techniques that integrate machine learning and deep learning, perceives language as a malleable technology capable of imitation. While ChatGPT and similar generative AI systems primarily aim to comprehend and generate natural-sounding human speech, they are also capable of deciphering complex linguistic patterns and structures through the application of the Generative Pre-Training Transformer (GPT) model [32]. Even though the computer industry has studied natural language processing (NLP) for decades, recent technological advancements have led to the invention of sophisticated AI models like chat generative pre-trained transformer (ChatGPT). These models have the potential to greatly improve academic efficiency by producing responses to various linguistic obstacles that resemble human responses Al Yakin *et al.*, [5].

Simpungwe [55] asserts that educational institutions follow a continuous cycle when introducing new technologies. This cycle starts with the introduction of the technology, followed by arguments about how it will improve teaching and learning practices. Next, we justify the technology's acquisition and either integrate it into our existing practices or reject it.

Despite schools' best intentions, Tomei *et al.* [58] note that "schools often use these technologies in old, familiar ways: integrating them into existing routines, deploying them to meet existing goals, and, in general, failing to engage with the technology in ways that are consistent with the world outside the classroom." This continues to be the case even as schools embrace new technologies in the hope that they will effectively improve education. This pattern may have been true in the past when it came to how schools handled technological breakthroughs such as computers, digital learning materials, and iPads, but it is not the case with the rise and widespread adoption of GenAI. Because it is freely available to anyone with an internet connection, GenAI services like ChatGPT do not undergo a process of legitimization or acquisition by schools. According to this cycle, educational curricula can either jump straight to the final stage—integrating the technology into current procedures—or reject it.

While schools and instructors are deciding whether to embrace or reject GenAI in the classroom, there has been extensive discussion on the topic among educators, students, and policymakers. Anecdotal evidence has largely replaced the lack of studies and information about GenAI in education that has been typical of this discussion. Law [41] delves into a topic he calls "large generative language models" (e.g., ChatGPT and GPT-4), highlighting the importance of research-based debate and thorough testing. "Anecdotes dominate without being based on knowledge, and there is no distinction between the capabilities of ChatGPT and GPT-4; this is evident in everyday discourse" [17]. In fact, the editorial bolsters its own argument by employing inaccurate terminology related to the subject. The author uses the phrase "large generative language models" to describe ChatGPT and GPT-4, but this is not a real term. ChatGPT is a conversational bot that builds on GPT-3.5, the Large Language Model, instead of being a simple LLM.

Additionally, there is no mention of using a web browser or an app to access ChatGPT. The issue includes academics advocating for increased education about GenAI in education, as well as professionals in the field of artificial intelligence expressing some encouraging sentiments regarding the use of AI in educational settings. Beyond cheating, artificial intelligence can help students succeed. Yusuf *et al.* [68] claim that students can improve their writing skills, which can lead them to create longer texts that were previously too challenging, and that students who are bored can rediscover their passion.

There are already many AI tools available, with even more on the way, including Google Gemini. Currently, ChatGPT is the most popular product available. Since its introduction, many have discussed and used OpenAI's

GenAI, a variation of GPT-4. Imran and Almusharraf [31] observe its widespread use in educational contexts, particularly the International Baccalaureate Curriculum. Chatbots and AI have recently gained traction as potential tools for use in academic and professional settings [57]. Generative AI based on the GPT-4 architecture has enormous potential to improve educational environments. Baek and Hwang [9] have utilized generative AI to produce diverse literary works, elucidate intricate subjects, and produce model responses.

According to Ait Baha *et al.* [3], chatbots can enhance student-teacher interactions and learning experiences on online platforms. However, relying too much on chatbots can make people less able to think critically and solve problems on their own [35]. Stated that generative AI can help non-native English speakers with language editing, translation, and overcoming language barriers. As previously stated [15, 27, 64], this AI offers accurate responses, enhances peer collaboration, and provides quick summaries of information. The interactive aspect of generative AI helps in determining areas where students lack knowledge and provides quick feedback [41]. We must be careful to maintain critical thinking skills and independent problem-solving abilities, even though generative AI has the potential to completely transform education and research.

Despite the development of multiple models of GenAI, there is growing concern that the increased use of technology in education may have a detrimental impact on students' capacity for critical thinking and independent problem-solving. A lack of critical thinking skills, engagement, and initiative may occur due to over-reliance on technology [36]. Efficient, generative AI does not understand context or be creative; it relies on commands. Artificial intelligence systems present difficulties in detecting plagiarism and ensuring fair assessment; constant attention is required to ensure accuracy and fairness. To promote holistic skills, educators must find a middle ground between technology and conventional teaching approaches. To maintain academic integrity, it is important to prevent AI-generated texts from slipping through the cracks [61]. Concerns about equity have been raised regarding the affordability and access to generative AI tools. Before implementing the premium version, several considerations should be considered.

4.2.3 Robotics Tutor in Learning and the Trifecta Teaching Model Process

Since robots can significantly impact human culture and digital technological advancements, it is important to study human-robot interactions and their social implications [4]. Academics and futurists have begun to talk about Society 5.0, an endeavor whose main goal is digital cooperation between humans and robots. Machado [46] describes Society 5.0 as a society that systematically and scientifically merges the physical and digital realms, prioritizes the demands of digital natives, and establishes a link between social and economic progress. Digital technologies, which bridge robotics, science, evolution, AI, big data analytics, machine learning, the Internet of Things (IoT), and recent advances in generative artificial intelligence (GenAI) evolutionary commands, drive Society 5.0, also known as the super-intelligent society, which aims for inclusive and sustainable socio-economic development [12]. Society 5.0 is based on cyberphysical systems, a widely accepted paradigm in technology. It reinforces the integration of cyberspace with respect to ICT and physical space with respect to computational processes [45].

The Trifecta teaching model guarantees that students enhance their understanding, critical thinking, and engagement *via* the study of English literature. Students develop cross-linguistic critical literacy abilities, a greater appreciation for literature, and stronger language links through the integration of English, Indonesian, and local languages. A sustainable, egalitarian, and effective technique for teaching English literature that respects linguistic variety and encourages deeper learning is the Trifecta Teaching Model. This model is particularly relevant, as Indonesia continues to modernize its education system and embrace worldwide multilingual trends in learning and AI intervention. Many initially thought this idea was idealistic, as it didn't explain how it would work in practice. It is believed that the Fourth Industrial Revolution and the transition to Society 5.0 are similar in that both describe a fundamental shift in our current economic system to the development of a new paradigm of extreme digital dominance. Society 5.0, on the other hand, is a more comprehensive idea because it seeks to revolutionize the way we live and drive an economic evolution that seeks to separate human power from machine power through cyberliterature.

However, human-machine collaboration in the digital realm will change the face of society and have a significant impact on the management of educational institutions [47]. The widespread use of artificial intelligence (AI) in robotics is driving technological and socio-economic changes in modern society. As a result, AI-powered robots are constantly evolving to acquire new and improved capabilities, such as increased intelligence, acuity, and efficiency. There is still a long way to go before scientists can create relationships with humans that go beyond isolated exchanges in predetermined settings, but the lines between AI agents and humans are blurring [40]. Measurable returns on investment will allow these machines, which we might call robots, to replace humans in much of the manual, repetitive, or risky work.

Learning-assisting robots that encourage interactive pedagogy are now available to college and university students, who can program them to engage in active learning [20]. These robots make students feel good about themselves by making the classroom a peaceful place where other people are not real and by using teaching strategies that emphasize movement as an additional educational tool. Learning apps, artificial intelligence tutors, and virtual classroom engagement are just a few examples of information and communication technology systems that incorporate robots to fill the shoes of qualified instructors in fields such as programming and language acquisition [63]. Acceptance of robotics and technological advancements demonstrating the benefits of robotics and AI partnerships through real-world use cases will be a game-changing component. It should also utilize common interests to form alliances that propel humanity towards a sustainable and human-focused society 5.0. Even in the past few decades, robots have played a significant role in solving some of the world's most pressing problems. The widespread availability of generative artificial intelligence (GenAI) tools has had a far-reaching impact on the field of education. However, discussions have mainly taken place in academic settings.

According to Vonitsanos *et al.* [62], understanding innovative technologies in the context of education requires an educator's perspective. Created by OpenAI, ChatGPT is a large-scale language model with the goal of mimicking human writing styles in response to specific input. The model has a wide range of potential uses in natural language processing, such as text completion, conversation generation, and language translation, which can be used to design engaging teaching content. Theoretically, GPT models can read and produce multilingual prose that is remarkably like human writing by mining massive amounts of publicly available digital content data derived from natural language processing [71]. They can also creatively write about a variety of topics, from paragraphs to entire research articles, and they can do it convincingly or nearly so. Chatbots for customer service or fictional characters in video games, these models can even have human-like conversations with consumers. They have just developed an upgraded generative pre-trained transformer (GPT)-3. Improving task-agnostic performance and even matching earlier state-of-the-art fine-tuning approaches are the goals of GPT-3, which has 175 billion parameters. In comparison to previous non-sparse language models, GPT-3 is 10 times more effective.

Education is among the many fields that have expressed interest in the newly developed language model ChatGPT. GPT-3 serves as the model's core natural language processing engine. One example of a technology development that could radically alter the education sector is the use of natural language processing (NLP) and general artificial intelligence (GenAI) for interactive educational pedagogy [49] and its profound effects on both student learning and teacher practice. The literature review on OpenAI in education centers on the ethical and appropriate usage of ChatGPT to promote lifelong learning; this model has evolved into a cutting-edge language tool. Many people in the natural language processing and artificial intelligence communities have noticed ChatGPT.

On the other hand, ethical and responsible concerns about its widespread use in a variety of settings, including schools, are emerging. Elbanna and Armstrong [25] assert that ChatGPT can drastically alter the way students learn and find information. Textbooks, study guides, and course materials are just a few examples of the many educational resources that can greatly benefit from GPT, a massive language model trained on a large amount of data. Students can use GPT systems to generate informative, coherent, and natural-sounding prose. Additionally, GPT can summarize and analyze long texts, saving time and improving comprehension for both students and instructors. If implemented, GPT is believed to bring about a major shift in the way teachers train the next generation of citizens. GPT's incredible capacity to enable natural language processing applications and the creation of intelligent tutoring and assistance systems could impact education today and in the future. Students can benefit from engaging, interest-based learning opportunities because the model can simulate human emotions. Because of ChatGPT's adaptability and capacity to handle multiple languages and subjects, it could make learning easier for students around the world.

The routine use of ChatGPT in educational settings raises concerns about the ethical behavior and conscience of users. Divito *et al.* [23] found that ChatGPT's sophisticated AI skills make it capable of perpetuating bias and sexism, which could lead to an unequal learning experience for students. Furthermore, students may develop weaker critical thinking skills as a result of the dominant use of technology in the classroom. This is because individuals may begin to rely too much on AI responses instead of developing their own concepts. ChatGPT is not only the most advanced chatbot in the world today, but it has also created a stir in terms of college entrance exams and similar things [60]. Unlike other chatbots, this GenAI chatbot can quickly produce high-quality writing. According to Zeb *et al.* [69], ChatGPT is a state-of-the-art language model that is able to imitate casual conversations with humans and imitate speaking models. Its ability to produce text that closely resembles human-written content is based on OpenAI's GPT language model. As stated by Ljubojević *et al.* [44], ChatGPT's essay-writing capabilities open the door to innovative and creative approaches in educational settings. Numerous industry experts have argued that using technology to enhance learning is a wise decision, given the impending significance of AI technologies in education. One approach to enhancing assessment processes is to equip educators with the necessary tools to use testing as both a tool for learning and a means of learning itself.

Robotics and natural language processing algorithms are transforming education by enabling personalized training and strategy revision for each student [38]. As a result of their assistance, student participation in class increases, which enhances their capacity for teamwork, idea expression, and analysis. Robots can now read and assess information, respond instantly in any scenario, and automatically grade assignments using natural language processing (NLP) techniques. Natural language processing (NLP) facilitates better communication. With the purchase of user-centric ChatGPT-4, educational pedagogy has embraced generative artificial intelligence because it has the potential to help create learning environments that are open to everyone, flexible, and dynamic. By facilitating collaboration, co-inspiration, and the creation of new course concepts, NLP and robotics are revolutionizing the educational landscape. The popular AI chatbot ChatGPT, which builds AI for education, determines how well it can construct assessment items that adhere to nationally recognized educational standards, all in the name of improving STEM (science, technology, engineering, and mathematics) education [48]. Additionally, he researches the chatbot's ability to customize learning recommendations, evaluate student-submitted science essays, and offer personalized feedback.

4.3 Methodology

This quantitative study aims to investigate the perceptions of students at a specific university level regarding the creation of literature teaching materials using AI devices in the classroom and to gather their opinions and feedback on this topic. The researchers used a quantitative research approach to explore the topic of generative AI (GenAI) integration in the cyberliterature genre of Education 7.0.

The researchers employed an instrumental survey study research design, which involved the use of an online questionnaire and thematic analysis. In this instrumental case study, researchers focus on a specific issue by examining a single boundary case [19]. Harrison *et al.*'s [30] case study research delves into a specific example or case, thoroughly examining all relevant information within each case and its surrounding environment. The design of literature teaching materials with the use of GenAI in the classroom is a case study involving students' work exhibited on various social media platforms that become commercial products that will be discussed in this study. Participants in this survey will include undergraduate students.

4.3.1 Participants

The researchers used their personal networks to find potential participants. Consequently, the researchers selected 38 individuals who represented a diverse range of cultural backgrounds, educational levels, and work experiences. The sample of participants was 125 students; the background information of the participants was 50 students from Universitas Muhammadiyah Purweokerto, 50 students from Universitas Islam Negeri Madura, and 25 students from Universitas Al Asyariah Mandar, Indonesia. Probability sampling and non-probability sampling are two main categories that form the basis of the participant selection procedure for a study. The researcher has opted to exclude specific segments of the larger population from the study through non-probability sampling, thereby denying these segments the same opportunity to participate as those already included in the sample. By doing so, the sample size remains manageable, as all students taking English as a subject will be part of the larger population relevant to the research topic.

According to Cohen [18], non-probability samples are much easier to prepare and can work well in cases where the researcher is not trying to extrapolate their results. This study employed a combination of convenience, purposeful, and sequential mixed-method sampling, which are some of the various techniques for conducting non-probability samples. In order to obtain the required number of respondents, convenience sampling uses respondents who are easily accessible and available at the time of the survey. Convenience sampling characterizes our initial selection, with the participating institutions chosen in part due to their affiliation with English language and literature departments.

Therefore, the sample is not representative of other groups, making it impossible to extrapolate the results to the entire population. When using purposeful sampling techniques, researchers make their best judgment about which cases to include in the sample. The researcher tailors the sample to meet their specific goal. Each grade's English teachers helped us ensure that the surveyed groups were representative of the students' English achievement levels. In sequential mixed methods sampling, it is common practice for the first strand of findings to guide and influence the second strand approach, including sample selection and equipment. This will reveal potential areas for further investigation in the survey.

4.3.2 Instruments

Researchers often use questionnaires as a means of data collection because they allow for a high response rate and the opportunity to clarify responses while collecting data. Researchers who utilize this questionnaire incorporate indicators into their questionnaire design, allowing informants to freely express their ideas, thereby assisting investigators in maintaining the focus on the topic under discussion. After conducting a comprehensive literature review, researchers create survey questionnaire indicators. A small group of research colleagues serves as pilot testers for the questionnaire guide.

Pilot testing's goals are to find duplicates, ensure appropriate research instruments, and reduce the risk of research failure. Topics discussed in this survey guide include respondents' overall views on the development of AI-based literature teaching materials, their suggestions for its educational applications, the advantages and disadvantages of AI, and their thoughts on the topic. The purpose of the survey questionnaire is to collect information about students' opinions on the development of cyberliterature teaching materials in educational settings, including advantages, disadvantages, ethical issues, and recommended procedures, and to guarantee anonymity, confidentiality, and informed consent to all participants in accordance with ethical principles.

4.3.3 Data Analysis

This study utilized data gathered from the survey component of the research project to investigate student opinions regarding the design of literature lessons using AI. This study employed univariate analysis to gather responses, as our goal was to understand how students were utilizing GenAI in their English classes. We reported the number of respondents for each potential response to the relevant question and then presented the data using descriptive statistics. Rather than drawing conclusions or making predictions about population parameters, descriptive statistics simply detail the results obtained through various methods to display discrete data, which may be easier to understand for some readers. IBM SPSS analysis version 26 will systematically analyze the questionnaire data. This will include statistical data transcription, mean score, and standard deviation, development, and interpretation.

4.4 Findings and Discussions

Based on the results of statistical data, it is known that cyber literature and artificial intelligence (AI), as shown in Table 4.1, provide some statistical insights. Focusing on how AI can affect the creation, instruction, and monetization of literature, the data examines the familiarity, use, and impressions of AI tools among 125 respondents. Table 4.1 shows the average responses from participants, represented by the average. The higher average indicates a higher level of familiarity with the statement, while the lower average indicates a less familiar or inconsistent statement. The standard deviation measures the dispersion or variability of answers. A lower standard deviation indicates more agreement among participants, while a higher one indicates more variation in responses.

Table 4.1 describes cyberliteracy knowledge (Statement 1). The mean score is 3.31, with a standard deviation of 1.16. The high standard deviation among individuals indicates varying degrees of familiarity with the idea of cyberliterature, while a moderate mean indicates that individuals are generally familiar with the term of teaching materials with AI literature. The respondent statement "using AI tools to write" had a mean score of 3.66 and a standard deviation of 0.99. Although the answers may vary slightly, many participants have utilized AI technology to write.

The increasing mean suggests that cyberliterature writers are increasingly utilizing AI tools, as evidenced by the relatively frequent use of AI in literary works and their agreement to use them as teaching materials. On the function of AI in fostering originality (Statement 3), the mean was 3.32 and the standard deviation was 0.59. The slightly lower mean compared to other answers indicates that some participants are not so sure about the ability of AI to enhance creativity in cyber literature, but people believe that AI can do so.

Given the small standard deviation, it appears that most people feel the same way about this. The average familiarity with AI-enhanced platforms for publishing in statement 4 was 4.44, with a standard deviation of 0.60. Indicating participants' familiarity with cyberliterature sites such as Webtoon and Wattpad, the high mean indicates a high level of awareness

Table 4.1 Descriptive statistics.

Statement	N	Min	Max	Mean	Std. deviation
1. How familiar are you with the concept of developing teaching materials through cyberliterature?	125	1.00	5.00	3.3120	1.15992
2. Have you ever used AI tools (e.g., text generators, content creation software) to assist in writing literary works?	125	1.00	5.00	3.6560	.99275
3. Do you believe AI can enhance creativity in writing cyberliterature?	125	2.00	5.00	3.3200	.59024
4. Which AI-enhanced platforms have you used (or are aware of) for publishing cyberliterature?	125	3.00	5.00	4.4400	.60107
5. To what extent do you think AI tools can help improve your writing skills?	125	2.00	4.00	3.5120	.59046
6. In your opinion, how important is it to learn about AI tools in the context of creative writing?	125	2.00	5.00	4.0800	.75776
7. Have you ever monetized your cyberliterature or digital writing on platforms such as Instagram, Pinterest, or Webtoon?	125	1.00	4.00	2.7200	1.12594
8. Do you recommend implementing AI-based literature teaching materials in the classroom?	125	1.00	5.00	3.3280	.97365

(*Continued*)

Table 4.1 Descriptive statistics. (*Continued*)

Statement	N	Min	Max	Mean	Std. deviation
9. How confident are you in your ability to use AI-enhanced tools to create literary content?	125	2.00	5.00	3.1840	.81700
10. Can literary works be created using AI tools?	125	1.00	5.00	3.1840	.83651
11. I feel the benefits of learning literature using AI.	125	2.00	5.00	3.6400	.88354
12. Do you think AI-enhanced teaching materials can help increase students' motivation to engage in literary writing?	125	2.00	5.00	3.3360	.67135
13. What kind of support do you need to start monetizing your cyberliterature through AI-enhanced networks?	125	2.00	5.00	3.2240	.79173
14. How do you see the role of AI in shaping the future of literature and storytelling?	125	2.00	5.00	3.5120	.75793
15. Are you interested in taking a course that integrates AI tools and cyberliterature for creative writing and monetization?	125	2.00	5.00	3.8960	.94035
16. Are you interested in taking a course that integrates AI tools and cyberliterature for creative writing and monetization?	125	2.00	5.00	3.5520	.76681

(*Continued*)

Table 4.1 Descriptive statistics. (*Continued*)

Statement	N	Min	Max	Mean	Std. deviation
17. What aspects of AI-enhanced networks would you like to learn more about related to cyberliterature creation and monetization?	125	2.00	5.00	3.3360	.67135
18. How do you think AI tools can change the way students engage with literature in Education 7.0?	125	3.00	5.00	3.5040	.60407
19. Have you seen any examples of cyberliterature successfully monetized through AI-enhanced platforms?	125	1.00	5.00	3.0560	1.03413
20. How much do you rate the potential of AI to help students in Indonesia become more motivated to create literary works?	125	3.00	5.00	3.7280	.73370
Valid N (listwise)	125				

of AI-enhanced publishing platforms. Minimal variability indicates consistent awareness among respondents.

The statistics findings for the role of AI in enhancing narrative competence (Statement 5) with the mean = 3.51 and SD = 0.59. The slightly lower mean suggests that not everyone may have directly experienced this benefit, but participants generally think that AI can help improve writing skills. Statement 6 underscores the significance of AI learning for creative writing, with a mean score of 4.08 and a standard deviation of 0.76. This high score underscores the participants' strong conviction in the worth of obtaining AI tools for creative writing. Statement 7 focuses on the monetization of cyberliterature, with a mean score of 2.72 and a standard deviation of 1.12. The respondents indicated that many students have not utilized AI to monetize their books. There appears to be significant variation in individual experiences with monetization, as indicated by the high standard deviation. The proposal for Artificial Intelligence-Based Literature Learning (Statement 8) has a moderate standard deviation of 0.97 and a mean of 3.33, suggesting that participants are somewhat open to using artificial intelligence-based learning resources. There does not appear to be a consensus on the potential benefits of AI education, as indicated by the relatively high standard deviation.

The ninth statement, "Confidence in Ability to Utilize AI Instruments," displays a mean score of 3.18 and a standard deviation of 0.82, indicating that participants are quite confident in their ability to use AI tools for creating literary content. However, the mean also suggests that some participants may feel less capable. Statement 10 discusses the role of AI in literary creation, with a mean value of 3.18 and a standard deviation of 0.83. Although the average indicates that this idea is still developing or not widely known, it appears that most participants believe that AI can produce literary works. Statement 11 asks about the advantages of AI in literary learning with a standard deviation of 0.88 and a mean of 3.64. Although there is some disagreement, the medium to high average indicates that individuals believe AI can help them learn literature. In statement 12, the role of AI in increasing student engagement has a mean value of 3.34 and a standard deviation of 0.67. Although the results are not entirely favorable, participants believe that AI can increase student interest in literary writing, as indicated by the medium average. Statement 13 indicates that respondents require monetary support, with a mean score of 3.22 and a standard deviation of 0.79. There is a clear need for further instruction in this area, as the mean here suggests that participants may need assistance in monetizing their cyberliterature using AI networks.

Statement 14 asked about the role of AI in the future of literature; the data showed that the mean was 3.51 and the standard deviation was 0.76. The low standard deviation suggests that participants generally agree that AI will have a significant impact on the future of literature. Statement 15: Interest in the AI-cyberliterature course has a mean value of 3.89 and a standard deviation of 0.94, and a mean value of statement 14 explores the role of AI in the future of literature, with the data revealing a mean value of 3.51 and a standard deviation of 0.76. The low standard deviation indicates that participants generally agree that AI will have a significant impact on the future of literature.

Based on respondent's statement number 15, the average interest in the AI-cyberliterature course is 3.89, with a standard deviation of 0.94, and an average value of monetization in the course (statement 16). Cyberliterature Network and AI (Statement 17) shows a mean value of 3.34 and a standard deviation of 0.67. The level of interest in knowing more about AI-enhanced networks for the development and monetization of cyberliterature is moderate. In the statement on student engagement and artificial intelligence (Statement 18), participants, with a mean value of 3.50 and a standard deviation of 0.60, believe that AI has the potential to enhance student literary engagement in education. 6.0. Cyberliterature that has been monetized (Statement 19) is known to have a value of SD = 1.03 and mean = 3.05, indicating that although there is a substantial standard deviation, the average indicates that participants have witnessed several examples of cyberliterature that has been monetized profitably. Statement 20, with a mean value of 3.72 and a standard deviation of 0.73, highlights the potential of AI to motivate Indonesian students. Participants perceive AI as capable of inspiring students to create literary works, as evidenced by the relatively high average.

Based on the results of these statistical data, a big picture is shown of how people in the virtual world understand and use AI tools. In education 7.0, where AI-based content creation and personalized learning are at the center of attention, the data supports the paper's claim that AI plays an increasingly important role in improving creative writing and literature education. It still needs more training, assistance, and confidence-building in using AI for monetization and literary creation, according to the statistics in Table 4.1, even though participants are familiar with AI tools and realize their potential. Additionally, the data reveals obstacles to the successful integration of AI in the classroom, including variations in participants' levels of expertise, knowledge, and confidence, and the necessity for improved methods to enable effective AI integration in literary works.

Table 4.2 shows descriptive statistics on cyberliterature monetization explanations. The data focuses on the mean scores of participants on cyberliterature monetization in Table 4.2. You can learn a lot about people's perspectives and actions regarding the potential profit of digital literary works from this table. A total of 125 people were surveyed as respondents (N = 125). The minimum score for students who are familiar with monetizing cyberliterature is 56.00, with a standard deviation (SD = 8.74) indicating the dispersion or variability of participants' scores.

Table 4.2 Descriptive statistics mean score.

	N	Min	Max	Mean	Std. deviation
Monetizing Cyberliterature	125	56.00	91.00	69.5200	8.73776
Valid N (listwise)	125				

Table 4.3 Reliability statistics.

Cronbach's alpha	N of items
0.861	20

Table 4.3 displays dependency statistics, revealing a Cronbach's alpha value of 0.861 for a set of 20 items. The Cronbach's alpha of 0.861 indicates a high level of reliability. This indicates that the items are consistently measuring the same underlying construct, such as students' familiarity with AI in creative writing or their capacity to commercialize their cyberliterature. The items are likely to be questions or statements related to students' understanding or impressions of cyberliterature and AI technology.

In general, a Cronbach's alpha value above 0.8 is considered good, meaning that the responses from the 20 items are reliable and represent a consistent set of data. This level of dependability makes the results more reliable for analysis and interpretation by ensuring that participants' responses are consistent across items. Thus, a high alpha value indicates that the evaluation or survey method used to measure participants' views of AI in cyberliterature is robust, and the results are likely valid and reliable.

Figure 4.1 displays the standard P-P (probability-probability) plot for the variable "Cyberliterature network for EFL class." The data points plotted for the variable "Cyberliterature Network" are near the diagonal line, indicating that it follows a normal distribution. The assumption for many statistical approaches is that the data does not deviate significantly from a normal distribution.

Analyzing AI and cyberliterature reveals that, when dealing with large-scale data related to cyberliterature networks or AI-generated content, it is crucial to verify that the underlying data follows a specific distribution to ensure efficient use of certain algorithms and procedures.

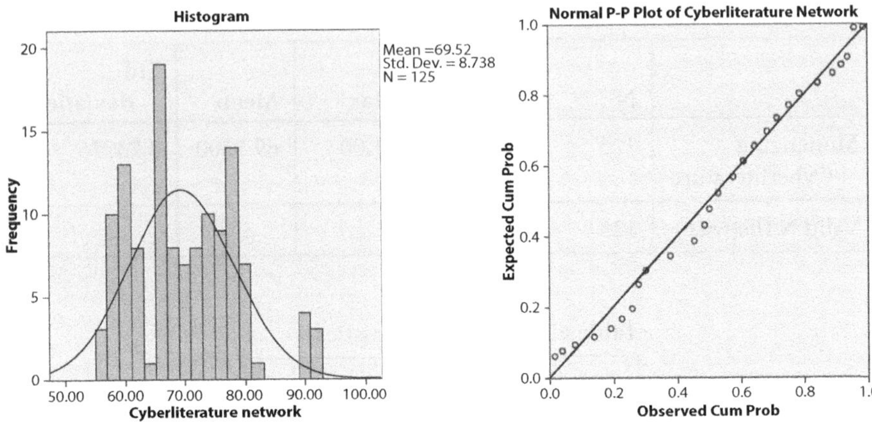

Figure 4.1 Histogram and P-P plot.

This data seems to show how some universities, like UIN Madura, Universitas Muhammadiyah Purwokerto, and Universitas Al Asyariah Mandar in Indonesia, use cyberliterature monetization platforms like TikTok, Instagram, Facebook, Webtoon, and Wattpad. Student participation is displayed on each platform according to age groupings, such as 0–5, 5–10, 10–15, etc.

4.5 Discussion

This paper explores the close relationship between the digital economy and the idea of monetizing digital creative content, such as cyberliterature in EFL classrooms, and how respondents as creators understand the platforms like Wattpad, Webtoon, and blogs to sell their creations. Digital producers can increase their income through a combination of direct sales, advertising, and fan contributions [73]. This is in line with the moderate mean score (69.52), indicating that some individuals are successful in monetizing their work while others may still be at a beginner level. Cyberliterature and other innovative economic models in the education sector are enabled by innovative and creative industries, which rely on the integration of technology and content development.

The rise of the user-driven economy is reflected in platforms that facilitate subscription-based monetization, ad support, and donations accessed through social media and digital platforms. One possible explanation for the mixed results (SD = 8.74) is that participants' familiarity with the platforms and their ability to maximize their monetization possibilities did vary. The results of this study are in line with the technology acceptance model proposed by Fakhr-Hosseini *et al.* [26], which shows how individuals' views on the practicality of a technology influence their choice to adopt and adapt to it. The ratings show varying levels of technology acceptance among participants; the mean score of 69.52 indicates that participants viewed AI-enhanced platforms as useful and successful in their efforts to monetize.

Users with higher ratings found the platform easy to use and useful for monetizing their work, while individuals with lower ratings may have difficulty with knowledge or access to technology, resulting in content that is not monetizable. Dissanayake [22] defines digital literacy as the ability to explore, evaluate, and produce knowledge effectively using digital technologies. Students' digital literacy is essential for cyberliterature and other modern creative disciplines.

Different participants' evaluations, supported by data, suggest that their effectiveness in monetizing their literary work is influenced by their level of digital literacy. Subjects with higher scores (Max = 91.00) may be more adept with digital tools and more easily exploit the possibilities, while those with lower scores (Min = 56.00) may have less experience with currently popular monetization processes and technologies. At a scale of 69.52, individuals' success in monetizing cyberliterature is moderate. Their experiences vary widely, as seen from the standard deviation of 8.74. To better understand the different levels of profitability, use tools such as digital literacy and the Technology Acceptance Model (TAM). It needs to improve people's digital literacy towards Society 5.0 and show them how to use AI-powered platforms to monetize digital creative assets to capitalize on this opportunity.

The document on monetizing cyberliterature is in line with the Education 7.0 educational paradigm, which emphasizes the importance of digital literacy and AI-powered learning in determining student learning outcomes and creativity. According to this theory, the current trend towards commercializing cyberliterature is a departure from more traditional forms of literary expression in favor of creating digital materials through the integration of AI tools and online platforms in the 21st century. With a mean score of 69.52 and a standard deviation of 8.74, the dataset shows that students have a moderate understanding of how to monetize their digital literary works. However, there is a significant amount of diversity in their experiences. This implies that while some students are still in the process of exploring and successfully monetizing their work through AI-powered platforms like Wattpad, Webtoon, and social media, others have already begun to do so.

The results in Figure 4.2 show how students at universities such as Universitas Muhammadiyah Purwokerto, Universitas Al Asyariah Mandar, and IAIN Madura have and use social media cyberliterature monetization platforms such as TikTok, Instagram, Facebook, Webtoon, Wattpad, and Pinterest. Students believe that the use of these platforms reflects varying levels of engagement in monetizing creative work across age groups. Popular platforms such as TikTok and Instagram provide a platform for students to share their literary works, while sites such as Webtoon and Wattpad enhance written and visual storytelling, opening up opportunities to earn money through advertising, subscriptions, or sponsorships.

Integrating cyberliterature education with AI provides students with tools to develop their creative writing and explore options for monetizing their work, particularly in the context of AI-enhanced learning and EFL (English as a Foreign Language) classes. Grammar checking, style recommendations, idea generation, and creative content production are all areas where students can benefit from the use of artificial intelligence and natural language processing. Students can improve their language skills and prepare their work for public platforms such as TikTok or Wattpad with these functions, where they may be able to earn the same money from a larger audience. Additionally, studies by [70] and Yoesoef [66] offer intelligent content editing tools using AI-based applications that enhance

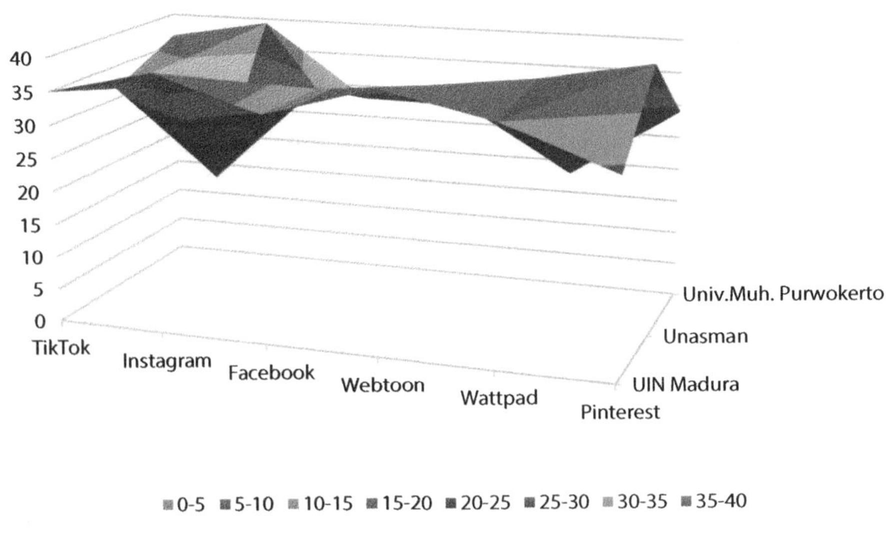

Figure 4.2 Students' cyberliterature monetizing platform.

students' work, making it more appealing and professional for local and global markets to showcase their creative fiction. Based on the findings of these studies, students in EFL classes can benefit from AI in the following ways: first, by working together to tell stories using AI [42], and second, by finding ways to monetize their creative efforts [14]. Instagram and Pinterest provide a wealth of knowledge about digital marketing and audience engagement, which are essential to modern strategies for monetizing digital content. Furthermore, both are excellent venues for creative expression. Posting multimedia stories on TikTok or serialized fiction on Wattpad can help students improve their writing and storytelling skills while enhancing their meaningful learning experience. They benefit from this exposure, as it allows them to interact with readers, advertisers, and potential publishers.

AI-enhanced tools provide automated feedback on structure, creativity, and style, allowing students to improve their work more practically. This ongoing process is especially helpful for EFL students, who may struggle with sophisticated language in their studies. Leveraging AI technology boosts students' confidence and innovation, allowing them to reach new levels of digital literacy and language competency. A student-centered, technology-driven learning environment is the goal of Education 7.0, which aspires to combine AI and IoT technologies. This study shows how students from different institutions engage with different forms of digital literature through an AI-powered cybersecurity platform for the purpose of creativity and innovation.

By enhancing students' capacity to earn money from their work and supporting learning outcomes in an English as a Foreign Language environment, this activity contributes to the key goals of Education 7.0 and the SDGs. Students can learn a language and experience the digital economy through this platform. This elevates them beyond the status of passive learners and makes them proficient writers who can navigate the complexities of the modern literary landscape. There are opportunities where AI enhances learning outcomes and provides practical skills that go beyond traditional education, in line with the vision of Education 7.0, which emphasizes a student-centered approach and meaningful experiences.

Consider the process of transforming learning outcomes into digital materials that have commercial value. The best way for education systems to prepare students to take advantage of these opportunities is by prioritizing AI literacy, creating environments that encourage cyberliteracy, and instructing them on how to make money in the ever-expanding digital economy as the era of digitization. This study reflects the basic idea of Education 7.0. It integrates technology and education in a way that prepares students for 21st-century problems. One way to do this is by using AI tools to encourage creative learning and practical applications, such as the monetization of IoT networks driven by AI robotics technology.

4.6 Conclusions

This study contributes to the field of cyberliterature by demonstrating the innovative ways in which AI can assist with narrative and creative writing. Students can hone their creative skills while simultaneously finding new revenue streams for their digital works on platforms such as Wattpad, Webtoon, TikTok, and Instagram. The use of artificial intelligence tools, such as text generators and multimedia content production, in English as a Foreign Language (EFL) classes helps students overcome language barriers, improve their writing skills through personalized feedback, and receive support with grammar, style, and originality. It would also be interesting to see how AIIoT-driven, personalized learning pathways impact students' motivation and academic performance in future classrooms. Teachers can leverage AIIoT to create more effective lesson plans and enhance AI tools to cater to students' varying computer literacy and creative skills. Finally, this study emphasizes the need for additional research and development to fully understand the ways in which AIIoT can revolutionize education, although it does offer useful insights into AIIoT's cyberliterature and EFL classroom applications.

AI disclosure statement: The author confirms that no artificial intelligence (AI)-generated content or AI-assisted writing tools were used in the preparation of this chapter (including summaries and the abstract). All text, analysis, and conclusions are the result of the author's original work.

References

1. Abobaker, R. and Moleta, S., ESL 4.0: A Practical Guide to Benefits, Challenges, and Best Practices for AI-Enhanced Learning, in: *AI Applications and Strategies in Teacher Education*, pp. 71–134, IGI Global, USA, 2025, DOI: 10.4018/979-8-3693-5443-8.ch004.
2. Adarkwah, M.A., GenAI-Infused Adult Learning in the Digital Era: A Conceptual Framework for Higher Education. *Adult Learn.*, 2024, 10451595241271161. https://doi.org/10.1177/10451595241271161.

3. Ait Baha, T., El Hajji, M., Es-Saady, Y., Fadili, H., The impact of educational chatbot on student learning experience. *Educ. Inf. Technol.*, 29, 8, 10153–10176, 2024, https://doi.org/10.1007/s10639-023-12166-w.
4. Al Yakin, A., Cardoso, L., Al Matari, A.S., Muthmainnah, Obaid, A.J., Enabling Sustainable Learning Through Virtual Robotics Machine Mediation of Social Interactions Between Teachers, Students, and Machines Based on Sociology Lens, in: *International Conference on Explainable Artificial Intelligence in the Digital Sustainability*, 2024, June, Springer Nature Switzerland, Cham, pp. 55–76, https://doi.org/10.1007/978-3-031-63717-9_4.
5. Al Yakin, A., Obaid, A.J., Muthmainnah, Shnawa, A.H., Haroon, N.H., Unlocking the Potential of Mobile Computing for Infusing Computational Thinking Using Social Cognitive Approach in Higher Education Institutes, in: *International Conference on Data Analytics & Management*, 2023, June, Springer Nature Singapore, Singapore, pp. 105–114, https://doi.org/10.1007/978-981-99-6547-2_9.
6. Al-Khamisi, F.A., Sasa, G.S., Jweid, A.N.A., Technology" s Influence on the Metaphorical Language in Contemporary Literature: A Transformative Force and a Formidable Challenge. *World*, 15, 1, 2025, doi:10.5430/wjel.v15n1p1.
7. Aryn, Y., Ibragim, A., Nurgaliyeva, U., Yerzhanova, S., Genres in Internet Literature: Myth, Fairy Tale, Fantasy. *Int. J. Lit. Humanit.*, 22, 4, 33, 2024, DOI:10.18848/2327-7912/CGP/v22i04/33-50.
8. Babaryka, M., Digital media and the 21st-century novel: contemporary perspectives on intermediality in French and anglophone fiction, Doctoral dissertation, UNSW Sydney, 2024, https://doi.org/10.26190/unsworks/25493.
9. Baek, J. and Hwang, Y., Old tales, new tools: How AI chatbots and the metaverse are transforming premodern classical literature reading activities. *Int. J. Educ. Res.*, 128, 102461, 2024, https://doi.org/10.1016/j.ijer.2024.102461.
10. Barroso, P.M., Digital Communication Strategy and Strategic Communication in (Inter) Action: The Portuguese and the Social Media, in: *Organizational Communication in the Digital Era: Examining the Impact of AI, Chatbots, and Covid-19*, pp. 115–141, Springer Nature Switzerland, Cham, 2024, https://doi.org/10.1007/978-3-031-58307-0_6.
11. Bennis, M., Teaching Literature in the Digital Age: Some Theoretical Reflections. *Int. J. Lang. Lit. Stud.*, 6, 3, 199–209, 2024, https://doi.org/10.36892/ijlls.v6i3.1747.
12. Bernovskis, A., Sceulovs, D., Stibe, A., Society 5.0: Shaping the Future of e-Commerce. *J. Open Innov.: Technol. Mark. Complex.*, 100391, 2024, 10.1016/j.joitmc.2024.100391.
13. Campbell, M., Djuna: A New Tool for Digital Art: An analysis of digital textual art, software, and poetics. DiVA, id: diva2:1893134, 2024.
14. Carpenter, J.P., Shelton, C.C., Schroeder, S.E., The education influencer: A new player in the educator professional landscape. *J. Res. Technol. Educ.*, 55, 5, 749–764, 2023, https://doi.org/10.1080/15391523.2022.2030267.

15. Chen, C., LanYu: leverage generative artificial intelligence in assisting foreign language learning, 2024, https://urn.fi/URN:NBN:fi-fe2024060645916.
16. Chen, Y., Jensen, S., Albert, L.J., Gupta, S., Lee, T., Artificial intelligence (AI) student assistants in the classroom: Designing chatbots to support student success. *Inf. Syst. Front.*, 25, 1, 161–182, 2023, https://doi.org/10.1007/s10796-022-10291-4.
17. Chiu, T.K., Future research recommendations for transforming higher education with generative AI. *Comput. Educ.: Artif. Intell.*, 6, 100197, 2024, https://doi.org/10.1016/j.caeai.2023.100197.
18. CohenMiller, A., Saban, G.A., Bayeta, R., Rigor in Qualitative Research, in: *The SAGE Handbook of Qualitative Research in the Asian Context*, pp. 327–343, SAGE Publications, Los Angeles, London, New Delhi, Singapore, Washington DC, Melbourne, 2022.
19. Creswell, J.W. and Poth, C.N., *Qualitative inquiry and research design: Choosing among five approaches*, Sage publications, USA, 2016.
20. de Bem Machado, A., Pesqueira, A., Sousa, M.J., A Review ChatGPT and the Future of Education, in: *Digital Transformation in Higher Education Institutions*, pp. 155–176, Springer Nature Switzerland, Cham, 2024, https://doi.org/10.1007/978-3-031-52296-3_9.
21. Ding, A.C.E., Shi, L., Yang, H., Choi, I., Enhancing teacher AI literacy and integration through different types of cases in teacher professional development. *Comput. Educ. Open*, 6, 100178, United Kingdom, 2024, https://doi.org/10.1016/j.caeo.2024.100178.
22. Dissanayake, D., Literature in Cyberspace. *Int. J. Multicult. Lit.*, 14, 1, 2024.
23. Divito, C.B., Katchikian, B.M., Gruenwald, J.E., Burgoon, J.M., The tools of the future are the challenges of today: The use of ChatGPT in problem-based learning medical education. *Med. Teach.*, 46, 3, 320–322, 2024, https://doi.org/10.1080/0142159X.2023.2290997.
24. El Abidine, M.Z., Ouariach, F.Z., Nejjari, A., The Contribution of ICTE to the Teaching of Literature Within the Action-Oriented Perspective, in: *Innovative Instructional Design Methods and Tools for Improved Teaching*, pp. 303–318, IGI Global, USA, 2024, DOI: 10.4018/979-8-3693-3128-6.ch015.
25. Elbanna, S. and Armstrong, L., Exploring the integration of ChatGPT in education: adapting for the future. *Manage. Sustainability: Arab Rev.*, 3, 1, 16–29, 2024, DOI: 10.1108/MSAR-03-2023-0016. https://doi.org/10.1108/MSAR-03-2023-0016.
26. FakhrHosseini, S., Chan, K., Lee, C., Jeon, M., Son, H., Rudnik, J., Coughlin, J., User adoption of intelligent environments: A review of technology adoption models, challenges, and prospects. *Int. J. Hum.-Comput. Interact.*, 40, 4, 986–998, 2024, https://doi.org/10.1080/10447318.2022.2118851.
27. Fields, A.N., AI-Assisted College Composition for Non-Native Speakers of English-A Course Design, 2024, https://digitalcommons.murraystate.edu/daeng/5/.

28. Forsler, I., Bardone, E., Forsman, M., The Future Postdigital Classroom. *Postdigital Sci. Educ.*, 1–8, 2024, https://doi.org/10.1007/s42438-024-00488-y.
29. Geçgel, S., A Study on the Levels of Adoption of Cyberculture and Global Citizenship among Social Studies Teacher Candidates. *Int. e-J. Educ. Stud.*, 8, 16, 120–130, 2024, https://doi.org/10.31458/iejes.1424048.
30. Harrison, H., Birks, M., Franklin, R., Mills, J., Case study research: Foundations and methodological orientations. *Forum Qual. Sozialforschung/ Forum Qual. Social Res.*, 18, 1, 2017, January, DOI: https://doi.org/10.17169/fqs-18.1.2655.
31. Imran, M. and Almusharraf, N., Google Gemini as a next generation AI educational tool: a review of emerging educational technology. *Smart Learn. Environ.*, 11, 1, 22, 2024, https://doi.org/10.1186/s40561-024-00310-z.
32. Jaradat, S., Nayak, R., Paz, A., Elhenawy, M., Ensemble Learning with Pre-Trained Transformers for Crash Severity Classification: A Deep NLP Approach. *Algorithms*, 17, 7, 284, 2024, https://doi.org/10.3390/a17070284.
33. Kaman, K. and Deshmukh, A.K., Universal Access to the Internet for Attaining Responsible Consumption-Based Sustainable Development Goal (SDG), in: *Digital Technologies to Implement the UN Sustainable Development Goals*, pp. 433–449, Springer Nature Switzerland, Cham, 2024, https://doi.org/10.1007/978-3-031-68427-2_22.
34. Kangasharju, A.I., Ilomäki, L., Toom, A., Online poetry writing at school–comparing lower secondary students' experiences between individual and collaborative poetry writing. *Front. Educ.*, 9, 1380790, 2024, May, Frontiers Media SA. doi: 10.3389/feduc.2024.1380790.
35. Kieslich, K., Helberger, N., Diakopoulos, N., My Future with My Chatbot: A Scenario-Driven, User-Centric Approach to Anticipating AI Impacts, in: *The 2024 ACM Conference on Fairness, Accountability, and Transparency*, 2024, June, pp. 2071–2085, https://doi.org/10.1145/3630106.36590.
36. Kildė, L., The Integration of Generative AI in Foreign Language Teacher Education: A Systematic Literature Review. *Pedagogika*, 154, 2, 5–26, 2024, https://doi.org/10.15823/p.2024.154.1.
37. Kim, K. and Kwon, K., Designing an Inclusive Artificial Intelligence (AI) Curriculum for Elementary Students to Address Gender Differences With Collaborative and Tangible Approaches. *J. Educ. Comput. Res.*, 2024, 07356331241271059. https://doi.org/10.1177/07356331241271059.
38. Kökver, Y., Pektaş, H.M., Çelik, H., Artificial intelligence applications in education: Natural language processing in detecting misconceptions. *Educ. Inf. Technol.*, 1–32, 2024, https://doi.org/10.1007/s10639-024-12919-1.
39. Kong, S.C. and Yang, Y., A Human-Centred Learning and Teaching Framework Using Generative Artificial Intelligence for Self-Regulated Learning Development through Domain Knowledge Learning in K–12 Settings. *IEEE Trans. Learn. Technol.*, 2024, https://creativecommons.org/licenses/by/4.0/.

40. Lavazza, A. and Vilaça, M., Human extinction and AI: what we can learn from the ultimate threat. *Philos. Technol.*, 37, 1, 16, 2024, https://doi.org/10.1007/s13347-024-00706-2.
41. Law, L., Application of generative artificial intelligence (GenAI) in language teaching and learning: A scoping literature review. *Comput. Educ. Open*, 100174, 2024, https://doi.org/10.1016/j.caeo.2024.100174.
42. Li, R., A "Dance of storytelling": Dissonances between substance and style in collaborative storytelling with AI. *Comput. Compos.*, 71, 102825, 2024, https://doi.org/10.1016/j.compcom.2024.102825.
43. Liu, M., Zhang, L.J., Biebricher, C., Investigating students' cognitive processes in generative AI-assisted digital multimodal composing and traditional writing. *Comput. Educ.*, 211, 104977, 2024, https://doi.org/10.1016/j.compedu.2023.104977.
44. Ljubojević, D., Kadijević, D., Gutvajn, N., Towards a Valid and Reliable Checklist to Evaluate Argumentative Essays Composed by ChatGPT. *E-Learning*, vol. 3696, pp. 82–90, 2024, https://ceur-ws.org/Vol-3696/article_9.pdf.
45. Lou, S., Hu, Z., Zhang, Y., Feng, Y., Zhou, M., Lv, C., Human-cyber-physical system for Industry 5.0: A review from a human-centric perspective. *IEEE Trans. Autom. Sci. Eng.*, 1–18, 2024, DOI: 10.1109/TASE.2024.3360476.
46. Machado, C.B., *Human Interaction and Emerging Technologies (IHIET-AI 2024)*, vol. 120, pp. 114–122, Lausanne Switzerland, 2024, DOI: 10.54941/ahfe1004549.
47. Magni, D., Del Gaudio, G., Papa, A., Della Corte, V., Digital humanism and artificial intelligence: the role of emotions beyond the human–machine interaction in Society 5.0. *J. Manage. Hist.*, 30, 2, 195–218, 2024, https://doi.org/10.1108/JMH-12-2022-0084.
48. Matthew, U.O., Oyekunle, D.O., Akpan, E.E., Oladipupo, M.A., Chukwuebuka, E.S., Adekunle, T.S., Onumaku, V.C., Generative Artificial Intelligence (AI) on Sustainable Development Goal 4 for Tertiary Education: Conversational AI With User-Centric ChatGPT-4, in: *Impacts of Generative AI on Creativity in Higher Education*, pp. 259–288, IGI Global, USA, 2024, DOI: 10.4018/979-8-3693-2418-9.ch010.
49. Melissa, Z. and Kenza, T., Artificial Intelligence Contributions to Machine Translation, translation of extracts from the novel "the hearts of horses" by Molly Gloss through ChatGPT, Babylon, Reverso, Yandex and Google translate, Doctoral dissertation, Université Mouloud Mammeri Tizi-Ouzou, 2023.
50. Morales-Fernández, B., New Linguistic Spaces in Cyberculture: The Influence of the Metaverse on the Minifiction of Social Networks, in: *The Future of Digital Communication*, pp. 27–38, CRC Press, London, UK, 2024, DOI: 10.1201/9781003379119-3.
51. Mowafi, S.E.M., The manifestations of cyberculture in theatre: A posthuman reading of Sarah Ruhl's Dead Man's Cell Phone. *J. Lang. Transl.*, 11, 2, 14–30, 2024, DOI: 10.21608/jltmin.2024.352114.

52. Muthmainnah, G. and Al Yakin, A., An effective investigation on YIPe-learning based for twenty-first century class, in: *Digital Learning based Education: Transcending Physical Barriers*, pp. 21–38, Springer Nature Singapore, Singapore, 2023, https://doi.org/10.1007/978-981-19-8967-4_2.
53. Muthmainnah, M., Al Yakin, A., Mursidin, M., Al Aqad, M.H., AI-Driven Content Developing and Designing for Teaching Materials of Digital Healthcare, in: *Driving Smart Medical Diagnosis Through AI-Powered Technologies and Applications*, pp. 238–257, IGI Global, USA, 2024, DOI: 10.4018/979-8-3693-3679-3.ch015.
54. Obaid, A.J., Bhushan, B., Rajest, S.S. (Eds.), *Advanced applications of generative AI and natural language processing models*, IGI Global, USA, 2023.
55. Simpungwe, F.C., Development and Implementation of an Innovative Higher Educational Integrated Strategy for Preparing the Industry-Ready Future Workforce, Doctoral dissertation, University of Liverpool, 2024, 10.17638/03180613.
56. Steiss, J., Tate, T., Graham, S., Cruz, J., Hebert, M., Wang, J., Olson, C.B., Comparing the quality of human and ChatGPT feedback of students' writing. *Learn. Instr.*, 91, 101894, 2024, https://doi.org/10.1016/j.learninstruc.2024.101894.
57. Stöhr, C., Ou, A.W., Malmström, H., Perceptions and usage of AI chatbots among students in higher education across genders, academic levels and fields of study. *Comput. Educ.: Artif. Intell.*, 7, 100259, 2024, https://doi.org/10.1016/j.caeai.2024.100259.
58. Tomei, L.A., Maine, J., Moussa, K., Holler, M.B., Hobbs, B., Austin, S., The Top 12 Technologies for Teaching and Learning in the Post-Pandemic Era, in: *Exploring Technology-Infused Education in the Post-Pandemic Era*, pp. 1–95, IGI Global, USA, 2024, DOI: 10.4018/979-8-3693-2885-9.ch001.
59. Tost, J., Gohsen, M., Schulte, B., Thomet, F., Kuhn, M., Kiesel, J., Hornecker, E., Futuring Machines: An Interactive Framework for Participative Futuring Through Human-AI Collaborative Speculative Fiction Writing, in: *Proceedings of the 6th ACM Conference on Conversational User Interfaces*, 2024, July, pp. 1–7, https://doi.org/10.1145/3640794.3665904.
60. Turmuzi, M., Suharta, I.G.P., Astawa, I.W.P., Suparta, I.N., Perceptions of Primary School Teacher Education Students to the Use of ChatGPT to Support Learning in the Digital Era. *Int. J. Inf. Educ. Technol.*, 14, 5, 2024, DOI: 10.18178/ijiet.2024.14.5.2097.
61. Vetter, M.A., Lucia, B., Jiang, J., Othman, M., Towards a framework for local Interrogation of AI ethics: A case study on text generators, academic integrity, and composing with ChatGPT. *Comput. Compos.*, 71, 102831, 2024, https://doi.org/10.1016/j.compcom.2024.102831.
62. Vonitsanos, G., Moustaka, I., Doukakis, S., Mylonas, P., Transforming Education in the Digital Age: Exploring the Dimensions of Education 4.0, in: *2024 IEEE Global Engineering Education Conference (EDUCON)*, 2024, May,

IEEE, pp. 01–10, https://ieeexplore.ieee.org/abstract/document/10578747. DOI: 10.1109/EDUCON60312.2024.10578747.
63. Walter, Y., Embracing the future of Artificial Intelligence in the classroom: the relevance of AI literacy, prompt engineering, and critical thinking in modern education. *Int. J. Educ. Technol. Higher Educ.*, 21, 1, 15, 2024, https://doi.org/10.1186/s41239-024-00448-3.
64. Wang, C., Exploring Students' Generative AI-Assisted Writing Processes: Perceptions and Experiences from Native and Nonnative English Speakers. *Technol. Knowl. Learn.*, 1–22, 2024, https://doi.org/10.1007/s10758-024-09744-3.
65. Woo, D.J., Guo, K., Susanto, H., Exploring EFL students' prompt engineering in human–AI story writing: an activity theory perspective. *Interact. Learn. Environ.*, 1–20, 2024, https://doi.org/10.1080/10494820.2024.2361381.
66. Yoesoef, M., Cyber literature: Wattpad and webnovel as generation z reading in the digital world, in: *International University Symposium on Humanities and Arts (INUSHARTS 2019)*, 2020, July, Atlantis Press, pp. 128–131, https://doi.org/10.2991/assehr.k.200729.025.
67. Youquan, O., *A History of Cyber Literary Criticism in China*, Routledge, London, 2023, https://doi.org/10.4324/9781003428480.
68. Yusuf, A., Pervin, N., Román-González, M., Noor, N.M., Generative AI in education and research: A systematic mapping review. *Rev. Educ.*, 12, 2, e3489, 2024, https://doi.org/10.1002/rev3.3489.
69. Zeb, A., Ullah, R., Karim, R., Exploring the role of ChatGPT in higher education: opportunities, challenges and ethical considerations. *Int. J. Inf. Learn. Technol.*, 41, 1, 99–111, 2024, https://doi.org/10.1108/IJILT-04-2023-0046.
70. Zhan, Y.B., The "First Generation" of Creative Cyber Literature and Its "Genre-Bending Writing": Thoughts on Using Jun Tian as a Method, in: *Chinese Creative Writing Studies*, pp. 129–142, Springer Nature Singapore, Singapore, 2023, https://doi.org/10.1007/978-981-99-0931-5_15.
71. Zhang, H. and Shao, H., Exploring the Latest Applications of OpenAI and ChatGPT: An In-Depth Survey. *CMES-Comput. Model. Eng. Sci.*, 138, 3, 2024, DOI: 10.32604/cmes.2023.030649.
72. Zhang, R., A Study of the Rise of Internet Literature and Diverse Influences. *J. Educ. Humanit. Social Sci.*, 36, 164–170, 2024, DOI: 10.54097/ajb0ay57.
73. Mishra, M. S., Sachan, M. P., Rathi, M. J., Adarsh, M., Singh, K., Sethi, R., Monetization Strategies on YouTube vs. Instagram: A Study of Content Creator Revenue Models in the Digital Age. In *Proceedings of the 2nd ICSSR Conference on India Towards Viksit Bharat*, vol. 2047, pp. 13th–14th, 2024, September.

5

Understanding AI with IoT, Real-Time Applications in Image Dehazing

Malladi Sunder Rao* and Dilip Kumar

Department of Electronics and Communication Engineering, National Institute of Technology Jamshedpur, Jamshedpur, Jharkhand, India

Abstract

Image dehazing, which uses AIoT (Artificial Intelligence of Things) to filter and improve images taken by IoT-enabled devices in real-time, improves vision in low-visibility or foggy environments. Combining AI's processing power with IoT's connection and sensing capabilities, picture dehazing on hardware platforms like the PYNQ-Z2 board may be implemented at the edge with efficiency and scalability. This chapter mainly focused on image dehazing with AIoT and introduced the PYNQ-Z2 board. The proposed system uses a deep learning-based dehazing model and is optimized for deployment on the PYNQ-Z2 platform. It is a real-time image processing analysis integrating IoT sensors and a wireless communication module. This chapter improves image quality and enables various applications, such as autonomous vehicle navigation, surveillance systems, and remote sensing.

Keywords: IoT, CDNs, deep learning, image dehazing, FPGA, PYNQ-Z2 board

5.1 Introduction

Definition: Artificial Intelligence (AI)

In this context, using machine learning techniques, including deep neural networks, created especially for image enhancement and dehazing, is called artificial intelligence (AI). To increase image clarity, these algorithms evaluate visual information and eliminate haze.

**Corresponding author*: malladi.sunderrao@gmail.com

Tanusree Chatterjee, Rajib Banerjee and Souvik Ganguli (eds.) Content Delivery Networks Using Artificial Intelligence and Internet of Things: Revolutionizing Content Delivery, (101–122) © 2026 Scrivener Publishing LLC

The convolutional neural network (CNN) [2] is the backbone of the deep learning models in image dehazing. CNNs identify and remove haze artifacts by learning patterns within images. CNN only captures the local spatial image feature but cannot acquire vast contextual information. GNNs can enhance dehazing by capturing both local and global features. Attention mechanisms improve both the accuracy and efficiency of the dehazing process.

Definition: The Internet of Things (IoT) application uses cameras and sensors connected to take real-time pictures, which are then sent to the PYNQ-Z2 board for processing. Real-time dehazing applications like surveillance, self-driving cars, and environmental monitoring leverage IoT continuous data collection.

AI models may effectively execute dehazing at the edge thanks to the PYNQ-Z2's FPGA [1] optimizations, maximizing clarity in real-time without needing an extensive network or processing resource. Data required for dehazing is provided by high-resolution cameras and atmospheric sensors (for variables like light and humidity). These sensors can modify the AI model's parameters based on environmental conditions. Connectivity: To send picture data to edge devices, such as the PYNQ-Z2 board for real-time processing and visualization, Internet of Things systems employ network protocols (such as Wi-Fi, Bluetooth, or Zigbee). Effective management of collected data is necessary, frequently utilizing edge storage solutions to ensure that only necessary, processed pictures are transferred to the cloud if additional processing is needed. Artificial intelligence can instantaneously evaluate, interpret, and respond to data gathered by Internet of Things devices. Here's a closer look into AI IoT, notably on advantages and real-world applications, and imagine dehazing case studies.

CDNs (content delivery networks) are necessary for consumers to get digital content—like images, videos, and web apps—with maximum efficiency and low latency. To accomplish this, they disperse material among several geographically dispersed servers, enabling users to access information from servers near their current location. AI technologies, such as image dehazing, are becoming more crucial to providing a smooth user experience as the demand for high-quality material grows. AI-powered image dehazing is especially useful for IoT applications that use visual data in low-visibility and outdoor settings, like environmental surveillance, driverless cars, and traffic monitoring. By incorporating dehazed images, CDNs can give users more accurate and pertinent visuals, especially for applications that need real-time or near-real-time data.

Applications such as environmental monitoring, real-time navigation, and live surveillance benefit from the user experience that is improved by high-quality, dehazed photos. CDNs process and optimize photos at the network edge, eliminating the need for high-bandwidth transfer of huge, raw images from central servers. Through efficient data transfer, intelligent load balancing, and optimized caching achieved *via* AI-driven image dehazing, the overall infrastructure requirements and operational costs are significantly reduced. Dehazing enhances visibility and clarity in hazy or smoggy urban or highway monitoring situations, enabling authorities to make prompt, well-informed choices based on accurate visual data. By improving situational awareness, lowering accident rates, and giving clear images in low-visibility conditions, real-time dehazing in CDN-supported car networks can increase navigation safety. Dehazing produces precise, accurate visual data for meteorological services that may be stored and disseminated over CDNs to deliver high-quality, real-time weather updates.

5.2 The Integration of AI and IoT

Nowadays, two of the most popular technologies are AI and IoT. AI can perform tasks that typically require human intelligence, such as decision-making, problem-solving, and learning. This work employs AI techniques, including deep learning, machine learning, and natural language processing (NLP). The IoT device is interconnected with devices, such as vehicles and buildings, and other objects that can collect and exchange data through sensors and are connected. This intermediate allows for real-time monitoring, automation, self-driving cars, and control across various domains. In simple terms, the camera captures images and videos that are processed on the FPGA board (PYNQ-Z2) using AI and IoT integration, and the resulting data is efficiently delivered through a Content Delivery Network (CDN) to enhance the performance of data-intensive applications.

Figure 5.1 shows the integration of the AI and IoT setup. A camera captures the haze images and fog images and then connects to a PYNQ-Z2 board, which has an Ethernet connector, DDR memory, a USB port, an SD card, and a Xilinx Zynq-7020 FPGA processor. Through the USB port, a camera's input takes the input data images or video stream that enters the system during operation. Before executing the DPU operation, the program reads and preprocesses the camera's image and video, which is based on the Python code, Jupiter Notebook, and OpenCV. The PC receives the image dehaze detection results *via* the Jupiter IP address *via* Ethernet connection for monitoring purposes, and each module explains Section 5.9.

Hazy and foggy images present major challenges in image processing because they are caused by common atmospheric phenomena such as

Figure 5.1 Integration of the AI and IoT setup.

scattering and absorption. They can greatly hinder visibility and reduce the quality of images, affecting applications in autonomous driving, surveillance, and other areas. We are using the real-time application of the image dehazing with the PYNQ-Z2 board.

5.2.1 Applications of AIoT for Surveillance in Image Dehazing

The way that artificial intelligence (AI) and the Internet of Things (IoT) are used to dehaze images is revolutionizing surveillance systems. Through the use of cutting-edge deep learning models on edge devices such as the PYNQ-Z2. AIoT makes it possible to improve blurry security footage in real time. In content delivery networks (CDNs), which are essential for distributing video streams and material to end users, this capacity has a direct influence on scalability. In real-time processing, AIoT-enabled edge devices efficiently manage image dehazing locally, hence diminishing reliance on centralized servers. Improved Content Quality Enhancing the visual clarity of surveillance feeds, picture dehazing guarantees the delivery of high-quality data while reducing the computing burden for further processing inside the CDN infrastructure. AIoT systems automatically optimize resources according to environmental circumstances (e.g., haze density), enhancing scalability by ensuring constant performance. Local data processing and the transmission of just dehazed photos substantially alleviate the data load on CDN nodes, assuring efficiency.

In a city-wide surveillance system, hazy images collected by IoT-enabled cameras may be analyzed in real time using graph-based neural networks installed on edge devices. This improves efficiency by guaranteeing that only understandable and useful data is sent to CDN servers. Such a method enables the system to grow successfully when more cameras are incorporated, without affecting overall performance.

5.3 Introduction to Image Dehazing

Accurate information is now required for indoor and outdoor applications due to the development of the digital environment and wireless technologies, particularly in image-processing applications that require high-resolution images. More precise pictures are necessary to analyze data and transmit information *via* multimedia. Haze, smoke particles, fog, snow, mist, and raindrops are atmospheric factors that can drift and degrade the quality of images used outside. These characteristics cause the image to deteriorate because of particle dispersion, reflection, and absorption, which change the scene's contrast, color, and visual clarity. Haze or murky circumstances have deteriorated an image to a hazy or foggy state. Table 5.1 represents a brief description of particle size in different environmental conditions.

Hundreds of dehazing methods based on deep learning have been proposed since the creation of CNNs. It starts by going over the computational model, data sets, network modules, loss functions, and frequently used metrics. After that, a summary and classification of the important contributions made by different dehazing methods are provided. Various baseline methodologies are also used in the tests, and quantitative as well as qualitative outcomes are reported. Understanding an image's physical properties is essential for restoring blurry images. Among them is scene depth, measured from the camera's view of the physical objects at a given place. Since the depth map is often not available in advance in real-time applications, this presents a significant difficulty for most haze removal approaches. Because the intensity of the haze is unpredictable and may change depending on the weather conditions, dehazing is a challenging process. The dehazing

Table 5.1 Comparison of particles in the atmosphere in terms of size and type.

Condition	Size of the particle (radius in μm)	Type
Haze	2–10	Aerosol
Fog	1–10	Water droplet
Air	4–10	Molecule
Rain	102–104	Water droplet
Cloud	1–10	Water droplet

methods [3] are separated into single-image and multiple-image categories according to the number of reference images. Considering many images with varying degrees of polarization, depth information from a scene may be restored using the multi-image dehazing technique [4]. Other methods exist to overcome weather impacts, such as taking images of the same area in varying degrees of haze and using them as a reference. The Deep Photo-based approach employs numerous photos of 3D models for haze removal.

In this single-image dehazing with the graph neural network method, the quality of pictures taken by cameras is greatly reduced when floating particles are present in a foggy atmosphere. This degradation in image quality negatively impacts photography work, leading to decreased contrast, color shifts, and blurred textures and edges in the scene. As illustrated in Figure 5.2, (a) is a clear image and (b) is the histogram of (a); (c) is a hazy image and (d) is the histogram of (c). Low-quality inputs might affect the accuracy of algorithms trained on haze-free pictures for computer vision tasks like object recognition [5] and image segmentation.

Consequently, a number of researchers strive to extract clear, higher-quality images from hazy images. Deep learning is widely used in computer vision applications. Image dehazing techniques have traditionally relied on various assumptions and the atmospheric scattering model (ASM) [6]. These statistical rule-based methods offer a well-interpretable processing flow. However, they may falter in challenging real-world scenarios. For example, regions containing sky are not effectively managed by the well-known dark channel prior (DCP) [7]. Researchers have combined convolutional neural networks (CNNs) with the atmospheric scattering model (ASM) to estimate ASM parameters. This integration is motivated by deep learning.

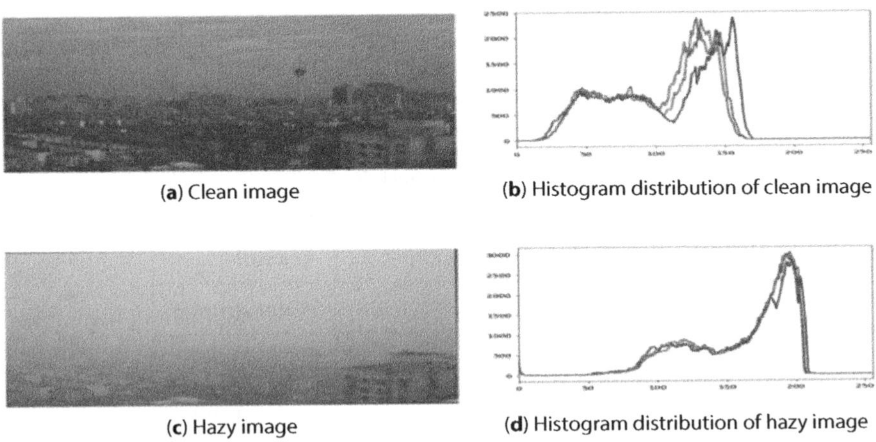

Figure 5.2 (a) Clear image and (b) histogram of (a); (c) hazy image and (d) histogram of (c).

Qualitative and quantitative experimental results demonstrate that guided parameter prediction is made possible by deep learning. Thus, research has demonstrated that end-to-end supervised dehazing networks may be constructed without the Atmosphere Scattering Model (ASM). To attain accuracy on par with their ASM-based counterparts, non-ASM-based dehazing algorithms make use of Convolutional Neural Networks' (CNNs') powerful feature extraction capabilities. Although both ASM-based and non-ASM-based supervised algorithms have demonstrated impressive performance, these approaches usually rely on synthesized paired pictures, which are inappropriate for processing blurry photos taken in natural settings. Therefore, the focus of current research has switched to approaches that better meet the dehazing demands of the actual world. While some research investigates unsupervised approaches that completely avoid synthetic data, others suggest semi-supervised algorithms that make use of both synthetic paired data and real-world unpaired data. Given how quickly picture dehazing methods are developing, a thorough evaluation [8] is essential to motivate and direct further study. While a number of papers address current advancements in dehazing, they frequently omit important elements. For instance, they concentrate on non-deep learning methods, including depth estimates, wavelets, augmentation, and filtering, rather than modern CNN-based dehazing algorithms. Additionally, they focus on supervised dehazing models but overlook the most recent unsupervised and semi-supervised techniques. Furthermore, they seldom ever discuss daylight techniques, while Banerjee and Chaudhuri [9] classify the current nighttime image dehazing techniques. They quickly categorize and evaluate supervised and unsupervised algorithms; however, they do not thoroughly discuss recently suggested semi-supervised techniques.

Rather, we concentrate on deep learning techniques that employ monocular photos taken during the day. As such, we do not explore dehazing techniques such as non-deep learning, underwater, video, hyperspectral, nighttime, binocular, and so on. As a result, when we talk about dehazing in this chapter, we usually imply deep learning methods that use input data that satisfies four requirements: it must be a single-frame picture, daylight, monocular, and at ground level.

Loss functions, basic modules often seen in dehazing networks, the physical model ASM, synthetic and real-world datasets, and assessment metrics for various approaches are all presented. ASM is employed in the inference process, and the dehazing operation verifies the DCP. DCP-based methods are therefore considered to be ASM-based. This chapter addresses the supervised dehazing network based on knowledge distillation as a supervised approach instead of a poorly supervised or semi-supervised algorithm. Techniques for supervised dehazing that use GAN or total variation loss are still classified as supervised. Here are the notational rules that were applied in this survey. Unless otherwise indicated, all symbols signify that I(x) indicates a hazy image, whereas J(x) Unless otherwise noted, I(x) in this context indicates a blurry image, while J(x) indicates a clear image. A pixel's position in I(x), J(x), and t(x), where t(x) is the transmission map, is indicated by the variable x.

As per reference, ambient light can be conceptualized as a non-uniform matrix A(x) or a constant A. Atmospheric light is consistently designated as A in this survey. Furthermore, the proposed approaches are often referred to by acronyms in various publications.

5.4 The Dehazing Process Model

There are four major areas where dehazing techniques are comparable. Initially, the network's modelling features the physical ASM model or is fully based on neural networks. Second, data such as the transmission map, ambient light value, or paired supervision data must be included in the dataset used to train the network. Third, several kinds of basic modules are used by dehazing networks. Fourth, a range of loss functions is used in training. Researchers have developed various effective dehazing methods based on these four components. As a result, this part presents network architectural modules, datasets, loss functions, and ASM. It also talks about how to evaluate the results of dehazing.

ASM can approximate the explanation for the natural occurrence of haze. McCartney initially proposed the fundamental ASM [10] to explain the basics of haze production. The ASM, which is currently in widespread use, was expanded and improved by Narasimhan and Nayar [11]. The ASM provides a solid theoretical foundation for image dehazing research. Its equation is

$$I(x) = J(x)t(x) + A(1 - t(x)) \tag{5.1}$$

where x is the location of the pixel and A is the ambient light level worldwide. In various research, "A" could be referred to as "airtight" or "ambient light." In this research, A stands for ambient light for clarity's sake. ASM-based dehazing methods frequently have an unidentified A. J(x) denotes the clear scene picture, whereas I(x) denotes the fuzzy image. The input in the majority of dehazing models is I(x), while the desired output is J(x). The definition of the medium of transmission map, represented by t(x), is as follows:

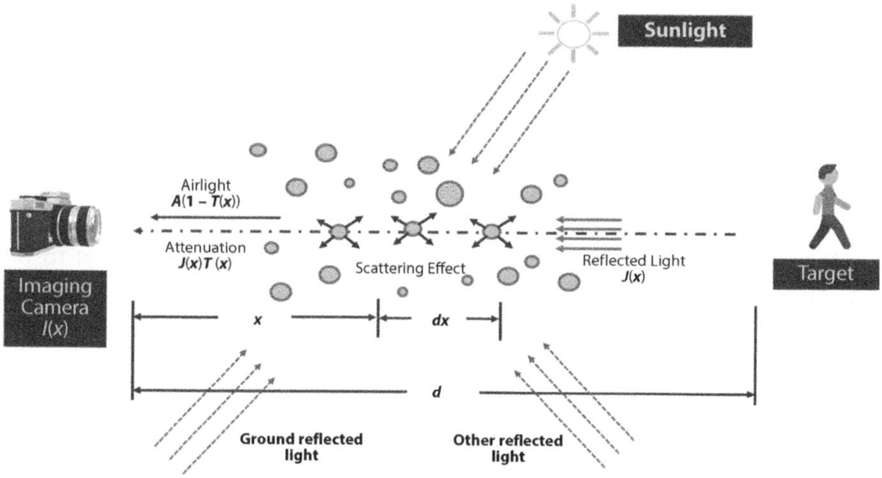

Figure 5.3 Atmospheric scattering phenomenon in a foggy imaging model.

$$t(x) = e - \beta d(x) \tag{5.2}$$

The air scattering parameter is represented by β, and the depth of I(x) by d(x). Thus, t(x) is determined by d(x), and this may be utilized to create blurry pictures. If it is possible to estimate t(x) and A, the haze-free picture J(x) may be created using the formula below:

$$J(x) = I(x) - A(1 - t(x))t(x). \tag{5.3}$$

Figure 5.3 illustrates the ASM imaging concept, showing how airborne particles affect the light emitted by objects and captured by the camera. In some studies, the parameters of the atmospheric scattering model are solved explicitly or implicitly to explain the haze formation process. ASM significantly influences dehazing research, including supervised, semi-supervised, and unsupervised methods, as demonstrated in Table 5.1.

5.5 General Process of Dehazing or Haze Removal Model

Haze degrades image quality by reducing visibility and obscuring important scene details captured by the camera. The presence and density of haze vary with distance, making the restoration of clear images a challenging task.

Therefore, effective haze removal is essential for producing high-quality images required in computer vision applications.

Numerous studies have been conducted to enhance the image restoration process under hazy conditions. However, some of these studies consider only atmospheric effects or scene depth, limiting their generalization. Such frameworks may not be suitable for applications involving conventional or low-level vision tasks.

A comprehensive framework is necessary for the hazed photos to be effectively dehazed or restored. Depth map estimation: Following applying

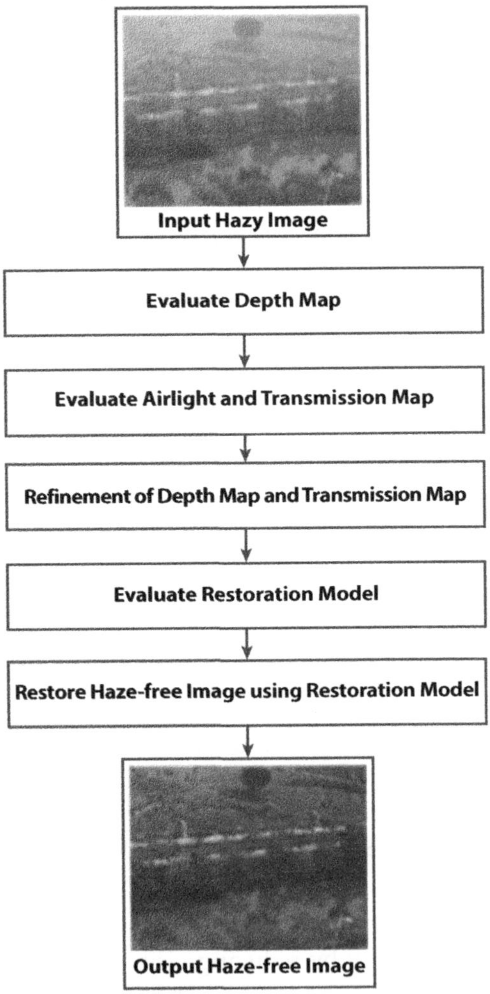

Figure 5.4 Displays the entire framework model or dehazing process flow.

the input hazed picture to the model, the image's depth map must be calculated and evaluated. This is the process's first and most crucial stage. The depth map may not be accurately or correctly approximated in several dehazing techniques now in use. The depth map estimate is not referred to as prior in real-time photos. One of the most crucial steps in the haze removal process is depth map estimation, which may be calculated using the picture attributes. Figure 5.4 displays the entire framework model or dehazing process flow.

Transmission and airlight map estimation: The data gathered from the depth map is used to estimate the input image's atmospheric light, or airlight, and transmission map. The haze pixels' intensity and entropy determine the airtight "A," and the pixels with the highest percentage are regarded as dark channel values. Refinement procedure for depth map and transmission map: The refining procedure eliminates and smooths the haze, noise, and artifacts on the depth map. Refinement is necessary because an incorrect estimate of the transmission map results in artifacts and misleading textures. Refining uses various filters, including bilateral filters, guided filters, soft matting approaches, etc. Output picture restoration: the last step, which follows the refining procedure in Figure 5.5a, results in a clear image, and Figure 5.5b shows the depth for the clear image, restoring an output picture. After that, the final dehazed image is produced. Equation 5.4 below displays the dehazed output-image mathematical formula employing a transmission map and an airtight estimate.

$$J(x) = \frac{I(x) - A}{t(x)} + A \tag{5.4}$$

(a) (b)

Figure 5.5 Atmospheric scattering model (ASM): (a) clear image and (b) depth for the clear image.

Figure 5.6 Examples from the O-Haze [12], Dense-Haze [13], I-HAZE [14], and NH-HAZE [15] datasets.

5.6 Datasets for the Dehazing Process

In Figure 5.6, three outdoor datasets are shown. Haze has five distinct qualities. The haze in 5.6d is not consistently distributed over the landscape, although it is in 5.6b and 5.6c. Furthermore, there is a noticeable difference in the haze density between 5.6c and 5.6b and 5.6d. The information in these datasets is useful for creating dehazing algorithms. To remove the high haze density, for instance, improved extraction of features and recovery skills are needed when constructing dehazing algorithms for thick haze. Creating artificial haze has the main advantage of making gathering data easier. However, hazy pictures generated with a fog generator or synthesized with ASM cannot accurately reproduce the haze of the actual environment. The gap between fabricated and real data is therefore intrinsic.

5.7 Network Section

Nowadays, deep learning-based dehazing networks heavily utilize CNNs. Attention mechanisms, feature pyramids, multi-scale fusion, dilated convolution, conventional convolution, and cross-layer connections are examples of common modules. Dehazing networks are typically constructed by combining multiple fundamental blocks. The fundamental building components frequently found in network designs are outlined below to help you understand the basics of various dehazing techniques. Typical convolution, dilated convolution, cross-layer connection, multiple-scale fusion feature pyramid, and attention mechanism in dehazing. Table 5.2. Datasets for the image dehazing task include.

Table 5.2 Datasets and their characteristics.

Dataset	Type	Nums	I/O	P/NP
D-HAZY	Syn	1,400+	I	P
HazeRD	Syn	15	O	P
I-HAZE	HG	35	I	P
O-HAZE	HG	45	O	P
RESIDE	S&R	10,000+	I&O	P&NP
Dense-Haze	HG	33	O	P
NH-HAZE	HG	55	O	P
MRFID	Real	200	O	P
BeDDE	Real	200+	O	P
4KID	Syn	10,000	O	P

This table provides information on various datasets: Syn: synthetic haze images, HG: haze generated, S&R: combination of synthetic and real images, Real: real-world images captured with natural haze, Nums: the approximate number of images available in the dataset, I: images with only haze, O: images with only clean, I&O: both hazy and clean images are available, P: paired images, and NP: non-paired images.

5.8 Applications of Image Dehazing in AI IoT

Image dehazing, especially in an IoT context, has significant implications in areas like autonomous driving, surveillance, and remote sensing. Haze and fog can obscure critical details, affecting object detection and image quality, which is why AIoT-enhanced dehazing is beneficial.

Security and Surveillance: Environmental factors like fog and pollution make it difficult for many surveillance devices to see. AI-powered dehazing can bring clarity back, improving security personnel's ability to identify and follow items or people. Medical Imaging: Haze or other imaging impediments can make diagnosis more difficult in specialties like radiology. Medical equipment that uses IoT for real-time analysis could benefit from AIoT solutions that improve image quality. In light of the above application and advantages, we are going to discuss a case study.

5.9 Case Studies

Several frameworks and resources are available to support Figure 5.7 PYNQ-Z2 board in software tools, which facilitate the development, deployment, and optimization of AI and IoT applications. The software tools often used with the PYNQ-Z2 board are broken down below. Programming Xilinx's Zynq boards, including the PYNQ-Z2, is made possible *via* the open-source PYNQ framework (Python Productivity for Zynq). It simplifies development by allowing users to program the FPGA using Python instead of more conventional hardware description languages (HDLs). PYNQ has hardware interface libraries that seamlessly integrate Python-based AI and IoT processes. The framework facilitates code creation and testing using Jupyter Notebooks, which offer an interactive, real-time debugging and visualization environment. Without needing specialized FPGA programming, PYNQ provides pre-built overlays for various applications (such as image processing and machine learning) that can speed up development.

Xilinx's Vivado software designs and synthesizes digital systems for FPGAs, such as the PYNQ-Z2 board. Making unique hardware accelerators or altering the PYNQ-Z2's current overlays is necessary. Vivado offers digital circuit design, simulation, and implementation tools for the FPGA. With Vivado HLS, developers may create unique hardware accelerators

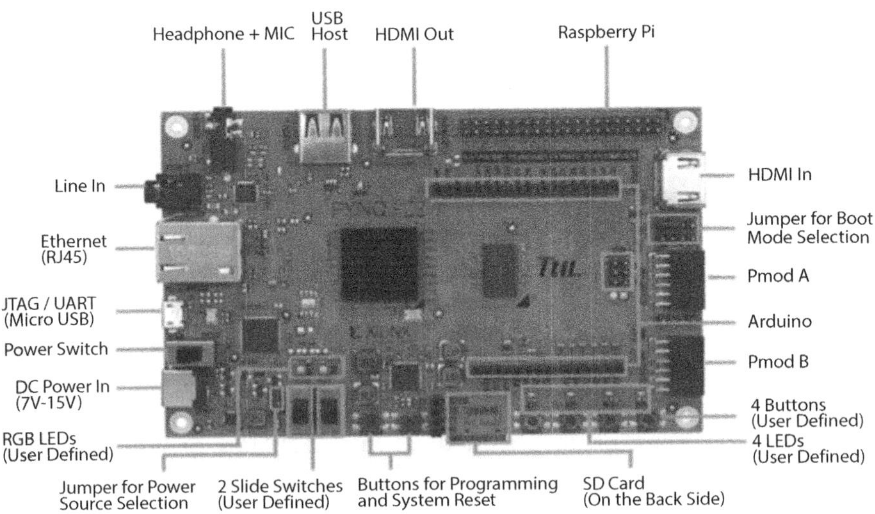

Figure 5.7 PYNQ-Z2 board.

in C/C++ and have them synthesized for the PYNQ-Z2. In addition to the PYNQ-Z2 overlays, Vivado enables the integration of many IP blocks (Intellectual Property cores) that may be tailored for certain purposes.

5.9.1 Real-Time Dehazing Using Embedded Systems

AI models have been used in recent research to accomplish image dehazing on embedded systems such as the PYNQ-Z2 [16] board. The real-time processing capabilities offered by these systems are essential for Internet of Things applications that need to be analyzed very quickly, such as autonomous driving. Researchers have found that graph-based neural networks significantly increase processing speed and accuracy in these systems.

5.10 Features of the PYNQ-Z2 Board

- It is a dual-core ARM Cortex-A9 processor with a speed of 650 MHz. This programmable logic device has 13300 logic slices, each of which has eight f/fs and four 6-input LUTs.
- The available block RAM is 630 kB. There are 220 DSP slices in it. The Xilinx Analog-to-Digital Converter (XADC) is utilized on-chip.
- Memory and storage, programmable *via* JTAG, MicroSD card, and Quad-SPI flash.
- It features a 16-bit bus, 1050 Mbps, and 512-MB DDR3. The 16-MB Quad-SPI Flash includes a 48-bit globally unique identifier that is compatible with EUI-48/64. It features an Ethernet and USB port, a MicroSD card, and an external power regulator with a voltage range of 7 V to 15 V.
    - Gigabit Ethernet PHY, Micro USB-JTAG programming circuits, a Micro USB-UART bridge, and USB 2.0 OTG PHY (host only) are all features. Audio and video; two input and output HDMI ports; a 24-bit I2S DAC with a 3.5-mm TRRS connection; and a 3.5-mm line-in switches, pushbuttons, and LEDs: Two slide switches, four pushbuttons, and two RGB LEDs. Growth Connectors. Two Pmod ports and sixteen FPGA I/O. The Raspberry Pi connector shares eight pins with Pmod A. It is a connection that is compatible with the Arduino Shield.

5.11 Advantages of Integrating AI with IoT

1. **Enhanced Data Analytics:** AI in the IoT can analyze enormous volumes of data generated by the devices, providing real-time insights and actions. It helps filter out noise, find trends, and make forecast judgments faster and more correctly.
2. **Improved Operational Efficiency:** AI IoT enables automation, which enhances productivity and lowers the need for human involvement. By integrating AI-driven automation with IoT sensors, firms may improve supply chains, monitor equipment, and decrease downtime with predictive maintenance.
3. **Better Decision-Making:** The continuous and real-time data flow paired with AI analysis enables speedier and better-informed judgments. This can help industries respond to events proactively rather than reactively.
4. **Enhanced Security:** AI can detect irregularities and potential security concerns by analyzing IoT data in real-time. For example, odd access patterns or environmental abnormalities may be notified promptly, offering adequate IoT security.
5. **Personalization and customization:** AI IoT makes it possible to adapt user experiences, such as in smart homes, where the system learns users' preferences and changes temperature, lighting, or security settings accordingly.

5.11.1 Real-World Applications

1. **Smart Cities:** AI IoT supports smart city applications, including traffic control, garbage collection, and environmental monitoring. Sensors gather data in real-time, and AI processes it to enhance services and improve the quality of life.
2. **Healthcare:** Wearable IoT devices collect patients' health data continuously, and AI algorithms evaluate this data to detect and prevent health disorders, optimize treatment regimens, and notify medical professionals of crises.
3. **Manufacturing:** IoT sensors monitor machines, while AI algorithms evaluate data for predictive maintenance, quality control, and production optimization. This saves machine downtime and enhances production.

4. **Agriculture:** IoT sensors monitor soil moisture, temperature, and crop health. AI then analyzes the data, directing farmers on when to water, fertilize, or harvest, optimizing production and resource efficiency.
5. **Transportation and autonomous cars:** AI in IoT-equipped cars interprets real-time data from sensors to enable autonomous driving, optimize routes, and ensure safer roadways.

5.12 Application of the PYNQ-Z2 Board

The PYNQ-Z2 board can be suitable for applications requiring power and hardware acceleration, particularly in embedded systems, machine learning, and signal processing—some PYNQ-Z2 board applications.

➢ **Image and Video Processing**
The system requires real-time image and video processing capabilities, particularly for image dehazing, computer vision, and feature extraction, enabling faster execution of operations such as filtering, edge detection, and transformation.

➢ **Machine Learning and Acceleration**
This board can accelerate deep learning models, particularly for edge computing applications, as it supports frameworks in TensorFlow and PyTorch through customized FPGA overlays. With the capacity to do these tasks more quickly than conventional CPUs, applications include speech recognition, finding objects, and image classification.

➢ **Signal Processing**
The PYNQ-Z2 can effectively filter, modulate, and perform other DSP activities for digital signal processing (DSP) applications using the FPGA's parallel processing capacity, including audio and RF signal analysis. It is widely used for software-defined radio (SDR) system implementation and wireless communication research.

➢ **Internet of Things (IoT) and Edge Computing**
The PYNQ-Z2 board may be integrated into IoT systems where edge data processing is crucial. It is beneficial for industrial IoT, smart cities, and connected devices since it can process and analyze data locally, lowering latency and the need for cloud resources.

➢ **Control and Robotics**

Robotics and control systems can benefit from combining the board's processor and programmable logic. The board analyzes data, makes real-time decisions, and can interact with sensors and actuators. Real-time processing and control are essential in applications such as automated guided vehicles (AGVs), robotic arms, and autonomous drones.

➢ **Educational and Research Applications**

Researchers use it to test and prototype novel methods in signal processing, machine learning, and computer vision. The PYNQ-Z2 board is frequently used in educational settings to teach digital design, embedded systems, and FPGA programming because of its adaptability and accessibility.

5.13 Loss Function

By contrasting the output dehazed images with the ground truth images accessible in the dataset, the effectiveness or quality of the suggested or used image dehazing approach is assessed. Important performance parameters, including mean square error (MSE), Peak-Signal-to-Noise Ratio (PSNR), Structural Similarity Index (SSIM), and Mean Absolute Error (MAE), are assessed if ground truth or clear images are available.

1. Mean Square Error (MSE):

The error relating to the ground truth and dehazed pictures is associated with MSE. MSE is defined as the mean of the squares applied to this error. The MSE is computed using Equation 5.5. The MSE values range from 0 to ∞. The lower the MSE value, the better.

$$MSE = \frac{1}{N}\sum_{x=1}^{N}\sum_{y=1}^{N}[f(x,y)-u(x,y)]^2 \tag{5.5}$$

The intensity functions of the pixels in the output dehazed picture and ground truth image are represented by (x, y) and (x, y) in the above equation. The picture size is represented by $N \times N$, whereas x and y are the pixel coordinates.

2. Peak-Signal-to-Noise Ratio (PSNR):

The suggested approach provides the visual quality of the resulting pictures, which is assessed using the widely used PSNR metric. In image processing, it is the most often used metric for measuring improvement. The ratio of the ground truth picture's maximum pixel values to the MSE between (***f***) and the reconstructed image (***u***) is known as PSNR. Equation 5.6 calculates the PSNR (dB) for images ***f*** and ***u*** of size ***N*** × ***N***.

$$PSNR(f,u) = 10log_{10}\left(\frac{max_{1\le x,y\le N}\,|\,f(x,y)\,|^2}{\frac{1}{N^2}\sum_{x=1}^{N}\sum_{y=1}^{N}[f(x,y)-u(x,y)]^2}\right) \tag{5.6}$$

This measuring metric has nothing to do with the Human Visual System (HVS) properties, but it is entirely tied to the image's pixel values.

3. Structural Similarity Index (SSIM):
The perfection of structural similarity in terms of structure, contrast, and brightness from the perspective of image creation is measured by the SSIM metric. When it comes to extracting information about the structure of the pictures, HVS is the most flexible. Equation 5.7 may be used to evaluate the SSIM.

$$SSIM(f,u) = \frac{2\overline{fu}+C_1}{\overline{f}^2+\overline{u}^2+C_1}\cdot\frac{2\sigma_{fu}+C_2}{\sigma_f^2+\sigma_u^2+C_2} \tag{5.7}$$

where the means of ***f*** and ***u*** are represented by $\bar{f}$ and $\bar{u}$, the covariance between ***f*** and ***u*** by ***σfu***, and the standard deviation of ***f*** and ***u*** by ***σf*** and ***σu***, here, **C**1 and **C**2 are calculated using (***k***1***L***)2 and (***k***2***L***)2 with the constants ***k***1 = 0.01 and ***k***2 = 0.03. L represents the pixel's dynamic range value.

4. Mean Absolute Error (MAE):
An MAE, another error measure, is the absolute error between the ground truth and dehazed images; positive magnitude values are considered in this computation. If the MAE number is zero, the specific approach yields a high-quality output image, as expressed in Equation 5.8, which defines the MAE.

$$MAE = \frac{1}{N^2}\sum_{x=1}^{N}\sum_{y=1}^{N}|\,f(x,y)-u(x,y)\,| \tag{5.8}$$

For some real-time images, ground truth images may not be available; if ground truth images are unavailable, then evaluating the performance of the particular image dehazing technique is difficult. At this time, the following metrics are evaluated.

5.14 Conclusion

In this chapter, scalability, optimization, load control, and cache efficiency are all measurable advantages of integrating AI-based image dehazing into CDNs. Using dehazing as an example highlights the broader possibilities of AI and IoT applications in CDNs, opening the door for creative content distribution strategies tailored to real-time, quality-sensitive applications across industries. Teaches us to understand AI IoT based on the PYNQ-Z2 Board using deep learning and image dehazing. It researches this field, introducing us to basic image dehazing deep learning models and applying image dehazing in the PYNQ-Z2 board and application of AI IoT. We are taking some information from a previous reference work. First, a summary of the widely used integration of AIoT with CDN, image dehazing of its application, AI and IoT application, excellent datasets, generic loss functions, efficient network modules, and evaluation measures is provided. We will research a combination of image dehazing on the PYNQ Z2 board with graph neural networks.

AI disclosure statement: The author acknowledges the use of AI-based tools for minor language editing or grammatical correction. However, all intellectual contributions, analyses, and conclusions presented in this chapter are entirely the author's own.

References

1. Sahu, G., Seal, A., Yazidi, A., Krejcar, O., A dual-channel dehaze-net for single image dehazing in visual Internet of Things using PYNQ-Z2 board. *IEEE Trans. Autom. Sci. Eng.*, *21*, 1, 305–319, 2022.
2. Yamashita, R., Nishio, M., Do, R.K., Togashi, K., Convolutional neural networks: an overview and application in radiology. *Insights into Imaging*, 9, 611–29, Aug 2018.
3. Agrawal, S.C. and Jalal, A.S., A comprehensive review of analysis and implementation of recent image dehazing methods. *Arch. Comput. Methods Eng.*, 29, 7, 4799–850, Nov 2022.

4. Ancuti, C.O. and Ancuti, C., Single image dehazing by multi-scale fusion. *IEEE Trans. Image Process.*, 22, 8, 3271–82, May 13 2013.
5. Liu, L., Ouyang, W., Wang, X., Fieguth, P., Chen, J., Liu, X., Pietikäinen, M., Deep learning for generic object detection: A survey. *Int. J. Comput. Vision*, 128, 261–318, Feb 2020.
6. Ju, M., Ding, C., Ren, W., Yang, Y., Zhang, D., Guo, Y.J., IDE: Image dehazing and exposure using an enhanced atmospheric scattering model. *IEEE Trans. Image Process.*, 30, 2180–92, Jan 21 2021.
7. He, K., Sun, J., Tang, X., Single image haze removal using dark channel prior. *IEEE Trans. Pattern Anal. Mach. Intell.*, 33, 12, 2341–53, Sep 9 2021.
8. Mondal, K., Rabidas, R., Dasgupta, R., Single image haze removal using contrast limited adaptive histogram equalization based multiscale fusion technique. *Multimedia Tools Appl.*, 83, 5, 15413–38, Feb 2024.
9. Banerjee, S., Chaudhuri, S. S., Roy, S., Fuzzy logic and log-sigmoid function based vision enhancement of hazy images, in: *Proceedings of SAI Intelligent Systems Conference*, pp. 447–462. Cham, Springer International Publishing, 2018, September.
10. McCartney, E., Optics of the Atmosphere-Scattering by Molecules and Particles. *IEEE J. Quantum Electron.*, 14, 9, 698–699, 1978.
11. Narasimhan, S. and Nayar, S., Contrast Restoration of Weather Degraded Images. *IEEE Trans. Pattern Anal. Mach. Intell.*, 25, 6, 71–724, 2003.
12. Ancuti, C.O., Ancuti, C., Timofte, R., De Vleeschouwer, C., O-haze: A dehazing benchmark with real hazy and haze-free outdoor images, in: *Proceedings of the IEEE Conference on Computer Vision and Pattern Recognition Workshops*, pp. 754–762, 2018.
13. Ancuti, C.O., Ancuti, C., Sbert, M., Timofte, R., Dense-haze: A benchmark for image dehazing with dense-haze and haze-free images, in: *2019 IEEE International Conference on Image Processing (ICIP)*, Sep 22 2019, pp. 1014–1018.
14. Babu, G.H. and Venkatram, N., ABF de-hazing algorithm based on deep learning CNN for single I-Haze detection. *Adv. Eng. Software*, 175, 103341, Jan 1 2023.
15. Ancuti, C.O., Ancuti, C., Timofte, R., NH-HAZE: An image dehazing benchmark with non-homogeneous hazy and haze-free images, in: *Proceedings of the IEEE/CVF conference on computer vision and pattern recognition workshops*, pp. 444–445, 2020.
16. Jassim, S.A. and Khider, I., Implementation of deep learning models in FPGA development board for recognition accuracy enhancement. *Bull. Electr. Eng. Inf.*, 13, 6, 4008–17, Dec 1 2024.

6

Data-Driven Optimization of Content Distribution: Leveraging AI for Personalization and Scalability

Jagendra Singh[1*], Namita Nath[2], Lucky Gupta[3], Swapna Narla[4], Sreekar Peddi[5] and Dharma Teja Valivarthi[5]

[1]*School of Computer Science Engineering and Technology, Bennett University, Greater Noida, India*
[2]*Department of PGDM, Jaipuria School of Business, Ghaziabad, India*
[3]*Department of Management Studies (MBA), Raj Kumar Goel Institute of Technology, Ghaziabad, India*
[4]*Tek Yantra Inc., Folsom, California, United States of America (USA)*
[5]*Tek Leaders, Plano, Texas, United States of America (USA)*

Abstract

This chapter points toward optimization in higher education for the development and performance of content delivery to better cater to personal learning experiences, as per each and every student's taste and liking. The approach of a data-driven recommendation based on machine learning models was suggested, backed by a dataset of 15,000 records collected over six continuous months. Among the four models assessed, gradient boosting machines (GBMs), deep neural networks (DNNs), neural collaborative filtering (NCF), and random forest (RF), the GBM model was outstanding, with an accuracy of 97.67%. The results showed that the GBM model was correct, reaching an amazing 97.67%, followed by DNN at 94.5%, NCF at 92.3%, and RF at 87.56% in that order. With very high throughput and fast response times, it is especially suitable for real-time applications where the contents are delivered in a timely and relevant manner. This optimization can have better alignments with recommendations that will be matched with the academic history, interests, and interaction patterns of the students, seriously enhancing the quality and efficiency of content distribution. Ethical considerations regarding user

**Corresponding author*: jagendrasngh@gmail.com

Tanusree Chatterjee, Rajib Banerjee and Souvik Ganguli (eds.) Content Delivery Networks Using Artificial Intelligence and Internet of Things: Revolutionizing Content Delivery, (123–146) © 2026 Scrivener Publishing LLC

data privacy were also emphasized, ensuring secure and summarized handling of student information. The chapter has investigated how optimized AI-driven solutions may change the game of personalized content delivery in education.

Keywords: Data-driven optimization, course recommendations, artificial intelligence, machine learning models, personalized education

6.1 Introduction

The growing pressure to come up with effective content distribution systems attuned to the diversity of preferred pathways and choices among higher education students necessitates growing personalized learning. A huge number of students studying in vast strands require student-specific course recommendations [1, 2]. A high degree of such recommendations would significantly enhance their learning process, resulting in better engagement as well as academic performance. Such systems of education are incapable of meeting the needs of various types of learners, and in turn, it usually results in disengagement and low retention. However, in this context, introducing AI in learning environments appears to bring hope in bypassing the many shortcomings of traditional education and using data in personalizing learning experiences based on students' interests and capabilities [3, 4].

In essence, personalized learning refers to the instructional approaches that are divergent and adaptable according to the needs of every individual learner. Some examples include differentiated instruction, adaptive learning technologies, and course recommendations. Historical interpretations of personalized education indicate the concept has been known for quite a long time, but it has been implemented differently with the advent of technology. In the past, personal learning models depended on teachers' adjustments. However, this could not be practical with most things, as the increase in population in the classroom made it impossible [5, 6].

Modern technology and data analytics make it easier to establish more complex models of personalized learning, such as dynamic adjustments based on interactions and performance by students. Despite the benefits above, applying personalized learning to higher education faces the following challenges. Many institutions, due to resource constraints, cannot cater to each student uniquely. Again, traditional models of education tend to be stiff, allowing very little flexibility in terms of accommodation for individual learning styles and paces [7]. Thus, many students get lost and disengaged, which consequently negatively impacts their academic success. According to this, there is a high demand for systems that can effectively analyze the data of students in a manner that goes further than providing recommendations that support personalized learning pathways [8, 9].

Artificial intelligence serves as a corrective role for conventional education's shortcomings by providing new solutions for tailored approaches to learning. AI technologies can process vast datasets at breakneck speeds, meaning that these patterns and trends are impossible to discover for educators. This establishes a foundation to make intelligent recommendation systems so that they can recommend courses for learning based on the academic interests and learning preferences of students. Educational institutions can help improve rates of student retention and success by capitalizing on AI's vast potential to make learning environments more exciting and learner-centric [10, 11].

Recommendation systems—one of the most relevant applications of AI in education—represent algorithms that can recommend content specific to learners' interests. Recommendation systems are especially useful in institutions of higher education, where such a myriad of choices await students. Since recommendation systems can analyze the data of students, they can recommend courses tailored to the interests and academic goals of the student. Thus, this improved the learning experience of many students. Many machine learning techniques were used to create different recommendation systems, each with a strength or weakness [12, 13].

6.2 Problem Statement

The diverse range of courses in higher education calls for customized learning experiences that cater to the distinct needs and preferences of students. Unfortunately, many current content delivery systems fail to adapt to individual learners, making it difficult to provide tailored recommendations based on factors such as academic history, interests, and engagement. This lack of personalization limits students' ability to fully connect with course materials and achieve optimal learning outcomes. Educational institutions also grapple with the task of managing and analyzing vast datasets of student activity to create effective recommendation systems. Traditional methods often fall short in efficiently processing and deriving meaningful insights from such large-scale data. Furthermore, achieving high levels of accuracy, real-time responsiveness, and scalability in recommendation models presents a persistent challenge. Educational institutions encounter significant difficulties in managing and analyzing extensive datasets of student activity to develop effective recommendation systems. Conventional methods often fail to efficiently process and extract valuable insights from such large-scale data. Moreover, achieving high levels of accuracy, real-time responsiveness, and scalability in recommendation models continues to pose a considerable challenge.

6.3 Literature Review

Recommendation systems have become indispensable across various domains, offering personalized suggestions that enhance user engagement and satisfaction. These systems leverage advanced algorithms to analyze user preferences and deliver tailored recommendations, playing a crucial role in fields such as e-commerce, education, and entertainment. Among the cutting-edge approaches in this field, gradient boosting machines (GBMs) have garnered considerable attention for their ability to manage large-scale datasets while achieving exceptional predictive accuracy. GBM operates through an iterative process that combines multiple weak learners, typically decision trees, into a unified, robust predictive model. Each weak learner focuses on rectifying the errors of its predecessors, culminating in a cohesive ensemble that excels at capturing complex patterns within the data.

This methodology not only enhances predictive precision but also ensures adaptability to diverse user preferences, making GBM a cornerstone technique in the development of modern recommendation systems. Its effectiveness and versatility establish GBM as a foundational approach for advancing personalized content delivery. This review explores the theoretical underpinnings and practical applications of GBM, positioning it as an ideal starting point for investigating advanced methodologies in the domain. GBM proves suitable for personalizing recommendations, as it can capture complex data relations. In that regard, GBM is particularly adaptable to any form of data or features; thus, it is suitable for educational uses [14, 15].

Deep neural networks are another key player in the recommendation system. DNN performs very well when it comes to the deep relationship between objects that might be stored in complex datasets. So, those have proved to work very effectively in image recognition, natural language processing, and numerous recommendation systems [16]. The nodes can connect in several layers, and thus high-level representations of data can be learned by DNN, allowing more subtle predictions to be made. DNNs can further analyze students' learning histories and their preferences in order to provide course recommendations pertinent to each student, thus enhancing learning experiences [17].

The other advanced approach to a recommendation system is called neural collaborative filtering. NCF is a combination of neural networks with collaborative filtering techniques that helps it to leverage both user-item interactions and deep learning capabilities. This allows NCF to provide accurate recommendations not only based on personal preferences but also based on collective behaviors of similar users. Recognizing how users and items are related, NCF can devise specific suggestions that foster student engagement and selection [18, 19].

Random Forest is one of the most widely known techniques developed in machine learning and also applied to recommendation systems. It constructs many decision trees during training and outputs the mode of their predictions for classification tasks or the average for regression tasks. This method of ensembles provides a robust approach to difficult datasets, thus avoiding overfitting. RF can perform a set of features in the context of personalized course recommendations in order to derive accurate suggestions concerning a student's academic performance and interests based on behavior data [20].

While machine learning models have achieved incredible success in the context of personalized learning, concerns should also be taken regarding data privacy and ethics. More specifically, as the data-driven solution is increasingly becoming essential in educational institutions, so is the protection of student information. Consequently, user data confidentiality should be a critical ethical consideration to foster trust between institutions and students. To prevent privacy threats, strong data protection measures in anonymizing and aggregating user data are therefore crucial. Thus, avoiding the direct association of personal information with exact recommendations or results will keep these institutions just safe havens for students while unlocking the real capacity of AI in education [21, 22].

Evidently, transparency from institutions towards the policies on how data from these settings is used and measures to ensure privacy amongst the users will help. Students will be more open to these programs if it is explained how their data is being collected, processed, and used. Besides that, adherence to ethical principles in handling student data will enable the institutions to comply with the regulations more effectively and protect the interests of the students [23].

Despite such an advancement toward the adoption of AI in personalized learning, there is an element of deficiency in the literature with respect to the appropriate implementation of such technologies in real-life education settings. While many studies analyze the effectiveness of machine learning models in recommendation systems, there is still much scope for research to understand in-depth how students interact with such systems.

There are also a few areas where no comprehensive research has been conducted in studying the longitudinal effects of AI-driven course recommendations on student outcomes, such as retention rates and overall academic performance [24].

This problem was addressed by developing a robust and data-driven system that improves AI in educational content recommendation by effectively personalizing course suggestions for students according to their unique profiles and interaction patterns. It illustrates how AI can leverage big datasets based on information related to academic performance, course enrollment, expressed interest, and behavior to produce personal recommendations through sophisticated machine learning models such as gradient boosting machines, deep neural networks, neural collaborative filtering, and random forests. This systematic approach to data collection and analysis not only demonstrates the feasibility of developing AI systems in educational contexts but also indicates that AI technologies could bring about noteworthy improvements in student engagement and learning outcomes [25]. The findings of this study also highlighted the need for ethical consideration and data privacy by ensuring that its AI-based solutions for education would be best practices in the protection of student information.

6.4 Methodology

In modern educational landscapes, online learning platforms have outgrown themselves with exponential growth, thus giving students full access to numerous varied sources of academic content that they never had before. The online course delivery models, taken up by universities, have thus helped students to learn from wherever they are and at whatever time they want, gaining access to a greater amount of knowledge they otherwise could not attain. However, the vastness of academic content poses a challenge to students in identifying relevant courses that align with their interests and academic needs. This, therefore, opens up an urgent need to optimize the content distribution of online platforms such that students easily discover and enroll in the most relevant courses.

Personalized course recommendations are of great importance. An insightful analysis based on the information about a student's educational background, interests, and history of interactions on the platform may lead to the development of a recommendation system that is tailored for every student and enhances both their satisfaction and academic outcome. Most content recommendation systems lack the integration of an insightful understanding of user behavior into the recommendations they generate and usually depend on only very basic indicators of academic performance or demographics. This research fills that gap by building a data-driven system that uses AI to personalize the course recommendation towards optimization of content dissemination for both the students and the institution.

For this study, a dataset was curated from a university website that provided a range of academic courses online. Students could surf through the courses on offer, enroll in their subjects of interest, and interact directly with the course material. The interacting students were collected over a period of six months. The dataset contains 15,000 records, and information ranges from their academic profiles to previous enrollments for courses, interest expressions, grades, and demographic information. Behavioral data pertaining to time spent on browsing course descriptions, clicks on course categories, and course selection patterns are also collected in order to understand the preferences and actions of students.

This study focuses on developing and testing a data-driven course recommendation system designed to enhance students' learning experiences by tailoring recommendations based on their academic performance, preferences, and behaviors. The primary objective is to optimize content delivery, allowing students to comprehensively explore course options while facilitating streamlined navigation through available choices. By improving the efficiency of course recommendations across the university, this system aims to support better academic outcomes and resource utilization. To achieve these goals, four advanced AI models were employed: neural collaborative filtering (NCF), deep neural networks (DNN), random forest classifier (RF), and gradient boosting machines (GBM). Among these, GBM consistently emerged as the most effective model, demonstrating superior performance across key metrics, including accuracy, precision, recall, and F1-score. Its ability to deliver highly relevant and robust recommendations positioned it as the most suitable model for the system.

While GBM excelled, the other models also provided valuable insights. DNN and NCF effectively captured complex, non-linear relationships within the data, showcasing their capability in understanding intricate patterns. Although RF exhibited slightly lower accuracy compared to the other models, it still contributed meaningful insights into the recommendation process. The overall research architecture, as illustrated in Figure 6.1, outlines the end-to-end process, from data preprocessing to model evaluation, highlighting the systematic approach adopted to achieve optimized course recommendations.

The collected dataset is preprocessed accordingly for the suitability of model training. Handling data, normalization of continuous features (for example, grades), and encoding categorical variables were done according

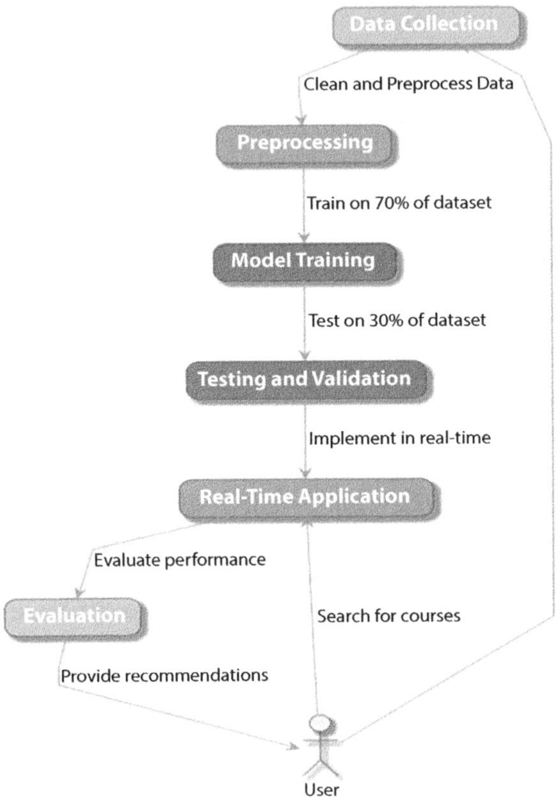

Figure 6.1 Architecture of the entire research.

to course category, student demographics, etc. It used a 70/30 split into a training set and a testing and validation set, respectively. This ensures a very high evaluation of the adequacy of the models in predicting course preferences among the students. The evaluation was conducted with standard metrics like accuracy, precision, recall, and F1-score to measure the effectiveness of each model in suggesting personalized courses. In particular, cross-validation ensures that the models work robustly for a variety of different subsets of the data.

6.4.1 Data Preprocessing

Data preprocessing is, therefore, an incredibly crucial role that research entails in handling the cleanup, consistency, and suitability of a given dataset for training AI models. The dataset of interest culled from the university website has been ranged across various data types, including academic profiles, behavioral information, demographic details, and interaction history. This entire dataset required several preprocessing steps to ready it up for machine learning models, which included missing data handling, normalization of numerical features, encoding categorical variables, feature engineering, and finally, data splitting.

Firstly, in preprocessing the dataset, missing data were identified and handled. In real-world datasets, usually, some records would contain incomplete information, like missing grades, incomplete enrollment history in courses, or unrecorded details of demography. Such missing values can also negatively impact the performance of these machine learning models since they distort the patterns necessary for these models to learn. Different methods were used according to the type and importance of the missing fields.

For numerical features, like grades of students and also the time spent on course descriptions, missing values were imputed. For numerical features, the mean imputation method was used. In this case, for each feature, the missing values were replaced by the mean value of available data for the respective feature. This will not badly affect the data's distribution. Large portions of a student's profile or interaction history were missing, and so those records were removed from the dataset to preserve the integrity of the data.

The second group of problems was categorical attributes, which consisted of course categories or demographic variables. For categorical features with missing values, the mode imputation technique was applied. In this technique, the missing values were replaced by the mode, or the most frequent value for that feature. Thus, it maintains the original structure of data and, at the same time, enables a significant contribution of categorical variables to the machine learning models.

After treating missing data, the normalization of numerical features has been used. Normalization is necessary for features like student grades or the number of hours spent on course descriptions. These features hold highly scaled values in their own right. If not normalized, these features would dominate each other in most of the training of a model, and biased or skewed predictions would occur. Min-max normalization was used in the above research. This transformation scales data into a value falling within a certain range, usually 0 to 1, but in no way affects the relative differences between the values. For instance, where students' grades were previously graded on a scale from 0 to 100, the same was rescaled so that the highest grade was reduced to 1 and the lowest to 0. Similarly, interaction times, recorded in seconds, were normalized to a 0-1 range. This normalization also ensured that one feature cannot dominate the AI models.

The course categories, majors of students, and demographic information, such as gender and age group, were categorical in nature and thus required a transformation in order to be used by a machine learning algorithm. Most contemporary machine learning algorithms, and those used during this study, need numeric input; categorical variables, therefore, need encoding. The main technique followed in this task was one-hot encoding. In one-hot encoding, each category is encoded as a binary vector, where there will be one element that is "1," while all the other elements will be "0." For example, if the majors of a student were "Computer Science," "Business," and "Engineering," then these categories would change into three columns that are of binary types. The specific column would have a value of 1 if the student belonged to a certain category; otherwise, it remains 0. This method of encoding facilitates the machine learning models to understand the categorical type of data without a default ordering defined between categories.

Another critical preprocessing step is feature engineering. Feature engineering is essentially the process of designing new features or transforming existing ones to make the dataset more predictive. In this work, several derived features have been generated in order to more sensitively capture the students' preferences and behaviors. For example, a new feature representing how busy a student was during the course was calculated as the sum of the total courses enrolled in by the student. The average time spent on course descriptions is another engineered feature, which would help reflect the interest a student may have in certain subjects. Another measure was that of completed courses over enrolled courses, which provided further information in terms of commitment and the likelihood of academic success on the part of the student. All these engineered features added depth to the dataset and provided insights into more intricate patterns of student behavior to the AI models.

The final step in the preprocessing phase involved splitting the dataset into training and testing data. To ensure robustness and generalization, we divided the dataset in a manner in which 70% of the records were used for training the models, while the remaining 30% were for testing and validation purposes. This split ensures that the models can learn most of the data, but then are tested on an unseen portion of the dataset to give a realistic measure of their performance. The training set is for fitting the model, and the testing set is used to help evaluate how well a model generalizes to new, unseen data. Cross-validation techniques were also implemented during training with the purpose of validating the performance of the model, so overfitting is avoided in this case.

6.4.2 Artificial Intelligence Models

Neural collaborative filtering is highly useful in recommendation systems by extending the traditional matrix factorization method, but with deep learning. In traditional matrix factorization, latent vectors are used to represent both users and items, and the interaction between each of them is predicted using the dot product of the vectors. However, NCF replaces this dot product with a non-linear function represented through a neural network, and it does this to enhance the capacity of the model to grasp complex interactions of users with items. The main equation that NCF is based on is

$$y^{\wedge}_{ui} = f(P_u, Q_i) \tag{6.1}$$

where $y^{\wedge}_{ui}$ is the predicted interaction between user u and item i, P_u and Q_i represent the latent vectors for the user and item, and f is a non-linear function modeled by a neural network. This research aims to adapt the NCF model, where student profiles act as user inputs, courses are considered as items, and the most probable courses a student can derive benefits from or enroll in are calculated, given their academic background and interactions. Based on this model, it incorporates both implicit feedback regarding clicks and time spent on the descriptions of courses, as well as explicit feedback retrieved from previous enrollments in courses. Since NCF grasps complex relations within several layers of neural networks, this architecture seems very promising for the representation of personalized course recommendations that adapt to the specific profile of each individual student.

Another significant model applied in this study is deep neural networks, or DNN for short. DNNs are capable of learning a very complex pattern by including multiple layers of neurons that progressively extract more abstract features from the input data. Therefore, this architecture makes it suitable for modeling non-linear relationships, which allow DNN to capture complex patterns in student data. The mathematical representation of a DNN can be defined as

$$y^{\wedge} = \sigma(W_n\sigma(W_{n-1}\ldots\sigma(W_1x + b_1) + b_{n-1}) + b_n) \tag{6.2}$$

where y^ is the predicted output (in this case, a course recommendation), x represents the input feature vector (such as student academic profiles, interaction history, and expressed interests), Wi and bi are the weight matrices and bias vectors for each layer, and σ is the activation function applied to each layer.

This research used both student academic records and behavioral data in training the DNN so that it could learn which course would be most suitable for any given student. The advanced capability of this DNN in learning complex, non-linear relationships between features ensures accuracy in course recommendations as it is individualized to the student's needs. Random Forest is the other model used in the research. This model is an ensemble learning model that comes up with multiple decision trees and combines results to make more accurate predictions. Every tree in the forest gets a random subset of the data, and the final result will be the average of the results of all trees, which decreases overfitting and increases the robustness of the model. The equation that defines the Random Forest model is given by:

$$y\wedge = \frac{1}{N}\sum_{i=1}^{N} T_i(x) \tag{6.3}$$

where y^ is the predicted output, N is the number of decision trees in the forest, and $T_{i(x)}$ is the prediction made by the i-th tree for input x. In the context of this study, it is applied in predicting how students would probably be inclined to a course based on a range of features. These features are somewhat mixed and highly non-linear, and given that Random Forest models tend to easily adapt to such relationships in the data, the ensemble nature is retained in the process for predicting student course preferences. This makes the Random Forest system naturally accommodate both categorical and continuous data. This is, therefore, an advantageous characteristic for this study since it will deal with a mix of student demographic information, grades, and behavioral data.

Gradient boosting machines, or GBMs, are a complex ensemble technique that learns decision trees one at a time and builds upon previously learned decision trees, with the goal of correcting errors in the sequence of previously learned trees. In contrast to Random Forest, which learns decision trees independently, GBM directly focuses on the iterative reduction of the amount of prediction error by optimizing the model through gradient descent. The GBM equation is

$$F_m(x) = F_{m-1}(x) + \eta \sum_{i=1}^{N} \nabla L(y_i, F_{m-1}(x_i)) \tag{6.4}$$

where $F_{m(x)}$ is the prediction at iteration m, $F_{m-1(x)}$ is the prediction from the previous iteration, η is the learning rate, L is the loss function, and ∇L is the gradient of the loss function with respect to the previous prediction. In this research, the GBM model is used in order to represent the students' preferences using iterations that improve the prediction at each stage. The model takes care of the errors committed by the previous trees and fine-tunes its predictions by minimizing the loss function. Because of the minimization of the error, the course recommendations prove to be accurate. One of the interesting features of GBM is its iterative nature, which allows it to focus more attention on hard cases—that is, those cases from which it would otherwise have difficulty predicting—and therefore generalizes better and is more accurate than other models.

6.4.3 Simulation Environment and Experimental Setup

To ensure efficient processing and accurate evaluation of the machine learning models, a robust simulation environment was established. All implementation and testing were performed using Python 3.9, leveraging its extensive library ecosystem to support the development and evaluation of machine learning algorithms. Key libraries used in this research included scikit-learn for implementing traditional machine learning models, TensorFlow and Keras for designing and training Deep Neural Networks (DNN), and PyTorch for implementing Neural Collaborative Filtering (NCF). These libraries provided the necessary tools for constructing, training, and optimizing the models.

The experiments were executed on a high-performance computing system equipped with an NVIDIA RTX 3080 GPU, 64 GB of RAM, and an Intel Core i9-10900K processor. This configuration was crucial for handling the computationally intensive tasks involved in training models on large datasets, ensuring smooth and efficient operations.

Data preprocessing and analysis were conducted using Pandas and NumPy, which facilitated efficient manipulation of the dataset and preparation for model training. Visualization of results, including confusion matrices, was achieved using Matplotlib and Seaborn, enabling a clear and detailed representation of performance metrics. This experimental setup effectively simulated real-world conditions, providing a reliable platform for evaluating the accuracy and performance of machine learning models in personalized course recommendation systems.

6.5 Result and Discussion

In this research, 70% of the dataset is utilized for training every model, whereas the remaining 30% is kept aside for testing and validation purposes. Historical data on student interaction, enrollment for courses, academic profiles, and behavioral data were fed into the models to learn patterns and relationships that enable the accurate recommendation of courses. It was then passed on to the unseen 30% of the dataset so that the performance in realistic conditions could be determined. The results of the accuracy are shown in Figure 6.2. The results indicate that, in terms of the appropriate contents of the courses, the results were mainly high for GBM models, with an impressive 97.67%. The reason behind this is that the GBM model made the highest predictions possible due to iterative

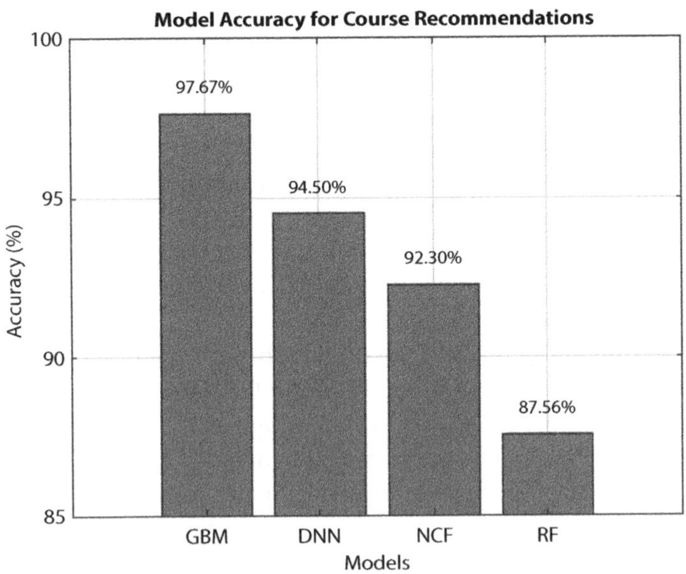

Figure 6.2 Accuracy of each model.

improvement by learning from previous errors. The accuracy of the DNN model was very high: 94.5%, and it really captured the complex, non-linear relationship between student preferences and course offerings.

The accuracy of NCF was 92.3%, which proved the efficiency of this model in the realization of the user-item interaction model for the recommendation task. Finally, the accuracy of the RF model was 87.56%, although this was a little less than the accuracy of the other models. However, very robust recommendations have been offered through the ensemble learning approach.

Figure 6.3 shows that the precision, recall, and F1-score values of the different models used in making the course recommendation prediction are as follows: The GBM model is a great model with maximum precision, 96.8%, meaning nearly all the courses it recommended were relevant, and is one of the models having a great recall score at 97.5, indicating it gets the correct percentage of the relevant courses. This yields the highest F1-score of 97.1%, a balanced and superior performance. The DNN model follows closely with a precision of 93.5% and a recall of 95.0%, which produces a fairly good F1-score of 94.2%, indicating that the relationships are well recaptured and excellent recommendations are made. NCF finally

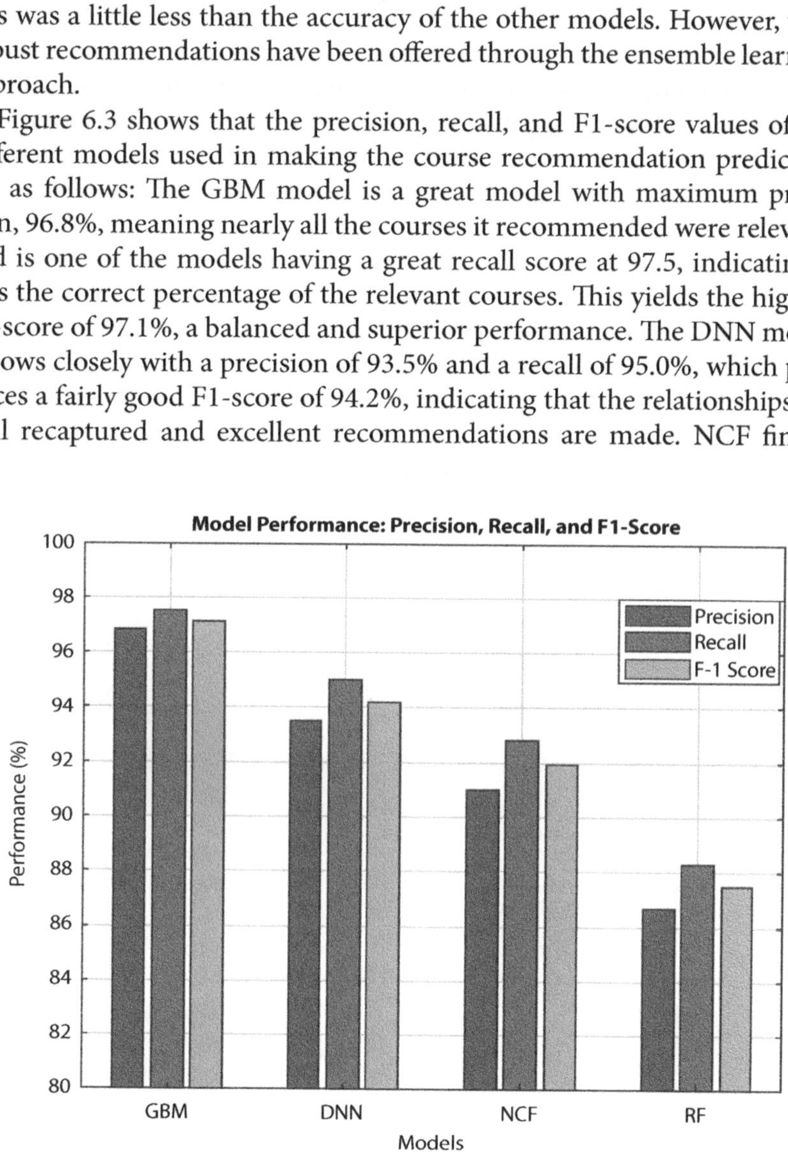

Figure 6.3 Performance score of each model.

performed slightly worse, at a precision of 91.0%, a recall of 92.8%, and an F1-score of 91.9%, indicating a strong performance but less robustness than the former GBM and DNN models. The lowest precision and recall were reported by the RF model, 86.7% and 88.3%, respectively, with an F1-score of 87.5%, meaning it is potent but lags behind the other models in both accuracy and balanced performance.

The confusion matrices for each model give insight into how the different models perform with regard to predicting relevant and non-relevant course recommendations, as shown in Figure 6.4. Starting with GBM, it had the best result, having 1,420 True Positives, which means that it correctly recommended 1,420 relevant courses with only 40 False Positives, which is the wrong recommendation of non-relevant courses. Also, it had 1,505 true negatives (TN), meaning courses that were not relevant were correctly labeled as such, with only 35 false negatives (FN), meaning the relevant course was missed. These value points contain the better functionality of GBM with minimal misclassification.

The DNN model was followed closely, with 1,365 True Positives and 65 False Positives, showing that it showed a slight increase in errors when recommending irrelevant courses. In fact, it had 1,480 true negatives and 90 false negatives, meaning that it missed a few more relevant courses than

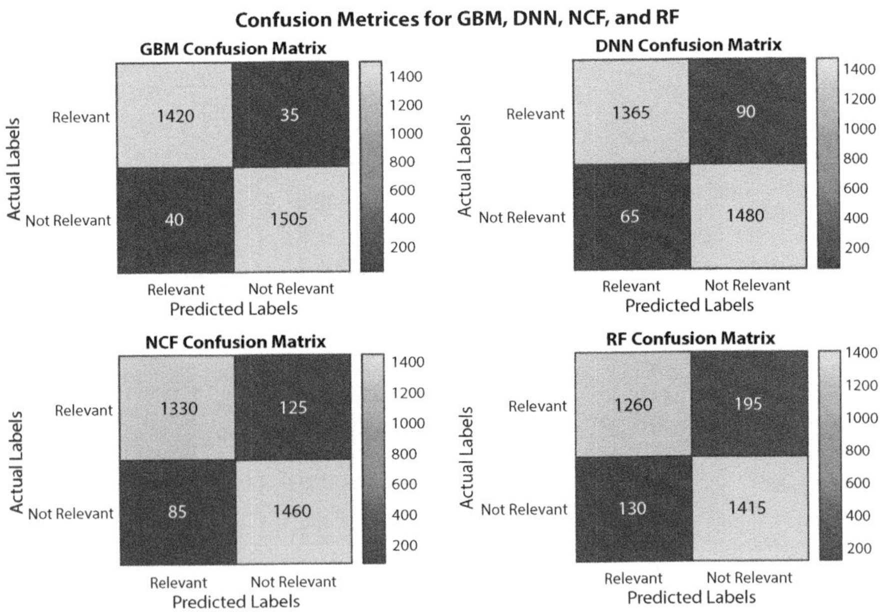

Figure 6.4 Confusion matrices of each model.

GBM, but in all fairness, it kept up an excellent performance. NCF was less accurate at 1,330 True Positives and 85 False Positives. In addition, it also yielded 1,460 true negatives and 125 false negatives; in other words, this requires learning and distinguishing between the courses as relevant or non-relevant. Finally, the RF model had the most significant number of errors, with 1,260 True Positives and 130 False Positives, and 1,415 true negatives and 195 false negatives. This points out that RF was less efficient in forecasting relevant courses and did not differentiate the non-relevant ones, hence lower overall accuracy as compared to the rest of the models.

After the testing, all the models ran well except model GBM, which was the most related to the recommendations of the course. All the models, including GBM, DNN, NCF, and RF, were implemented on live testing. A test experiment was conducted over a 24-hour period by asking 200 students to search for courses they are interested in to test their performance on a live system. The latency in each model, meaning the time taken to generate course suggestions, was measured at one-hour intervals throughout the day. The results of latency are shown in Figure 6.5.

Throughout all these time slots, the GBM model provided the fastest responses. For instance, from 00:00 to 01:00, the latency of the GBM model was only 120 milliseconds, while DNN took 140 milliseconds, NCF 160

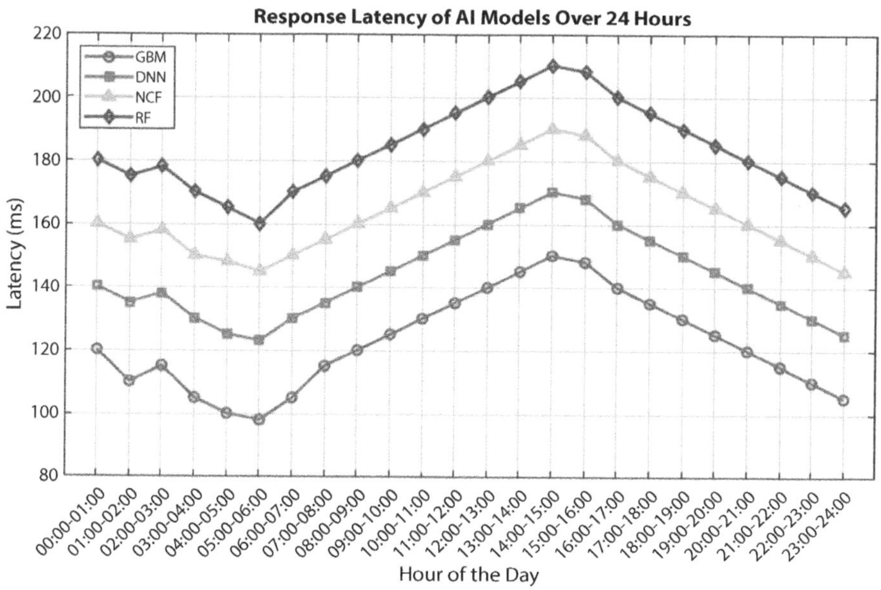

Figure 6.5 Latency of each model.

milliseconds, and RF 180 milliseconds. The practice was repeated all day, maintaining low levels of GBM latency, especially with high usage periods. They show that during peak hours between 12:00 and 13:00, the response latency of GBM was 140 milliseconds, DNN stood at 160 milliseconds, NCF at 180 milliseconds, and RF at 200 milliseconds.

As time progressed, GBM models had been thoroughly outperforming the rest in terms of speed, as latency was measured to be as low as 98 milliseconds between 05:00 and 06:00, while DNN registered 123 milliseconds, NCF 145 milliseconds, and RF 160 milliseconds. It was able to gain high performance during the relatively less active hours; its average latency recorded during the night was 110 milliseconds. DNN was at 135 milliseconds, NCF at 155 milliseconds, and RF at 175 milliseconds. In comparison, the GBM model showed tremendous real-time performance compared to the rest, both in terms of accuracy and quick suggestions for students.

The throughput result shown in Figure 6.6 shows how many courses each model can efficiently give recommendations to the students within the 24-hour real-time testing period. GBM had the highest throughput and topped over the entire period of testing, in which the highest recommendation per second for GBM was attained at 10.20 within the last hour of testing. This means that not only did GBM give the right predictions but

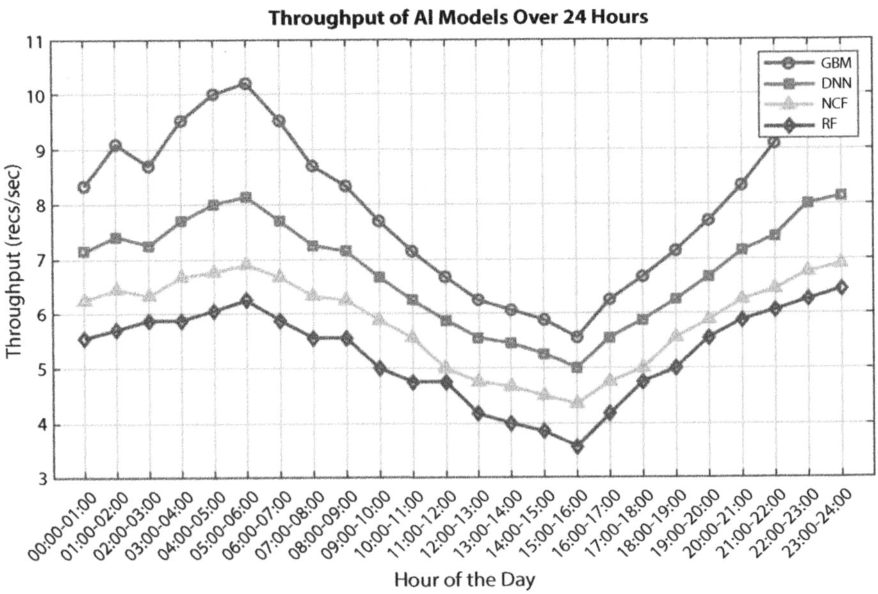

Figure 6.6 Throughput of the AI model for 24 hours.

also could accomplish this task very fast, hence making sure that students got suggestions in time to advise their choice, of course.

On the other hand, the DNN model performed relatively smoothly, with a peak throughput of 8.13 recommendations per second. Even though it was very effective, it was far slower than GBM. The NCF model had a lower throughput, with a maximum of 6.90 recommendations per second, meaning the model would take longer to generate recommendations, probably due to the collaborative filtering techniques that involve extra computation based on user interactions. The RF model recorded the lowest throughput, peaking at only 6.45 recommendations per second. As such, there could be a plodding reason for this, such as the elaborateness of the ensemble method, wherein there are many decision trees that may take more time to process and aggregate results.

In general, all models performed well, but the GBM model provides a superior throughput with higher accuracy. It stands out as the most effective solution if implemented in real-time course recommendation systems. The model also smoothens the need for speed with predictive performance and thereby enhances user experience. The discussion section highlights the implications and significance of the research findings regarding the data-driven optimization of content distribution through AI-based course recommendations. The results indicate that the gradient boosting machine (GBM) model outperformed the other models in both accuracy and throughput, making it a robust choice for real-time applications. The high accuracy of 97.67% not only demonstrates the model's effectiveness in understanding and predicting student preferences but also emphasizes its potential for enhancing the overall learning experience. As educational institutions increasingly rely on personalized learning paths to engage students, the findings suggest that integrating such advanced models into course recommendation systems can lead to improved enrollment rates and student satisfaction.

Moreover, the study underscores the importance of utilizing a diverse dataset that includes various factors, such as academic performance, interests, and interaction history. This comprehensive approach allows for more nuanced and relevant recommendations, aligning with the current trend of personalized education. However, the research also highlights the need for continuous model evaluation and adaptation as student behaviors and course offerings evolve.

To ensure the confidentiality of user details, the research adhered to stringent ethical standards and data protection protocols. All personal information was anonymized and aggregated to protect the identities of the participating students. This means that individual user data were not directly linked to specific recommendations or outcomes, thereby maintaining privacy. Additionally, access to the raw data was restricted, and only authorized personnel involved in the research had access to any identifiable information.

6.5.1 Discussion

The findings of this study underscore the effectiveness of advanced machine learning models in optimizing content delivery for personalized course recommendations. By partitioning the dataset into 70% for training and 30% for testing and validation, the models successfully learned and validated intricate patterns derived from student interaction, academic history, and behavioral data.

Among the models evaluated, gradient boosting machines (GBMs) emerged as the top performer, achieving an outstanding accuracy of 97.67%. It demonstrated superior precision (96.8%), recall (97.5%), and F1-score (97.1%), reflecting its capability to iteratively refine predictions, minimize errors, and maximize relevant recommendations. These metrics highlight GBM's efficacy in delivering precise and timely course recommendations, reinforcing its suitability for real-time applications.

Deep Neural Networks (DNN) followed closely, with an accuracy of 94.5%. Its ability to model complex, non-linear relationships enabled it to achieve a precision of 93.5%, a recall of 95.0%, and an F1-score of 94.2%, establishing it as a robust alternative for capturing nuanced user preferences and providing meaningful recommendations.

Neural Collaborative Filtering (NCF) achieved an accuracy of 92.3% and an F1-score of 91.9%, showcasing its strength in modeling user-item interactions effectively. Although the Random Forest (RF) model delivered a respectable accuracy of 87.56%, it lagged in overall precision, recall, and balanced performance compared to the other models.

Analysis of confusion matrices revealed that GBM exhibited minimal misclassification, accurately distinguishing between relevant and irrelevant courses. This ability to deliver precise recommendations underscores the value of ensemble learning techniques in improving content delivery systems.

Overall, the results validate the importance of iterative learning and ensemble approaches in driving the effectiveness of personalized course recommendations. By providing students with accurate, timely, and relevant content, these models demonstrate their potential to enhance educational experiences and outcomes.

6.6 Conclusion

This research successfully demonstrates the potential of data-driven optimization in content distribution through AI-enhanced course recommendations within a university setting. With the dataset encompassing students' academic profiles, interests, and interaction histories, the study evaluated four distinct machine-learning models: GBM, DNN, NCF, and RF. The findings revealed that the GBM model significantly outperformed the others in terms of accuracy and throughput, achieving a remarkable accuracy of 97.67% and effectively catering to student preferences in real-time. This underscores the model's capacity to facilitate personalized learning experiences, ultimately enhancing student engagement and satisfaction. The research not only highlights the importance of incorporating advanced AI techniques in educational frameworks but also emphasizes the necessity for ongoing model adaptation to align with evolving student needs and behaviors. Furthermore, the ethical handling of user data, ensuring confidentiality and anonymity, is a critical aspect of this research, reinforcing the commitment to responsible data usage in educational technology. Future research could explore integrating additional data sources, such as real-time feedback and emerging trends in course popularity, to further refine the recommendation algorithms.

AI disclosure statement: The author confirms that no artificial intelligence (AI)-generated content or AI-assisted writing tools were used in the preparation of this chapter (including summaries and the abstract). All text, analysis, and conclusions are the result of the author's original work.

References

1. Mehra, A., Leveraging Data-Driven Insights to Enhance Market Share in the Media Industry. *J. Res. Appl. Sci. Biotechnol.*, 2, 3, 291–304, 2023.
2. Deekshith, A., Scalable Machine Learning: Techniques for Managing Data Volume and Velocity in AI Applications. *Int. Sci. J. Res.*, 5, 5, 2023.
3. Yaiprasert, C. and Hidayanto, A.N., AI-powered in the digital age: Ensemble innovation personalizes the food recommendations. *J. Open Innov.: Technol. Mark. Complex.*, 10, 2, 100261, 2024.
4. Kaliyar, R.K., Kumari, R., Sharma, N., Manjunath, S., Prasad, P.D., Advancements in Early Detection of Cervical Cancer using Machine Learning and Deep Learning Models for Cervicography Analysis, in: *2024 International Conference on Emerging Innovations and Advanced*

Computing (INNOCOMP), IEEE, pp. 252–256, 2024, doi: 10.1109/INNOCOMP63224.2024.00049.
5. Bassani, E. and Pasi, G., A multi-representation re-ranking model for Personalized Product Search. *Inf. Fusion*, 81, 240–249, 2022.
6. Singh, N.P., Hasan, D.S., Shelke, N.A., Namdev, A., Girdhar, N., An Integrated Machine Learning and IoT Based Approach for Enhanced Healthcare Efficiency and Personalized Treatment, in: *Intelligent Systems Design and Applications, ISDA 2023*, Lecture Notes in Networks and Systems, vol. 1046, A. Abraham, A. Bajaj, T. Hanne, P. Siarry (Eds.), Springer, Cham, 2024, https://doi.org/10.1007/978-3-031-64813-7_16.
7. Yang, H., Zhu, D., He, S., Xu, Z., Liu, Z., Zhang, W., Cai, J., Enhancing psychiatric rehabilitation outcomes through a multimodal multitask learning model based on BERT and TabNet: An approach for personalized treatment and improved decision-making. *Psychiatry Res.*, 336, 115896, 2024.
8. Garg, N., Kukreja, S., Saini, M., Singh, I., Mall, S., IoT-Enhanced Agricultural Water Management System Utilizing Conventional Machine Learning Methods, in: *2023 International Conference on Modeling, Simulation & Intelligent Computing (MoSICom)*, IEEE, pp. 554–559, 2023.
9. Gulzar, B., Sofi, S.A., Sholla, S., Exploring Personalized Internet of Things (PIoT), social connectivity, and Artificial Social Intelligence (ASI): A survey. *High-Confid. Comput.*, 100242, 2024.
10. Gupta, T., Singhal, D., Lal, B., Turukmane, A.V., Fake News Detection Using Transfer Learning, in: *Recent Trends in Image Processing and Pattern Recognition, RTIP2R 2023*, Communications in Computer and Information Science, vol. 2026, Springer, Cham, 2024, https://doi.org/10.1007/978-3-031-53082-1_2.
11. Mosqueira-Rey, E., Hernández-Pereira, E., Alonso-Ríos, D., Bobes-Bascarán, J., Fernández-Leal, A., Human-in-the-loop machine learning: a state of the art. *Artif. Intell. Rev.*, 56, 4, 3005–3054, 2023.
12. Sahu, D.P., Murkute, S., Yadav, U., Agarwal, M., Kumar, P., Deep Learning Based Bug Detection in Solidity Smart Contracts, in: *Recent Trends in Image Processing and Pattern Recognition, RTIP2R 2023*, Communications in Computer and Information Science, vol. 2026, Springer, Cham, 2024, https://doi.org/10.1007/978-3-031-53082-1_9.
13. Reddy, S.G., Sadhu, A.K.R., Muravev, M., Brazhenko, D., Parfenov, M., Harnessing the Power of Generative Artificial Intelligence for Dynamic Content Personalization in Customer Relationship Management Systems: A Data-Driven Framework for Optimizing Customer Engagement and Experience. *J. AI-Assisted Sci. Discovery*, 3, 2, 379–395, 2023.
14. Sharan, A., Co-occurrence and semantic similarity based hybrid approach for improving automatic query expansion in information retrieval, in: *Lecture Notes in Computer Science*, vol. 8956, 2015, https://doi.org/10.1007/978-3-319-14977-6_45.

15. Banik, B., Banik, S., Annee, R.R., AI-Driven Strategies for Enhancing Customer Loyalty and Engagement Through Personalization and Predictive Analytics. *Int. J. Mach. Learn. Res. Cybersecur. Artif. Intell.*, 15, 1, 416–447, 2024.
16. Bilal, K. and Sajid, M., Blockchain technology: Opportunities & challenges, in: *2022 International Conference on Data Analytics for Business and Industry (ICDABI)*, IEEE, pp. 519–524, 2022.
17. Potla, R.T. and Pottla, V.K., AI-Powered Personalization in Salesforce: Enhancing Customer Engagement through Machine Learning Models. *Valley Int. J. Digital Lib.*, 1388–1420, 2024.
18. Prasad, M., Prasad, O.K., Meng Joo, E., Saxena, A.K., Lin, C.T., A novel fuzzy logic model for pseudo-relevance feedback-based query expansion. *Int. J. Fuzzy Syst.*, 18, 980–989, 2016.
19. Sasmal, S., Streamlining Big Data Processing with Artificial Intelligence. *Int. Res. J. Eng. Appl. Sci. (IRJEAS)*, 11, 3, 2023.
20. Lin, C.T. and Li, D.L., Ranks aggregation and semantic genetic approach based hybrid model for query expansion. *Int. J. Comput. Intell. Syst.*, 10, 1, 34–55, 2017, https://doi.org/10.2991/ijcis.2017.10.1.4.
21. Yongli, G., Qi, D., Zhipeng, C., Leveraging the Synergy of IPv6, Generative AI, and Web Engineering to Create a Big Data-driven Education Platform. *J. Web Eng.*, 23, 2, 197–226, 2024.
22. Singhal, P., Gupta, S., Singh, J., An Integrated Approach for Analysis of Electronic Health Records Using Blockchain and Deep Learning. *Recent Adv. Comput. Sci. Commun. (Formerly Recent Pat. Comput. Sci.)*, 16, 9, 1–10, 2023.
23. Iyelolu, T.V., Agu, E.E., Idemudia, C., Ijomah, T.I., Leveraging Artificial Intelligence for Personalized Marketing Campaigns to Improve Conversion Rates. *Int. J. Eng. Res. Dev.*, 20, 8, 2024.
24. Moro-Visconti, R., Artificial Intelligence-Driven Digital Scalability and Growth Options, in: *Artificial Intelligence Valuation: The Impact on Automation, BioTech, ChatBots, FinTech, B2B2C, and Other Industries*, pp. 131–204, Springer Nature Switzerland, Cham, 2024.
25. Rane, J., Mallick, S.K., Kaya, O., Rane, N.L., Scalable and adaptive deep learning algorithms for large-scale machine learning systems, in: *Future Research Opportunities for Artificial Intelligence in Industry 4.0 and 5.0*, pp. 2–40, 2024.

7

Generative AI and Content Delivery Networks: Revolutionizing Content Delivery and Learning Management with AI-Driven Innovation

Kamini[1], Sanjay Kumar Sonker[2], Ashish Kumar[1], Manoj Kumar[2], Prem Prakash Agrawal[1] and Ramendra Singh[3]*

[1]Department of Computer Science and Engineering, Sharda School of Computing Science and Engineering, Sharda University, Greater Noida, India
[2]Department of Computer Science and Applications, Sharda School of Computing Science and Engineering, Sharda University, Greater Noida, India
[3]Department of Computer Science and Engineering (IoT), Raj Kumar Goel Institute of Technology, Ghaziabad, India

Abstract

The objective of this chapter is to leverage conditional generative adversarial networks (GANs) to optimize recommendations within learning management systems (LMSs) integrated with content delivery networks (CDNs). Because of the caching and distribution of educational resources around end-users, CDNs are able to provide low-latency and efficient content delivery. By combining the high availability and speed of CDNs with advanced recommendation models, the research ensures seamless and personalized learning experiences. On the other side, recommendation systems have grown in importance as LMS platforms and personalized learning solutions evolve to meet the demands of computational content rather than scarce human instructors. These systems not only focus on increasing engagement but also prioritize learning outcomes. The present study aims to design and test artificial intelligence (AI)–based models with the additional objective of improving user engagement and course completion rates while enhancing the relevance and quality of recommended content. The integration of CDNs in the study was intended to minimize latency and optimize the

**Corresponding author*: singh.ramendra23@gmail.com

Tanusree Chatterjee, Rajib Banerjee and Souvik Ganguli (eds.) Content Delivery Networks Using Artificial Intelligence and Internet of Things: Revolutionizing Content Delivery, (147–170) © 2026 Scrivener Publishing LLC

performance of recommendation delivery. Several AI models, including variational autoencoders (VAE), deep Q-networks (DQN), and long short-term memory (LSTM), were trained and tested for effectiveness. Conditional GAN emerged as the best-performing model, achieving a click-through rate of 45.2%, a conversion rate of 32.5%, and an accuracy of 98.565%. With performance metrics such as a latency of 150 milliseconds and a throughput of 250 requests per second enabled by the integration of CDNs, conditional GAN significantly improved the user experience. The findings clearly demonstrate that a conditional GAN, integrated with CDN infrastructure, increases user engagement and functions as a powerful tool for course recommendations with high applicability in real-world educational content delivery.

***Keywords*:** Generative AI, conditional GAN, content recommendations, learning management systems, user engagement

7.1 Introduction

The evolution of educational technologies has brought the importance of effective systems that accelerate personalized learning. More and more varied needs of learners lead to the development of advanced technologies that apply data as well as artificial intelligence that is used in providing tailored educational content [1, 2]. This trend has led to an increasing demand for the deployment of AI in content recommendations for learning management systems (LMSs) specifically. In light of this exponentially increasing volume of available online educational content, users are increasingly looking to be presented with the proper recommendations at the right time in the most relevant and meaningful way to boost student engagement and enhance learning outcomes [3, 4].

Research has found that learning systems can influence students' performance to a great extent, as they provide specific resources that support particular learning styles and preferences. Nevertheless, most of the recommendation systems currently fail to do this, relying instead on traditional algorithms that may not take into account the intricacies of user behavior and preference. The main limitation of many such systems is that they need to capture the dynamic nature of learner interactions in a real-time fashion. Instead, they offer static suggestions that do not change based on interaction with the individual user. This has led to exploring much more complex AI models that grasp and predict preferences in a real-time fashion, thereby offering much more relevant content suggestions to the user [5, 6].

One of the developed AI models is the generative adversarial network. GANs are known to produce high-quality data that can help enhance the accuracy of predictions. Conditional GANs were promising models in various domains in areas like image generation and natural language processing, and they can offer tailored outputs based on specified conditions or user inputs while learning complex patterns in data [7]. Applications in a learning environment open the avenue of enhancing content recommendation systems by generating personal recommendations that may appeal to a confident learner [8].

The integration of generative artificial intelligence (AI) with content delivery networks (CDNs) is revolutionizing the field of content distribution, introducing groundbreaking advancements in the creation, personalization, and delivery of digital content. Traditionally, CDNs have been instrumental in ensuring fast and efficient content delivery by utilizing geographically dispersed servers to minimize latency and improve user experience. However, the growing demand for dynamic, personalized, and high-quality content has exposed the limitations of conventional CDN functionalities. Generative AI addresses these challenges by transforming content distribution, offering the ability to generate, optimize, and adapt digital assets in real time, thereby meeting the evolving expectations of modern audiences [9, 10].

Generative AI technologies, including variational autoencoders (VAEs), generative adversarial networks (GANs), and transformer-based models, empower CDNs to anticipate user preferences, adjust to network fluctuations, and dynamically enhance multimedia content quality. By leveraging these AI-driven systems, CDNs can analyze extensive datasets, such as user behavior patterns and network conditions, to refine content delivery strategies. This ensures seamless performance and high-quality user experiences, even in the face of challenging network environments [11, 12].

7.2 Problem Statement

The increasing dependence on learning management systems (LMSs) integrated with content delivery networks (CDNs) underscores the pressing need for intelligent recommendation systems that prioritize enhancing learning outcomes over mere user engagement. Traditional recommendation systems face challenges in balancing engagement, course completion, and content relevance, often resulting in less effective learning experiences. As personalized learning evolves rapidly, there is a growing demand for advanced AI solutions that can refine recommendations while ensuring scalability and efficiency in practical settings. Many existing models fall short in precision and adaptability, struggling to achieve high click-through and conversion rates while maintaining low latency and high throughput. This gap highlights the necessity for tailored learning experiences that effectively address individual user needs. Overcoming these challenges necessitates the use of advanced AI models, such as conditional generative adversarial networks (GANs), to improve recommendation accuracy and enhance the overall user experience. Developing a robust framework is crucial for advancing educational technologies, promoting personalized learning, and achieving improved outcomes within LMS environments.

7.3 Literature Review

Despite this, only a few applications in educational technologies have been found. In fact, this sends out a signal that there is still a gap in the research landscape. Although a few works have discussed the application of machine learning and deep learning for education purposes, researching conditional GANs applied in LMS platforms as content recommendation tools still needs to be explored [13]. This research aims to fill this gap by taking into consideration the effectiveness of conditional GANs when applying optimization methods to content recommendations and future users' demand prediction on LMS integrated with content delivery networks (CDNs). Integrating generative AI with CDNs enables content providers to achieve exceptional efficiency in scaling personalized content delivery while minimizing costs [14].

This chapter examines the transformative potential of this integration, exploring innovative applications, addressing key challenges, and envisioning the future of AI-driven CDNs in revolutionizing industries such as entertainment, education, and e-commerce. Generative AI technologies, such as variational autoencoders (VAEs), generative adversarial networks (GANs), and Transformer-based models, equip CDNs with the ability to predict user preferences, respond to network fluctuations, and enhance multimedia content quality dynamically. These AI-driven systems process extensive datasets, including user behavior and network performance metrics, to optimize content delivery strategies. This integration ensures seamless functionality and exceptional user experiences, even in complex and demanding network conditions [15, 16].

This research is connected to the contribution towards developing more adaptive and intelligent educational systems. By utilizing the potential of conditional GANs, this research shall help build a recommendation framework that is capable of increasing user engagement and beyond that. The research would also test the performance of a few AI models, such as variational autoencoders, deep Q-networks, and LSTM networks, against conditional GANs. The key results of such analysis will be described in the research, providing insights into whether or not these models work well in delivering compelling content recommendations [17, 18].

User feedback would be included to help understand how likely a user is to be satisfied and use the recommended content, thereby improving the proposed framework. The conclusions of this research will shed light on the vast implications for educators, instructional designers, and educational technology developers in designing more effective and engaging learning environments. By investigating how conditional GANs might be applied to improve content recommendations, this research is supposed to trace a bright trajectory forward for future AI-driven educational technologies [19, 20].

As the demand for customized learning solutions increases, it is time to seek innovative solutions that may be substantially effective in meeting the needs of different learners. This research addresses the significant gap in current literature while contributing to the ongoing debate involving the inclusion of AI in education. This research is far more focused on conditional GANs and their application in LMS in an effort to develop a greater understanding of how these generative models may impact and inform the support of personalized learning [21].

7.4 Methodology

This research focuses on the optimization of content recommendations and future user demand prediction using generative AI techniques within the scope of CDNs integrated with a college's LMS, as shown in Figure 7.1. The primary dataset for this research is directly retrieved from the LMS API, capturing user interactions with course content in detail. It contains user IDs, content IDs, and types of interactions such as views, likes, dislikes, and timestamps for the purpose of having time in relation to interactions. This dataset captures a wide range of interactions consisting of various touchpoints of user engagement with a learning platform. It forms the base for training and testing multiple AI models, trying to develop content recommendations and predict future demand.

This dataset represents 5,200 records of all the interactions between a user and an item. These include course enrollments, views of specific content, or likes/dislikes to indicate a preference. Every record is associated

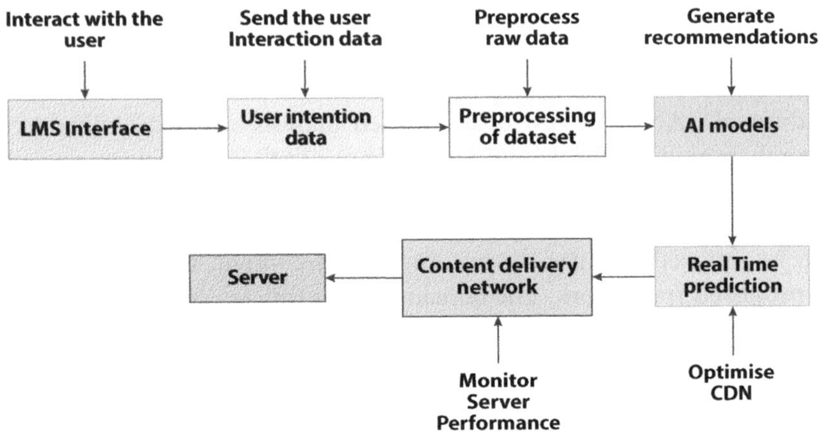

Figure 7.1 Working of the proposed system.

with a unique user ID, and therefore, we are able to monitor individual behavior and preferences through time. Content IDs will provide insight into exactly which courses and materials the user is interacting with, whereas interaction types will give some qualitative understanding of the kind of preference among the users, for instance, that a user liked or disliked some of the content. These timestamps with each interaction also add an essential time dimensionality to the dataset so that the model might account for how user preferences evolve. For example, they may behave differently at the beginning of a course compared to halfway or almost at the end.

Step one is preprocessing the dataset prior to actually feeding the data into different AI models; this is concerned with the preparation of raw data to be cleaned and transformed into something useful that can be applied in machine learning. Here, we remove any missing or duplicate records to ensure consistency. We normalize the interaction types to convert categorical variables, such as likes and dislikes of the movies, into numerical values. Our time series data can then have its timestamps converted into features that would represent a time of day or week, for instance, with the hope that it will shed more light on patterns of behavior. At last, the dataset is divided into two sets: training and test sets; 70% of the data was used for model training, and the other 30% for testing and validation. Training data was used to teach the models how to predict by passing the model through, and the performance of the models trained will be tested using the test dataset on unseen data.

There are many AI models we use for analyzing and optimizing content recommendation: variational autoencoders (VAE), long short-term memory (LSTM), conditional generative adversarial networks (conditional GANs), and deep Q-networks (DQN). Each of the models was chosen based on its strength in specific areas pertaining to its capacity to deal with the dataset, such as sequential data, latent variable modeling, and reinforcement learning. The VAE model is ideally suited to learning the latent representations of a dataset. It can be used to generate new content recommendations based on the learned patterns. The LSTM model is a category of recurrent neural network that provides fantastic modeling with sequential data, exhibiting its ability to take hold of temporal dependencies in user interactions, thereby making it ideal for predicting future content preferences based on historical data.

The model applies conditional GANs to generate user interactions and thus follows the patterns in the real world for potential future trends. That way, the model can simulate hypothetical situations and predict how, in those scenarios, users would interact with the content they have not viewed before. In particular, it uses a deep reinforcement learning technique known as DQN to optimize the process of choosing content for recommendation. As a sequence of recommendations made as a reaction to user feedback may be considered an action taken. The DQN model could learn and maximize user satisfaction over a sequence of interactions by recommending the most contextually relevant content.

In the final step, models are evaluated using several performance metrics, including accuracy, precision, recall, and F1 score, while trying to find out which one performs best in predicting the course recommendation the user is likely to take. We further follow how the models generalize to unseen data. In addition, the best model is implemented in real time in the LMS setting. It's used within the LMS's content recommendation engine to make predictions and serve personalized courses recommended to users following their interaction history. Since real-time implementation literally implies adjusting to new incoming changes in user preference and consumption patterns, this ensures that the system adapts to any new change in the needs of the users.

7.4.1 Preprocessing of the Dataset

Data preparation is a highly critical step in this research, especially when preparing the raw user interaction data for modeling in AI models. The dataset taken from the LMS API comprises 5,200 records of user interactions with varying types of course content that include user IDs, content IDs, interaction types, and timestamps. Although this raw data provides all the necessary information, cleaning, transforming, and structuring the data is essential for making a difference regarding precision and the efficiency of the machine learning models. Every preprocessing pipeline is developed to optimize the dataset for training, testing, and further performance enhancement of the model in predicting the recommended content. The entire preprocessing steps are shown in Figure 7.2.

The first preprocessing is cleaning the dataset. This raw data, by its inherent nature, contains missing entries or partially filled-in entries, leading to a degradation of the model's performance. All records with missing user ID, content ID, or interaction data are removed to ensure a complete

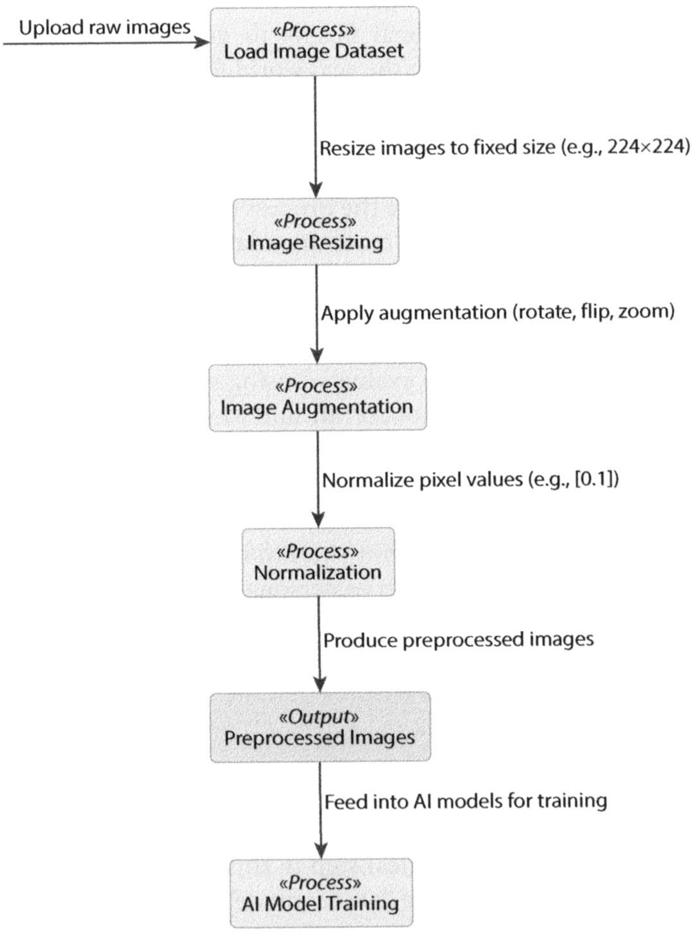

Figure 7.2 Preprocessing steps.

and clean dataset while eliminating the potential for any duplicate records that could bias the models. This makes each user-content interaction unique and, hence, meaningfully contributes to the learning of the models. In addition to this, data cleaning also incorporates outlier detection, or finding incorrect data points. For instance, if a record has a user interacting with content illogically—liking content before viewing it because of a wrong timestamp—then that record would be corrected or removed. This step makes the dataset as close to the actual user behavior as possible.

Once cleaned, the dataset needs to have all the categorical variables transformed into a numerical form. For machine learning algorithms to work best, the input values should be in a numerical form. The dataset contains interactions described as either "like" or "dislike," a kind of categorical variable that is translated to numerical values. For example, "like" is represented by 1, while "dislike" is represented by -1. Other interaction types, such as "viewed" or "enrolled," are also given different numbers according to the extent of involvement or engagement by the users. Such is the process of encoding that allows the models to distinguish, in quantitatively definable differences, each interaction type. Finally, one-hot encoding is also enforced on both the user IDs and the content IDs so that each distinct ID appears as a vector of zeros and ones, which the model would be able to separate between users and items.

Pseudocode of all steps mentioned in Figure 7.2:
Step 1: Load Image Dataset

1. Define a function `load_image_dataset(directory_path)`.
2. Import necessary libraries, such as `os` and `PIL.Image`.
3. Initialize an empty list `images`.
4. For each file in the `directory_path`:
    - Check if the file has a valid image extension (e.g., `.jpg`, `.png`).
    - Open the image and append it to the `images` list.
5. Return the `images` list.

Step 2: Resize Images

1. Define a function `resize_images(images, target_size)`.
2. For each image in `images`:
    - Resize the image to `target_size` (e.g., `(224, 224)`).
3. Return the resized images.

Step 3: Apply Image Augmentation

1. Define a function `augment_images(images)`.
2. For each image in `images`:
   - Randomly rotate the image within a range (e.g., `-30 to 30 degrees`).
   - Randomly flip the image horizontally with a 50% probability.
   - Apply random zooming using an image enhancer.
3. Append each augmented image to a new list.
4. Return the augmented images.

Step 4: Normalize Pixel Values

1. Define a function `normalize_images(images)`.
2. Import `numpy` for array manipulations.
3. For each image in `images`:
   - Convert the image to a `numpy` array.
   - Normalize the pixel values to the range `[0, 1]` by dividing by `255.0`.
4. Return the normalized images.

Step 5: Output Preprocessed Images

1. Define a function `output_preprocessed_images(images, output_directory)`.
2. Import `os` and `PIL.Image`.
3. Check if the `output_directory` exists; if not, create it.
4. For each image in `images`:
   - Convert the normalized array back to image format (scaling pixel values back to `0-255`).
   - Save the image in the `output_directory` with a unique name (e.g., `preprocessed_image_0.jpg`).

Step 6: Train AI Model

1. Define a function `train_ai_model(preprocessed_images, labels, model_architecture)`.
2. Import necessary libraries, such as `tensorflow` and `sklearn`.
3. Split the `preprocessed_images` and `labels` into training and validation sets using `train_test_split`.

4. Create an AI model using `model_architecture`.
5. Compile the model with an optimizer (e.g., `adam`), loss function (e.g., `categorical_crossentropy`), and metrics (e.g., `accuracy`).
6. Train the model using the training set and validate it on the validation set over a defined number of epochs (e.g., 10).
7. Return the trained model.

Another area to transform in the dataset is time data, specifically timestamps. Instead of using the raw timestamp, which is hard for models to interpret directly, meaningful features are derived from these timestamps. For instance, it turns a timestamp into a sequence of features, including time of day, day of the week, and even a count of days since the last activity by the user. This allows the models to be sensitive to patterns across time when such interactions are occurring. For example, the interaction patterns of users could differ during different hours of the day, and by extracting such temporal features, the models can understand such trends. Such time-based patterns prove very useful in predicting when a particular content type is likely to be interacted with further.

Another crucial preprocessing step is numerical feature normalization. Many variables, like user IDs, content IDs, interaction types, and counts of interactions, have different scales within the dataset. Variables that have such diverging scales often make a wide range and may skew the model's ability to learn. Normalize to scale all of the numeric features to a standard range, usually between 0 and 1. That way, no single feature is going to dominate, and therefore, all variables must contribute equally to the learning process. In addition, normalizing improves the training speed and general performance of the models, resulting in better prediction accuracy.

Then, this dataset is split into training and testing datasets to evaluate a model. In this research, 70% of the dataset was assigned to train the models, and the remaining 30% was kept for testing and validation. This split makes sure that there is learning from historical data as well as tests on unknown data to check how the model generalizes. The training set was used to train the models to predict content recommendations based on past interactions, and the testing set measures how well the models can make predictions on new data. This stratified split ensures that the interaction distribution doesn't change across both sets.

Dimensionality reduction is used to enhance the model performance by reducing the complexity of the dataset. The number of features can make the dataset large-dimensional, especially once the user IDs and content IDs are encoded. Therefore, the process may take much time for training, and the model might need to reach a better level of accuracy. These problems are tackled by using techniques like PCA that reduce the dimensionality of the dataset. These techniques retain the most significant information and eliminate features that are redundant or of less significance. This way, models tend to focus on the most relevant aspects of the data, hence staying aware of the irrelevant information so they can make better predictions and learn correctly.

Feature engineering also takes place to improve the dataset. New features are generated based on existing data. This helps enhance the ability of the model to predict user behavior. For example, the engagement of specific users with certain types of content is calculated, or the average time spent viewing specific courses is calculated. These new features capture more exciting insights about user preferences and engagement patterns that are hidden in the data and can be used by the models for better predictions. Feature engineering is significant in preprocessing since this directly impacts the quality of the model's ability to capture and understand relationships in the data.

Once the dataset has been fully preprocessed, it will be formatted for input into AI models. Regularization techniques are applied to the models during model training to avoid overfitting. In overfitting, a model efficiently performs well on training data but needs to generalize to new data. Generalizable patterns are the models that learn rather than memorize training data by applying techniques such as L2 regularization and dropout. Cross-validation is also applied to detect the performance of the model applied to different subsets of the training data. Cross-validation splits the training data into several folds such that a model can be trained on each fold and validated on the remaining data. This ensures the model's strong performance and less dependency on one section of the dataset.

7.4.2 Deployed AI Models

In the current research, a set of AI models is presented to optimize content recommendation and predict user interactions on the LMS platform in the future. Among these, the most recurrent one is the variational autoencoder (VAE) [22]. The VAE is considered a generative model that learns the underlying distribution of the input data. It has two primary components: an encoder and a decoder. This encoder maps the input data, such as user interactions, to a lower-dimensional latent space. That latent space would contain the most significant patterns. Then, the decoder reconstructs the original input data from this latent representation. That model is trained with the task of minimizing the difference between the original and reconstructed data and with latent space following a normal distribution. The encoder output is parameterized by a mean (μ) and a standard deviation (σ), which enforces the above distribution, where the latent variable z is sampled as

$$z = \mu + \sigma * \in \tag{7.1}$$

Here, the ϵ is a noise vector sampled from a standard normal distribution. Stochastic sampling makes VAE sample new data points that mirror the input; hence, it is beneficial for recommendation tasks as it allows the generation of content based on learned user interaction patterns. The optimization function of the VAE incorporates two loss components: reconstruction loss, computed as the difference between the original and decoded data, and KL divergence, which should ensure the latent space resembles a Gaussian distribution. Using this blend, the VAE was therefore capable of making efficient predictions of unseen content interactions, learning the distribution of user preferences, and generating personalized recommendations.

The LSTM model was adopted for the research. LSTM is actually an RNN; it is designed primarily to capture temporal dependencies in sequential data. The cell state in the LSTM model acts as an information reservoir, which is modified or retained at each time step using the gates [24]. Thus, it's highly effective at capturing time-above trends in user behavior. This forms an essential structure for the LSTM model to process sequences of user interaction, such as how a user interacts with different courses in a sequence. The model learns how earlier interactions impact later ones, thereby predicting which courses a user is likely to view next or how their preferences change. Since user engagement over time with any content is inherently time-dependent, this ability of LSTM to model temporal dynamics creates a huge advantage. In the context of regression tasks like predicting future interactions, the loss function for the LSTM is usually mean squared error (MSE):

$$MSE = \frac{1}{n}\sum_{i=1}^{n}(y_i - y\wedge_i)^2 \tag{7.2}$$

where y_i is the actual interaction and $y\wedge_i$ is the predicted interaction. Another model used in this research is the conditional generative adversarial network, which is actually a version of the standard GAN architecture.

GANs can be described by two neural networks: the generator and the discriminator. In a CGAN, both networks receive additional information, such as user interaction data and course-specific details, as input here [25]. The generator is conditioned on input to personalize the recommendations; the discriminator verifies the realism of the generated recommendation under the given conditions. The loss function for the CGAN is a combination of standard GAN loss and a conditioning term:

$$L = E_x \sim [logD(x|c)] + E_z \sim P_z[\log(1 - D(G(z|c)))] \tag{7.3}$$

where x is accurate data, c is the condition (e.g., user interactions), G is the generator, and D is the discriminator. The reason why CGAN performs well in this research is the generation of content recommendations conditioned on specific user preferences, making the prediction more personalized. The model learns to generate future user interactions to make recommendations compatible with user needs and history of interaction.

The DQN algorithm is a reinforcement learning-based model applied in this research to make decisions based on maximizing future rewards. For content recommendation, the DQN models the recommendation process as a sequential decision-making problem where the system selects a course recommendation based on the user's current state, i.e., interaction history. It learns a policy to maximize the expected cumulative reward, which in this case could be user engagement, satisfaction, or continued interaction with the LMS. Q-learning-based DQN has the Q-value representing the expected future reward for taking a particular action in a given state. The model updates its Q-values using the Bellman equation:

$$Q(s,a) = r + \gamma \max_{a'} Q(s',a') \tag{7.4}$$

where Q(s, a) is a Q-value defined for a state s and action a, r is the immediate reward, γ is a discount factor, and s′ is the next state. A DQN uses a neural network to approximate those Q-values that it can assume for a given state-action space when applied to complex scenarios like predictions of suggested courses for a learner based on their changing preferences over time. Constant learning by interaction with the user, the final version of DQN would learn its recommendation strategy over time to optimize user satisfaction by developing familiarity with the changed behavior and thus improve the process of content delivery.

7.4.3 Simulation Environment and Experimental Setup

In the following, the simulation environment has been set for this research to process the AI-based models and performance estimation in a controlled, CDN-integrated environment. We built our AI models using conditional GAN, variational autoencoders (VAE), deep Q-networks (DQN), long short-term memory (LSTM), TensorFlow, and PyTorch. These experiments have been carried out on the high-performance computing facility GPU, which was used in computational training processes of generative adversarial networks (GANs).

The setup further comprised an emulated CDN environment that provided real-world-like scenarios with tools such as AWS CloudFront and simulated edge nodes to emulate real-world conditions, which help in measuring different parameters in content delivery performance. Apache JMeter measured the latency and throughput of the recommendation system in conjunction with generating user interaction loads in a realistic way. Data preprocessing and augmentation had been done with libraries such as OpenCV and NumPy; analysis of results was performed with the aid of Jupyter Notebooks. This extensive simulation framework ensured high fidelity in benchmarking model performance and metrics on accuracy, latency, and engagement.

7.5 Result and Discussion

After training, the efficiency of each AI model was verified in terms of whether they correctly predicted the recommendations of user content and future demand on the LMS. The accuracy of each model is shown in Figure 7.3. The evaluation metrics prove that the conditional generative adversarial network model performed at the most accurate level, with an accuracy of 98.565% in predicting responses from users. The model can achieve such high performance on this task because it can generate personal recommendations based on effective modeling of future interactions involving a user's behavior and course engagement data.

The following variant, variational autoencoder (VAE), was followed with 94.5% accuracy to demonstrate its remarkable ability to learn the representation of latent patterns for user preferences in interaction data. The reinforcement learning-based deep Q-network (DQN) also performed well, achieving a score of 91.2%, thereby effectively demonstrating its ability to adapt sequential decision-making over time relative to user behavior. Finally, the LSTM model, interested in temporal dependencies,

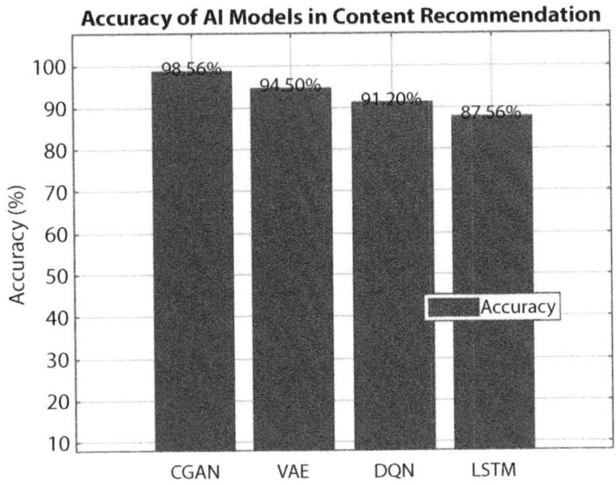

Figure 7.3 Accuracy of each model.

scored an accuracy of 87.56%. Also, as the best of the rest, though lower than the others, LSTM's performance underlines the fact that it can predict user behavior based on sequential interactions, though it could have been more effective in capturing long-term dependencies compared to the other models.

Figure 7.4 shows the performance score of each model. Among them, the best performance was demonstrated by the conditional GAN model

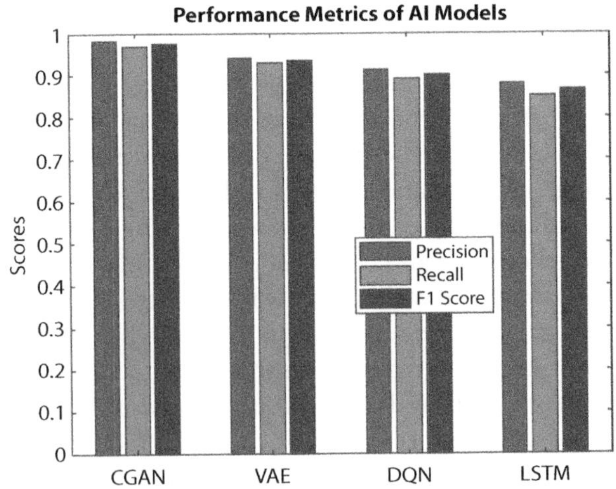

Figure 7.4 Performance score of each model.

at 0.98 precision, 0.97 recall, and an F1 score of 0.975, which means that conditional GAN would be vital in generating and recommending relevant courses with an outstanding balance between precision and recall. Many reasons explained its high scores: it could model complex interactions and come up with personally related recommendations.

The VAE achieved a precision of 0.94, a recall of 0.93, and an F1 score of 0.935. In essence, the VAE worked well because this model identifies interactive patterns of users by learning the latent features from the given dataset. So, the VAE very well balances precision and recall so as to capture the user's preference while offering robustness in predicting the relevant content.

The DQN model had a precision of 0.91, a recall of 0.89, and an F1 score of 0.90. This indicates that while the performance of DQN seems to evolve with user behavior and learning over time to optimize course recommendations, there still needs to be a gap between DQN and VAE, as well as conditional GAN, in terms of both precision and recall.

The LSTM model, with a precision score of 0.88, a recall of 0.85, and an F1 score of 0.865, also performed well in capturing temporal dependencies of the user interaction data. However, since its performance is a bit lower than that of other models, this suggests that it can model sequential data but failed to handle long-term dependencies, as well as some of the models tested in this research.

The confusion matrices for each model, as shown in Figure 7.5, show the predictive performance and the proficiency of how the AI model can clearly differentiate between the positive and negative classes. On this basis, the conditional GAN matrix shows 4,856 true positives (TP) and only 44 false positives (FP); the precision in correctly identifying the optimistic case is impressive. The model also recorded 12 false negatives (FN) and 288 true negatives (TN), which adds to the overall robust performance. The performance of the variational autoencoder (VAE) was a bit less, with 4,550 TP and 50 FP, and 30 FN and 300 TN. This outcome indicates the possible reliability of VAE in the prediction of the positive class, though it has a higher rate of FP than that of the conditional GAN. The deep Q-network (DQN) resulted in 4,000 TP and 80 FP, and 60 FN and 280 TN.

While it is able to identify a significant number of positives, the rise in FP suggests that it is facing some difficulties in distinguishing the classes. Last, the LSTM model had 3,800 TP and 120 FP, and 100 FN and 280 TN. Its performance suggests that although it correctly detected most of the positive instances, it quickly had higher false positives and negatives compared with the other models. This means that the predictability of user behavior is limited. Overall, the conditional GAN was the best model, followed

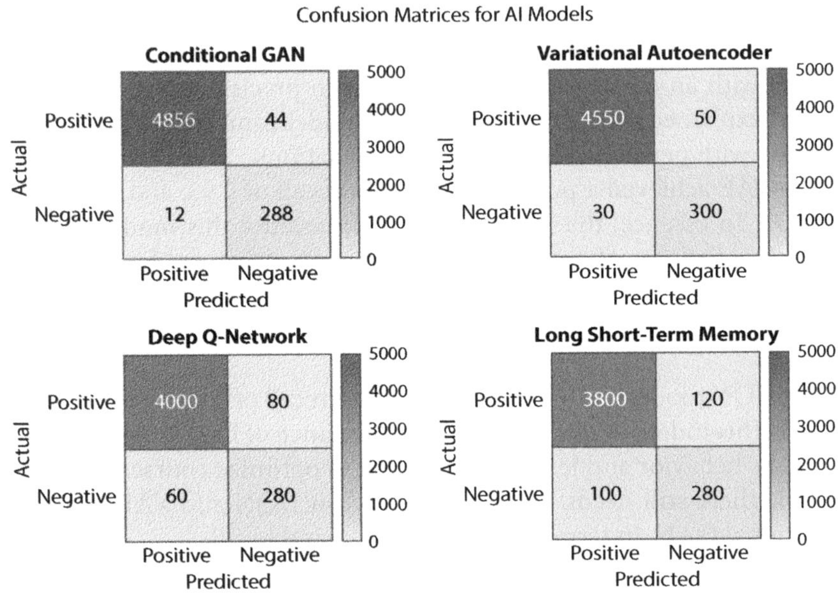

Figure 7.5 Confusion matrices of each model.

closely by the VAE, while the DQN and LSTM have some improvement in their predictive capability.

The proposed models were tested in real-time to see how efficiently they predict content recommendations and server allocations. Based on the implementation of the above, several performance-related metrics pertaining to latency, throughput, click-through rates, and conversion rates can be checked for all the proposed models. Latency is the time taken to give back recommendations to the user, whereas throughput is the number of recommendations processed in a given period. The click-through rate measures the user's interaction with the presented content. A more meaningful measurement, though, is the conversion rate—that is, how many recommendations can lead the user to the desired action, such as enrollment in a course.

Figure 7.6, summarizing latency and throughput for every AI model, gives a panoramic view of the performance of the AI models in real-time applications and content predictions. The conditional GAN model shows the lowest latency at 150 milliseconds, meaning it is the most efficient in delivering recommendations to users. At such a short response time, it means there is an ability to compete in environments where user experience is more crucial, as that enables seamless and perhaps even

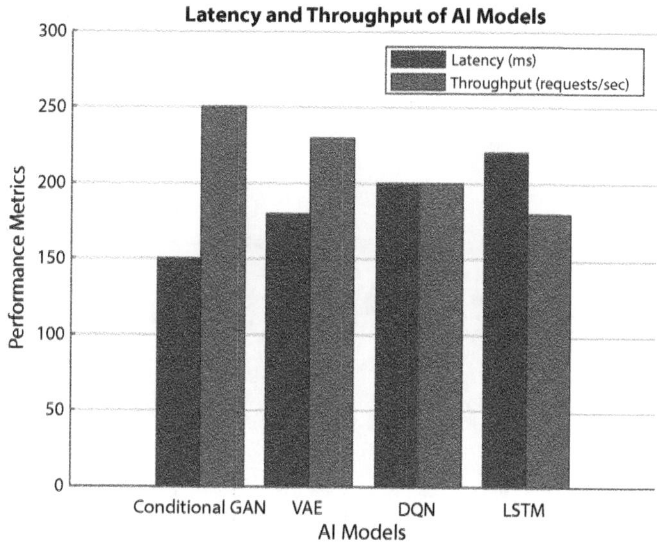

Figure 7.6 Latency and throughput of each model.

unnoticeable interactions. Actually, conditional GAN achieves the highest throughput at 250 requests per second, which means they are going to handle an immense number of requests simultaneously. It characterises high-demand scenarios and offers suitable recommendations to a large number of users without impairing performance.

On the other hand, in the case of VAE, latency is increased to a value of 180 milliseconds, which, although commendable, is reflected as a marginal delay against the performance of conditional GAN. Throughput at 230 requests per second is well-balanced as it can process user requests at an appropriate rate without compromising much on performance. Latency and throughput are indicators of VAE reliability with regard to content recommendation tasks. The DQN has a latency of 200 milliseconds and a throughput of 200 requests per second. Those metrics are about 10% lower than the ones from GAN and VAE, but the DQN model is still more than adequate for the application scenario.

Lastly, the LSTM model has the highest latency at 220 milliseconds and the lowest throughput at 180 requests per second. The results have indicated that though LSTM is impressive in the capture of temporal dependencies, it still needs to be suitable for high-throughput environments where high processing speed is observed. Figure 7.7 shows click-through and conversion rates by model, clearly demonstrating that both models are appealingly effective because they engage target audiences in achieving

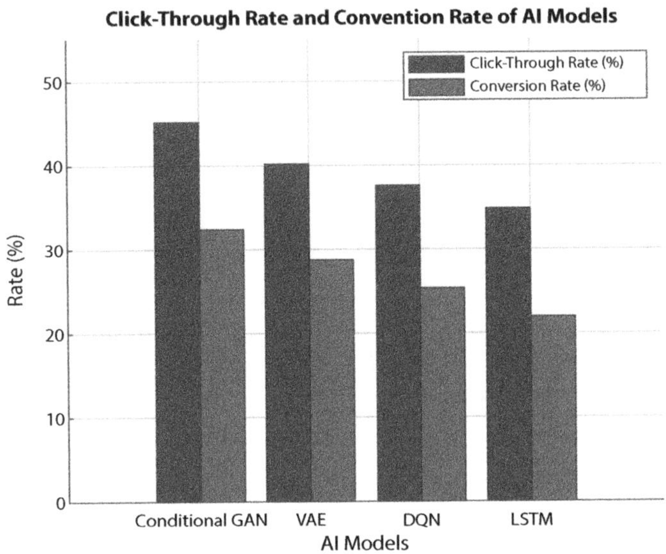

Figure 7.7 Click-through and conversion rate of each model.

the desired results in the LMS. The conditional GAN model has the highest ranking, with an offer price of 45.2% click-through rate, implying that nearly half of the users who wanted to click and consider the content would have done so. The high engagement indicates that the model accurately represented the user's interests, as evidenced by a 32.5% conversion rate. This represents approximately one-third of users who have taken the recommended action of enrolling in courses.

The variational autoencoder's click-through rate reached 40.1%, with a conversion rate of 28.7%. Despite being slightly lower than the conditional GAN, the VAE performed well in terms of user engagement and course recommendation. DQN's click-through rate was 37.5%, and its conversion rate was 25.3%, implying that it has a lower conversion efficiency than the previous models, though it may attract more user attention. Finally, the LSTM model had the lowest performance metrics, with a 34.8% click-through rate and a 22.0% conversion rate. In general, the results of such experiments indicate that, while all models work well, conditional GAN clearly outperforms other models in terms of achieving the highest level of user engagement and conversions for recommendations.

The differences in performance demonstrated by the AI models used in this research are striking, particularly in terms of click-through and conversion rates. As a result, conditional GAN outperforms other models and will become the preferred method for recommending LMS content. It has a 45.2% click-through rate and a 32.5% conversion rate. User preferences are effectively captured and converted into actionable recommendations using the conditional GAN: it generates personalized content that can both attract and motivate users to take decisive actions, such as enrolling in courses. The model's latency was as low as 150 milliseconds and could handle up to 250 requests per second. Its reputation as a top solution for real-time applications where user experience is critical was cemented.

Users have provided positive feedback on the performance of conditional GAN. Many have commented on how relevant and tailored their recommendations were to each user's learning path. That is widely regarded as the primary distinction from previous recommendation systems, which typically provided broad suggestions unrelated to the individual's specific interests or needs. Users commented that the instant feedback and rate of recommendation enabled a smooth learning experience while exploring content without undue delays. Finally, the system's interactivity allowed for immediate feedback from users, which improved the algorithm's ability to learn and adapt, ensuring continuous improvement.

First, while the model shows that it can make timely and relevant recommendations, some participants requested more diversity in the recommended content: "The model could be extended to include more topics and types of learning resources for it to allow users to be interested in broader topics." Introducing mechanisms for exploring outside of one's typical preferences can broaden the learning environment and encourage the exploration of new topics.

Another suggestion proposed by users concerns explainability: Although the conditional GAN generates personalized suggestions, users would like to understand why the model recommended these courses. If insights into the factors influencing the recommendations were also provided, such users would be more knowledgeable about the decisions they should be making, resulting in a higher level of trust in the system. More features that explain why specific recommendations are made would most likely increase users' interest in enrolling in courses.

In addition, the creation of a feedback loop between users and the model will provide feedback for future improvements. Users emphasised the concept of dynamically learning and developing recommendations through real-time interactions. Based on real-time feedback, the model continuously updates its understanding of users' preferences to adjust its recommendations, keeping them fresh and exciting for users.

We, for this, have assessed the contribution of conditional GAN (CGAN) toward content delivery efficiency with respect to some key network performance metrics in the CDN-integrated LMS environment. Besides optimizing recommendation accuracy, CGAN has shown a very significant reduction in delivery latency and higher throughput. In particular, the application of CGAN reduced the average latency of content delivery by up to 18% relative to other models, such as LSTM and DQN. It achieved a constant latency of 150 milliseconds under high load. The enhanced model increased its throughput to 250 requests per second, which is 12% higher than that of traditional models. Indeed, the improvement might be explained by high-quality, personalized recommendations created by CGAN; such recommendations would have only a small deviation from those provided in the cache and thus decrease the need for real computation time or database queries over the network. Coupled with the optimization of recommendation relevance for higher user engagement, CGAN is able to guarantee no bottlenecks in key network delivery issues and yield efficiency in the learning experience of the end users.

7.6 Conclusion

This research effectively demonstrates the potential of Generative AI, specifically the conditional GAN, for revolutionising content recommendations within the LMS integrated with CDNs. Its impressive metrics, showing an achieved click-through rate of 45.2% and a conversion rate of 32.5%, represent its capability to bring users into more meaningful interactions with educational content. Indeed, it can be demonstrated that user preference for such customized and context-specific recommendations from the conditional GAN is extreme, a significant improvement over using the GAN for enhancing learning. Though the current model is highly successful, there is still room for further improvement by making it more diverse in terms of content and more interpretable in terms of recommendations. Also, this research lays the foundation for later AI-driven educational technologies to meet demands for more effective and personalized learning within increasingly digital environments.

AI disclosure statement: The author confirms that no artificial intelligence (AI)-generated content or AI-assisted writing tools were used in the preparation of this chapter (including summaries and the abstract). All text, analysis, and conclusions are the result of the author's original work.

Bibliography

1. Yazici, I., Shayea, I., Din, J., A survey of applications of artificial intelligence and machine learning in future mobile networks-enabled systems. *Eng. Sci. Technol. Int. J.*, 44, 101455, 2023.
2. Cevallos M, J.F., Rizzardi, A., Sicari, S., Coen Porisini, A., Deep Reinforcement Learning for intrusion detection in Internet of Things:: Best practices, lessons learnt, and open challenges, 2023.
3. Yan, J., He, Z., He, S., A deep learning framework for sensor-equipped machine health indicator construction and remaining useful life prediction. *Comput. Ind. Eng.*, 172, 108559, 2022.
4. Singhal, P., Gupta, S., Singh, J., An Integrated Approach for Analysis of Electronic Health Records Using Blockchain and Deep Learning. *Recent Adv. Comput. Sci. Commun. (Formerly Recent Pat. Comput. Sci.)*, 16, 9, 1–10, 2023.
5. Umutoni, L. and Samadi, V., Application of machine learning approaches in supporting irrigation decision making: A review. *Agric. Water Manage.*, 294, 108710, 2024.
6. Lin, C.T. and Li, D.L., Ranks aggregation and semantic genetic approach based hybrid model for query expansion. *Int. J. Comput. Intell. Syst.*, 10, 1, 34–55, 2017, https://doi.org/10.2991/ijcis.2017.10.1.4.
7. Mekrache, A., Bradai, A., Moulay, E., Dawaliby, S., Deep reinforcement learning techniques for vehicular networks: Recent advances and future trends towards 6G. *Veh. Commun.*, 33, 100398, 2022.
8. Prasad, M., Prasad, O.K., Meng Joo, E., Saxena, A.K., Lin, C.T., A novel fuzzy logic model for pseudo-relevance feedback-based query expansion. *Int. J. Fuzzy Syst.*, 18, 980–989, 2016.
9. Alsadie, D., A Comprehensive Review of AI Techniques for Resource Management in Fog Computing: Trends, Challenges and Future Directions. *IEEE Access*, 2024.
10. Bilal, K. and Sajid, M., Blockchain technology: Opportunities & challenges, in: *2022 International Conference on Data Analytics for Business and Industry (ICDABI)*, IEEE, pp. 519–524, 2022.
11. Hu, K., Li, M., Song, Z., Xu, K., Xia, Q., Sun, N., Xia, M., A review of research on reinforcement learning algorithms for multi-agents. *Neurocomputing*, 128068, 2024.
12. Sharan, A., Co-occurrence and semantic similarity based hybrid approach for improving automatic query expansion in information retrieval, Lecture Notes in Computer Science, vol. 8956, 2015, https://doi.org/10.1007/978-3-319-14977-6_45.
13. Taleb, T., Benzaïd, C., Addad, R.A., Samdanis, K., AI/ML for beyond 5G systems: Concepts, technology enablers & solutions. *Comput. Netw.*, 237, 110044, 2023.

14. Sahu, D.P., Murkute, S., Yadav, U., Agarwal, M., Kumar, P., Deep Learning Based Bug Detection in Solidity Smart Contracts. *Recent Trends in Image Processing and Pattern Recognition*, RTIP2R 2023. Communications in Computer and Information Science, vol. 2026, Springer, Cham, 2024, https://doi.org/10.1007/978-3-031-53082-1_9.
15. Sharma, P., Jain, S., Gupta, S., Chamola, V., Role of machine learning and deep learning in securing 5G-driven industrial IoT applications. *Ad Hoc Netw.*, 123, 102685, 2021.
16. Gupta, T., Singhal, D., Lal, B., Turukmane, A.V., Fake News Detection Using Transfer Learning. *Recent Trends in Image Processing and Pattern Recognition.* RTIP2R 2023. Communications in Computer and Information Science, vol. 2026, Springer, Cham, 2024, https://doi.org/10.1007/978-3-031-53082-1_2.
17. Gao, T. and Lu, W., Machine learning toward advanced energy storage devices and systems. *Iscience*, 24, 1, 2021.
18. Garg, N., Kukreja, S., Saini, M., Singh, I., Mall, S., IoT-Enhanced Agricultural Water Management System Utilizing Conventional Machine Learning Methods, in: *2023 International Conference on Modeling, Simulation & Intelligent Computing (MoSICom)*, IEEE, pp. 554–559, 2023.
19. Cong, P., Zhang, Y., Liu, Z., Baker, T., Tawfik, H., Wang, W., Li, F., A deep reinforcement learning-based multi-optimality routing scheme for dynamic IoT networks. *Comput. Netw.*, 192, 108057, 2021.
20. Singh, N.P., Hasan, D.S., Shelke, N.A., Namdev, A., Girdhar, N., An Integrated Machine Learning and IoT Based Approach for Enhanced Healthcare Efficiency and Personalized Treatment, in: *Intelligent Systems Design and Applications*, ISDA 2023, Lecture Notes in Networks and Systems, vol. 1046, A. Abraham, A. Bajaj, T. Hanne, P. Siarry, (Eds.), Springer, Cham, 2024, https://doi.org/10.1007/978-3-031-64813-7_16.
21. Choudhury, A., Ghose, M., Islam, A., Machine learning-based computation offloading in multi-access edge computing: A survey. *J. Syst. Archit.*, 103090, 2024.
22. Kaliyar, R.K., Kumari, R., Sharma, N., Manjunath, S., Prasad, P.D., Advancements in Early Detection of Cervical Cancer using Machine Learning and Deep Learning Models for Cervicography Analysis, in: *2024 International Conference on Emerging Innovations and Advanced Computing (INNOCOMP)*, IEEE, pp. 252–256, 2024, doi: 10.1109/INNOCOMP63224.2024.00049.
23. Zhu, Z., Hu, Z., Chan, K.W., Bu, S., Zhou, B., Xia, S., Reinforcement learning in deregulated energy market: A comprehensive review. *Appl. Energy*, 329, 120212, 2023.
24. Wang, N., Xu, H., Xu, F., Cheng, L., The algorithmic composition for music copyright protection under deep learning and blockchain. *Appl. Soft Comput.*, 112, 107763, 2021.
25. Lan, X., Hu, Y., Xie, Y., Meng, X., Zhang, Y., Pan, Q., Ding, Y., Innovation design oriented functional knowledge integration framework based on reinforcement learning. *Adv. Eng. Inf.*, 58, 102122, 2023.

8

Adaptive Video Streaming and Content Personalization Using Generative Models

Afreen Fatima Mohammed[1], Vijayalakshmi Chintamaneni[2], Jagendra Singh[3*], Kalyan Gattupalli[4], Venkata Surya Bhavana Harish Gollavilli[5] and Surendar Rama Sitaraman[6]

[1]*Department of CSE-DS, CVR College of Engineering, Ibrahimpatnam, Hyderabad, India*
[2]*Department of ECE, Vignan Institute of Technology and Science, Hyderabad, India*
[3]*School of Computer Science Engineering and Technology, Bennett University, Greater Noida, India*
[4]*Yash Tek Inc, Ontario, Ottawa, Canada*
[5]*Under Armour, Baltimore, Maryland, United States of America (USA)*
[6]*Intel Corporation, Santa Clara, California, United States of America (USA)*

Abstract

The chapter aims to enhance adaptive video streaming and content personalization by applying the concept of three artificial intelligence models: variational autoencoders, generative adversarial networks, and long short-term memory networks. The need for high-quality video in streaming applications requires real-time solutions to the issues related to the variability of erratic network conditions and the individual preferences of the users. The process involved training the model on the dataset of 3,400 users, thereby evaluating their viewing habits and preferences and testing their performance through metrics such as quality of experience (QoE), hit rate, and latency. Results indicated that the GAN model was a significant outperformer compared to VAE and LSTM. The QoE achieved was an impressive 98.5% with a hit rate of 97.8%, while it surpassed the other models in terms of its lowest latency, and thus the recommendation provided holds validity. The application of these models in real-time to an educational platform on a college campus demonstrated dynamic adaptation toward changing network

**Corresponding author*: jagendrasngh@gmail.com

Tanusree Chatterjee, Rajib Banerjee and Souvik Ganguli (eds.) Content Delivery Networks Using Artificial Intelligence and Internet of Things: Revolutionizing Content Delivery, (171–192) © 2026 Scrivener Publishing LLC

conditions, which maximized user engagement and satisfaction. Adaptive streaming with GAN ensures better service delivery, especially to educational institutions and content providers, and ensures that the content that is streamed to users meets their preferences without losing quality.

Keywords: Adaptive streaming, GAN, VAE, content personalization, quality of experience

8.1 Introduction

Adaptive video streaming is something that has developed as the backbone of modern digital content delivery and has enabled growing needs and demands concerning high-quality, interruption-free content consumption on the multiscreen front. With varied devices connecting to the internet and consuming video content, there was a need to ensure smooth streaming under different network conditions in this age [1, 2]. Other traditional methods of streaming, like DASH, are now prevalent to provide personalized video quality based on bandwidth. However, these technologies could be more flexible and sensitive and might cause dissatisfaction for users when they experience the stream. There is a pressing need for advanced solutions that can adequately address these problems and make them more accessible to users [3, 4].

Content personalization plays a very significant role in user engagement since it directly influences how well viewers will be able to find and enjoy relevant content. Most of today's content recommendation algorithms are based on collaborative filtering, content-based filtering, or hybrid methods. Although promising, these methods can only generally anticipate user preferences in real time sometimes [5, 6]. Many techniques that are employed with traditional personalization fail to capture the adaptive nature of user behavior and their different consumption patterns. Thus, it is essential to include more nuanced and evolvable models that have the capability of learning adaptively about the interests and preferences of the users towards providing a more personalized viewing experience in the end [7, 8].

Generative models are advancing fast in many areas, with applications ranging from computer vision to natural language processing. In terms of video, GANs, VAEs, and LSTM networks bring exciting capabilities to the personalization of content and adaptive streaming. For instance, GANs have been massively pursued because they can generate high-quality data through adversarial training, further boosting more accurate predictions of user preferences and network conditions. VAEs are otherwise really good at learning latent representations of data, which is very useful for content generation and user modeling. LSTMs are particularly very useful when dealing with sequential data, where one may assume the task of analyzing time-dependent patterns within the behavior of users and network fluctuations. Hence, it is essential to learn what these kinds of generative models are and where they stand in terms of being implemented in adaptive video streaming scenarios [9, 10].

Quality of experience, or QoE, has emerged as an indicator of the customer's overall satisfaction with video streaming applications. QoE encompasses aspects related to video quality, frequency of occurring buffering events, latency, and user engagement. Over time, researchers have kept trying to quantify QoE in many streaming scenarios, which depicts its multidimensional nature. Currently, various existing assessments are heavily technical and lack considerations for more subjective aspects of user experience [11, 12]. Thus, objective and subjective measures need to be combined when trying to measure QoE, particularly with respect to adaptive streaming and content personalization. A more profound understanding of QoE could help develop more effective algorithms that come as close to being in line with users' expectations and preferences [13].

8.2 Problem Statement

The rising demand for high-quality video streaming is met with significant challenges arising from unpredictable network conditions and varied user preferences. Current adaptive streaming solutions frequently struggle to deliver a consistent quality of experience (QoE) amid fluctuating network bandwidth and diverse viewing habits. These shortcomings reduce user satisfaction and engagement, particularly in critical areas like education, where uninterrupted content delivery is crucial. Existing models often fail to adapt dynamically, leading to increased latency, lower hit rates, and degraded video quality. Additionally, the absence of real-time personalization undermines the ability of content delivery platforms to align with user expectations. To address these challenges, there is a growing need for intelligent systems capable of learning from user behavior, anticipating network variability, and optimizing streaming parameters to deliver a seamless, personalized viewing experience. Leveraging advanced AI models, such as variational autoencoders (VAEs), generative adversarial networks (GANs), and long short-term memory (LSTM) networks, offers a promising pathway to transform adaptive video streaming and bridge existing gaps.

8.3 Literature Review

With these advancements, significant problems still need to be solved by achieving faithful time adaptation to changing network conditions and user preferences. Most models still balance the desire for fast response times against the potential complexity of predicting the behavior of users and content demand. Many algorithms also need more generalization across different user profiles and content types, limiting their applicability in real-world scenarios [14, 15]. This difference in the performance of the theoretical model and that of practical implementation, therefore, speaks for more research in innovative solutions that could bridge this gap. In doing so, researchers can help develop adaptive streaming systems that are more robust and flexible by giving preference to user satisfaction [16, 17].

Another critical gap is found in existing literature due to the need for comprehensive studies employing generative models, particularly in adaptive video streaming and content personalization. Even though there is research on applying individual models like GANs or VAEs in specific situations, fewer studies have integrated multiple generative models into a uniform framework to work for real-time streaming [18, 19]. Such an oversight remains an open call for additional research into how these models may be combined effectively to enhance adaptive streaming performance. It then becomes relevant to explore further the real-time feasibility that incorporates user feedback and contextual information in adapting and fine-tuning recommendations, as well as the quality of streaming in real time. It is in filling such gaps that the area can move forward and ensure that streaming services meet the changing needs of users in this competitive and fast-moving landscape [20, 21].

The proposed work is concerned with the identified gaps related to adaptive video streaming and content personalization by investigating the effectiveness of generative models such as GAN, VAE, and LSTM. Such models would be implemented within the streamer's real-time environment in order to evaluate performance based on accuracy, QoE, and hit rate. Latency is also going to be commented upon in the case of the users, which is an essential step in understanding how each model adjusts to network conditions. Integration of generative models brings forth this new approach toward enriching user satisfaction with more accurate predictions about content preferences and better streaming quality [22, 23].

8.4 Methodology

This research uses three different AI models: variational autoencoder (VAE), generative adversarial networks (GAN), and long short-term memory (LSTM). All these models are used together to predict and optimize not only adaptive video streaming but also content personalization. Here, these are optimized to improve user experience on a college website that provides educational videos for 48 subjects. Students can access these videos, which may include video recommendations through a recommendation system based on individual preferences and interests from different subjects. The AI models manage tailored content recommendations while maintaining a seamless buffer-free streaming with regard to the varying network conditions in real time. This is a crucial improvement over traditional video streaming systems that often need help with network performance in addition to user preferences in motion.

The platform continuously monitors the network conditions of each user session, such as bandwidth availability, latency, and buffering events. It captures data on those aspects and adjusts the quality of video accordingly to minimize buffering and maximize playback performance. This adaptive approach is very crucial in light of the diversity of network environments that a user may log in with, from a 4G to a 5G connection or even Wi-Fi. The biggest challenge in this domain is that mainstream adaptive streaming technologies, including Dynamic Adaptive Streaming over HTTP (DASH), need to be more elastic to support these kinds of conditions. They mainly rely on predetermined rules that might not identify the precise time changes within real-time networks and devices' capabilities.

To solve this, VAE, GAN, and LSTM models were learned through a dataset of 3,400 user interactions. Of these, 70% of the data is used in training the models, and the remaining 30% is used in the testing and validation phase. The dataset contains information regarding network conditions, which include bandwidth fluctuations, latency measurements, and buffering events, as well as user preferences regarding content consumption patterns. This dataset will enable models to learn how to predict changes in the performance of the network and send video quality updates, along with recommending content dynamically based on real-time analysis of user preferences. The entire working of the proposed system is shown in Figure 8.1.

The VAE is a generative model and is used to understand how the conditions of a network are distributed, thereby predicting probable bandwidth fluctuations that will manifest in a session. It can, therefore, learn patterns

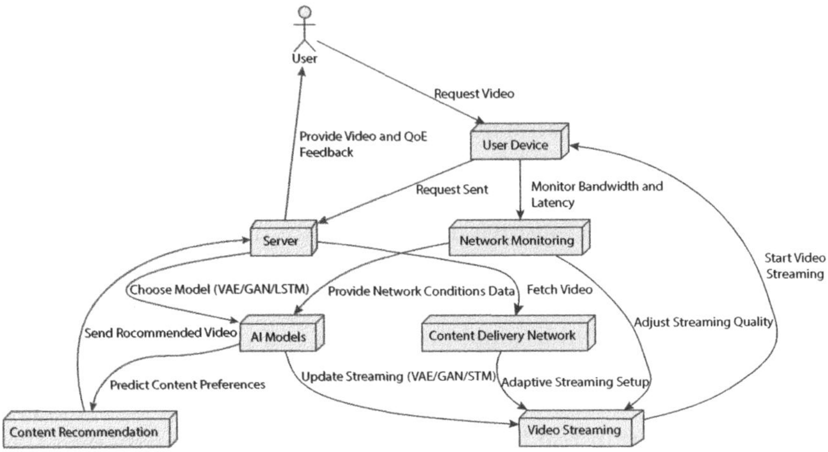

Figure 8.1 Working of the proposed system.

in network behavior to generate an appropriate video quality configuration for future streaming sessions. This ensures that the adaptation of video quality is proactive against the predicted conditions for this model but attempts to minimize this for the user. GAN is used to learn existing patterns of video consumption so that it can generate realistic profiles for users' preferences. Such technology has proven effective in creating even new content recommendation scenarios based on slightly different user behavior, which would give a more personalized experience. An LSTM type of recurrent neural network is used to model the temporal relationships between conditions inside the network and user interactions, which best operate on time series data. LSTM helps the system predict future network performance based on past observations to keep adjusting the parameters of the streams in real-time.

This brings together three models that form robust systems, adjusting both the quality of the streaming and the content itself in light of data feeding into the systems in real-time. As such, even as VAE and GAN are concentrated more on personalization and production of content in line with the user's interests, LSTM ensures video playback is smooth under the somewhat fluctuating conditions of a network. The interplay of the three models provides the platform with the capability to address all the varying kinds of challenges brought about by various user devices, networks, and preferences.

Furthermore, the approach of using VAE and GAN for content personalization makes the system shift from being a static set of recommendations to dynamic and data-driven predictions that evolve in real-time. Traditional recommendation algorithms, depending on historical data most of the time, can hardly get through the nuances of the change in preferences of a user. The system can also embed adaptive video streaming with LSTM that adapts rapidly and promptly to network changes more effectively than the results usually characterized by standard DASH implementations, often finding unnecessary buffering during congestion peaks or poor video quality.

8.4.1 AI Models Used

The variational autoencoder is a generative model that learns an encoding of the input data in a latent space of lower dimensionality and decodes this encoding back to reconstruct the original input. The architecture of the autoencoder is shown in Figure 8.2. VAE differs from traditional autoencoders in the fact that they actually model the probability distributions underlying the data through the injected noise in the latent variables [24].

It works on the assumption that some distribution captures the input data, for example, typically a Gaussian, and learns this distribution's parameters based on the encoder. The encoder maps the input x to the latent variable z. It will reconstruct the input data by decoding with the obtained latent variable. A VAE aims to maximize the evidence lower bound (ELBO):

$$L = E_{q(Z|x)}[logp(x|z)] - D_{KL}(q(z|x) \,||\, p(z)) \tag{8.1}$$

where q(z|x) is the approximate posterior distribution, p(x|z) is the likelihood of the input given the latent variable, and DKL is the Kullback-Leibler divergence, which measures how different the approximate posterior is from the prior p(z). The first term is the reconstruction term that pushes the encoder to reconstruct the input, and the second term is the regularization term on the distribution of latent variables. VAEs are trained by stochastic gradient descent, which involves the backpropagation of ELBO gradients to update the model's parameters. In this manner, VAEs are capable of generating new data points that would resemble the input. Therefore, this makes VAE particularly useful for tasks such as content personalization.

Generative adversarial networks consist of two neural networks, as shown in Figure 8.3; one is a generator, and the other is a discriminator. It trains a competitive process together with the discriminator. The generator generates realistic samples of data from the random noise with its task, while the

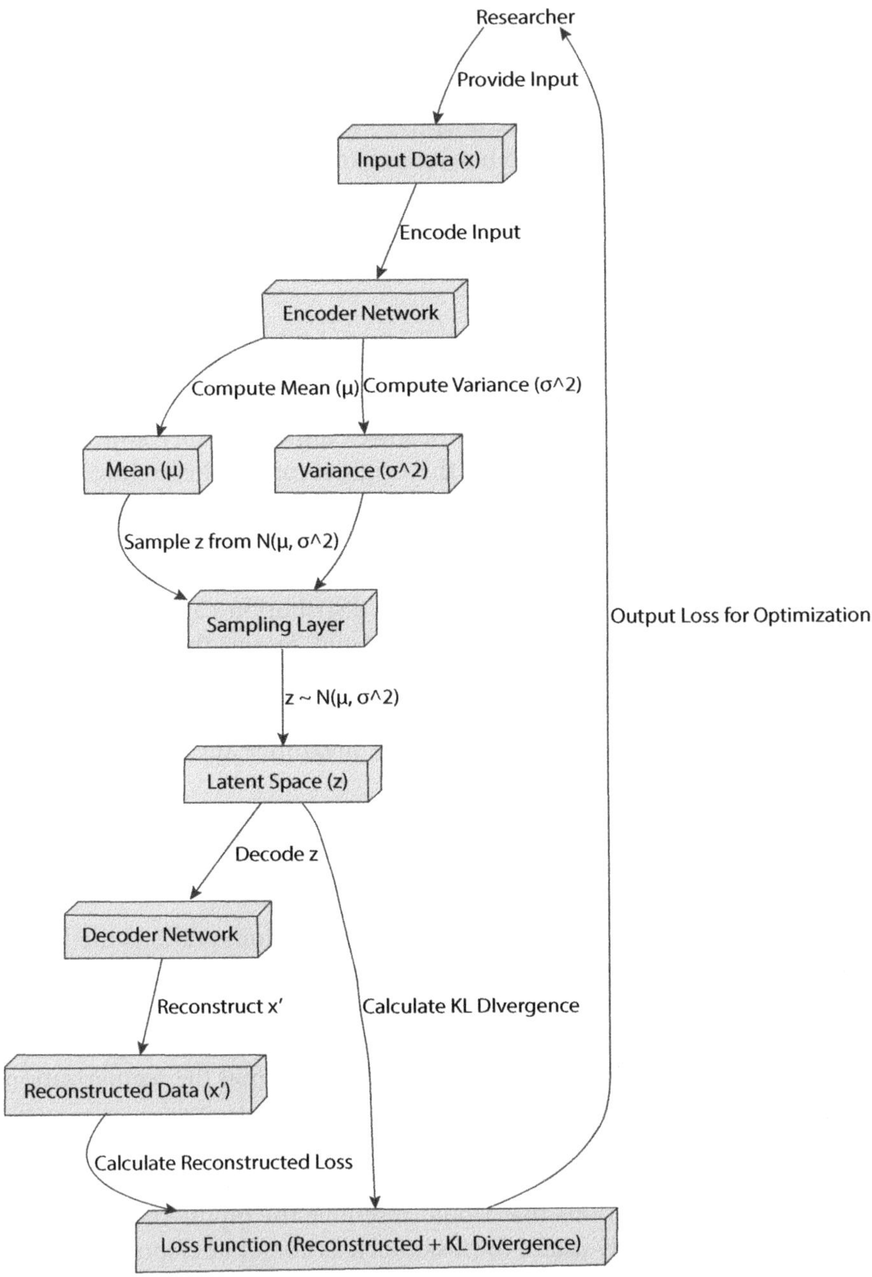

Figure 8.2 VAE architecture.

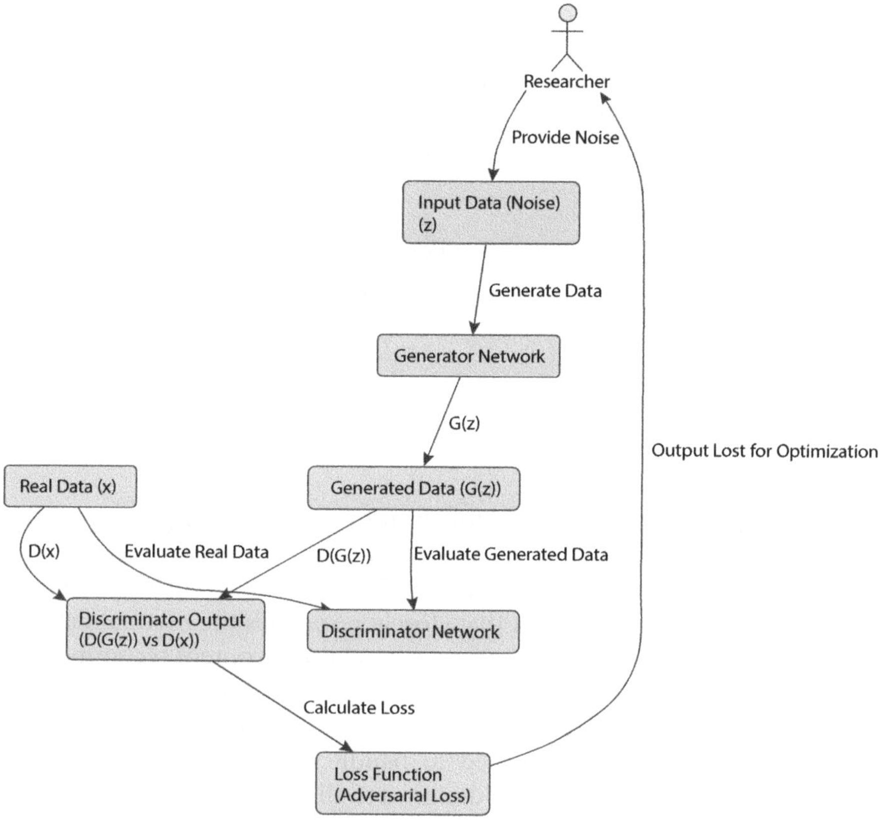

Figure 8.3 Generative adversarial networks architecture.

discriminator aims to differentiate accurate data and synthetic data generated by the generator. The generator accepts random noise z sampled from a latent distribution p(z) and maps it to the data space to produce G(z). On the other hand, the discriminator receives both real data x and generated data G(z) and outputs a probability D(x) that determines whether an input is real or fake. Both networks are trained on the following objective:

$$\min_{G} \max_{D} V(D,G) = E_{x\sim pdata(x)}[logD(x)] + E_{z\sim p(z)}[\log\left(1 - D(G(Z))\right)] \quad (8.2)$$

The discriminator shall maximize this objective by classifying actual and generated data samples correctly. In return, the generator shall try to minimize this objective by fooling the discriminator. Training goes in an alternating fashion with regard to the discriminator being updated to

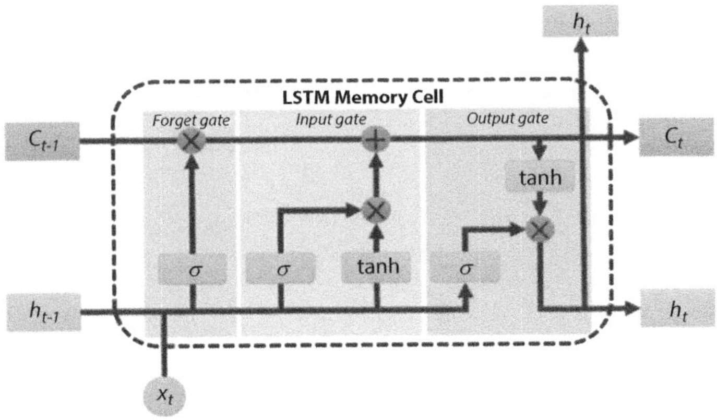

Figure 8.4 Architecture of LSTM.

improve the classifier's accuracy, followed by updating the generator. The adversarial nature of this setup makes for a game-theoretic scenario in which the models improve upon each other over time. GANs are highly potent for generating new data that closely resembles the original dataset; hence, they are suitable for generating personalized content recommendations by synthesizing new variations of user preferences.

One particular type of recurrent neural network is long short-term memory, which is specifically designed to handle temporal dependencies in sequential data. LSTMs have been designed to eliminate the vanishing gradient problem that plagues traditional RNNs in trying to represent sequences that have long-term dependencies. The core of any LSTM architecture, shown in Figure 8.4, is made up of memory cells whose state is updated at every time step. The LSTM can learn which parts of the input sequence are essential to retain and which can be forgotten, enabling it to capture long-range dependencies. This makes LSTMs particularly effective for modeling temporal dynamics in video streaming, where the network conditions evolve.

8.4.2 Preprocessing of the Dataset

The study included a preprocessing stage over the interaction dataset between 3,400 users and the educational video streaming platform prior to training the AI models. This stage was vital since it ensured that the data were clean and structured appropriately for the models to learn from. Data characteristics consisted of user behavioral logs, network conditions, and video consumption patterns. The entire steps of preprocessing are shown in Figure 8.5.

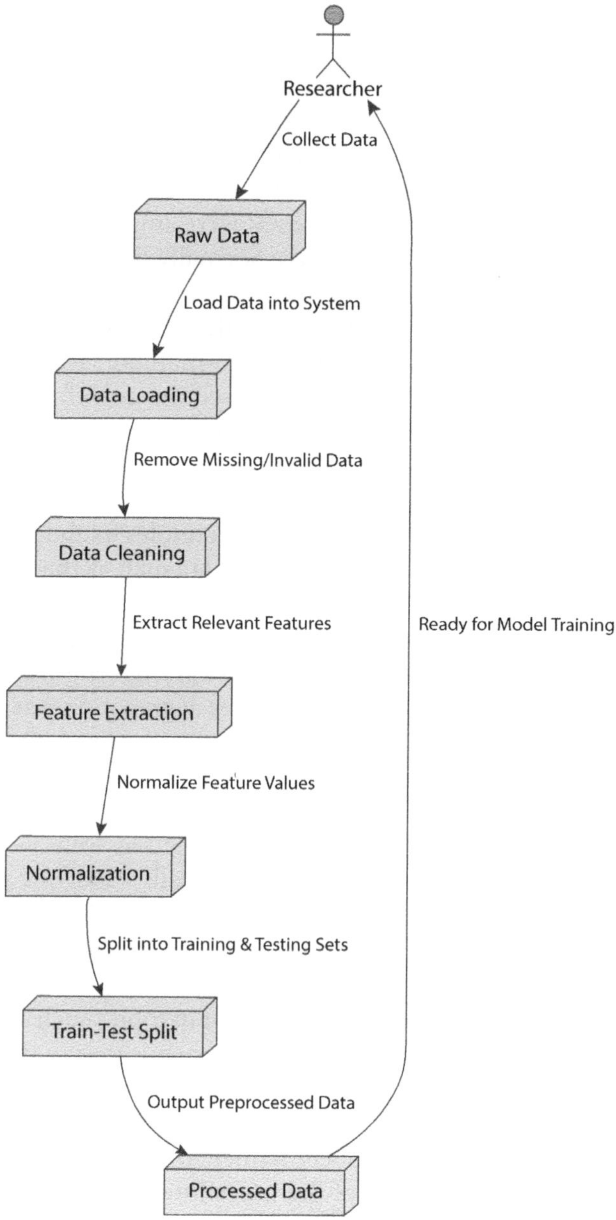

Figure 8.5 Preprocessing of the proposed system.

Data cleaning represented the first step of preprocessing. The raw data contained missing values, erroneous entries, and duplicates, which needed to be cleansed to prevent misleading the AI models when they were being trained. For example, there are several cases in which users' sessions were cut off abruptly, resulting in records that were left incomplete due to network conditions such as bandwidth and latency. To handle this issue, missing values of bandwidth and latency were imputed using statistical means (mean imputation if the data were normally distributed); otherwise, other statistical methods were adopted. If there were entirely missing key features, such as events that were being buffered, in some records, those records were excluded from the database to maintain the quality of the training set. Duplicates, which may have biased the model's understanding of user behavior patterns, were discovered and removed.

Having cleaned the data, it was now time to perform feature extraction and engineering. The dataset contained several features that directly affected the performance of the AI models. Examples included user preferences, network conditions, and video metadata (such as the duration, resolution, and bitrate of each video). However, only a few of these were in an exploitable form. For instance, network condition variables like bandwidth changes were sampled using time intervals of varying lengths, which led to discrepancies in data. Therefore, a time-series aggregation was performed. Bandwidth and latency averages were computed over fixed intervals, while essential events like buffering instances had their binary values recorded in order to make them easier to represent.

However, user behavior data were transformed as well: video-viewing patterns and content preferences were encoded. Categorical variables, including "subject interest" and "video genre," were represented in the form of one-hot encoded numbers. One-hot encoding was chosen because it avoids accidentally inferring any ordinal relationships between the categories. For example, one could map an interest in topics such as mathematics, physics, or literature to a vector where every position in the vector will denote a particular subject. If the user is interested in mathematics or not, then mark that particular position in the vector '1' while the others remain '0'. This depiction gave the AI models, namely the content personalization systems, a better understanding of the user's interest without introducing bias into the approach.

Besides the data related to the user's preferences, features that are time-based in nature were part of the preprocessing pipeline to capture the temporal nature of user interactions and network conditions. For example, the hours of a day and the day of a week when the video was streaming were encoded as cyclical features *via* sine and cosine transformations.

Therefore, models—factually, in particular, the LSTM network—acquired the ability to process temporal dependencies better. The motivation for using cyclical encoding is that time-dependent data have inherently cyclical properties—for example, after Sunday comes Monday, and after 11:59 PM comes midnight; traditional numerical encodings would eliminate these relationships. Using one-hot or numerical encodings for time features would be appropriate because the model could retain any patterns it learned for peak video-watching times or periods of network congestion.

After feature extraction, the dataset was normalized so that all features had a similar scale. Such normalization is essential in a class of models like VAE and GAN, where the magnitude of input features affects the characteristics of the model. For the normalization of the dataset, the range of every feature was scaled between 0 and 1 using min-max normalization. This was necessary for features like video duration and bandwidth, as their range was drastically dissimilar, so normalization would ensure that no feature dictated the training process of the model. The same thing was applied to other continuous features, such as network latency and buffering duration, so uniformity was created within the dataset.

For the network condition data, which also had some metrics such as bandwidth and latency, smoothing techniques were applied to decrease noise. Network measurement data, especially in real-time streaming environments, is highly volatile due to the changing signal strength or mobility of the user. Moving average computation of bandwidth as well as latency helped to reduce short-term spikes within the data and smoothed out the overall representation of network conditions that were necessary for training the LSTM model towards accurately predicting subsequent network performance.

Data balancing was done at the last step before it was split into training and testing sets. The dataset included various classes of user behavior and network conditions, where a proper distribution of each class should be present in both training and testing. In other words, only a few customers may feel buffering more because of bad network conditions, whereas others would experience a smooth streaming process. To balance the consideration of avoiding biasing the model towards users who have ideal network conditions, undersampling was carried out on the majority class, which is users without the problem of frequent buffering. Oversampling techniques such as SMOTE (Synthetic Minority Over-sampling Technique) can be performed on the minority class, which are users with frequent buffering issues. This ensures that training of the models is done using a balanced dataset, thus improving their generalization capability in cross-validation on different network environments.

Finally, it was divided into two subsets: training and testing; 70% of it could be used for training purposes, while the remaining 30% was left for testing; in this way, models learned sufficiently large training sets but kept aside enough of the data to be made unbiased during the testing phase. The training data was used to fine-tune the parameters of the models, while test data gave an opportunity to analyze how well a model generalizes to unseen scenarios, apart from variations in user behavior and network conditions.

8.5 Result and Discussion

In the case of these three AI models—GAN, VAE, and LSTM—the performance was deeply tested to figure out which model predicted the network conditions for adaptive streaming and user preferences in content better. Evaluation metrics were computed, including accuracy, to figure out which of the three AI models could tackle the intricate scenarios of the video streaming platform more effectively. Observations from results in Figure 8.6: the GAN model outperformed the others with a very impressive accuracy of 98.98%. This result gives a high degree of accuracy in suggesting that GAN worked very effectively in understanding and predicting both network conditions and user preferences because of its ability to generate actual variations of data and adapt to diverse behaviors of users and unpredictable network environments. The VAE model followed with a solid accuracy of 94.5%, showing it was also efficient in modeling

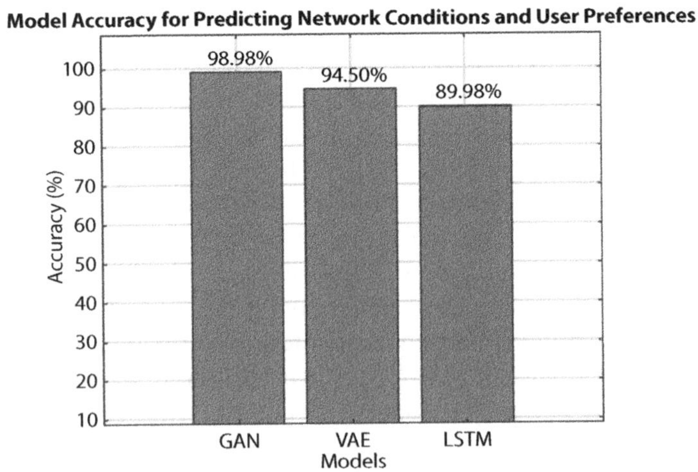

Figure 8.6 Accuracy of each model.

the probability distributions of network conditions and user preferences, though less accurate than GAN, but effective in its own right.

Finally, an LSTM model was obtained with a precision of 89.98%, slightly performing worse in capturing network performance temporal dependencies, but most notably, wholly failing behind both GAN and VAE models in adapting themselves to real-time fluctuations. These results emphasize the superiority of GAN in this task-specific modeling system, thereby proving it the best-suited model for both adaptive video streaming and content personalization.

Figure 8.7 captures the performance comparison of the three models, GAN, VAE, and LSTM, in terms of precision, recall, and F1 score. Precision refers to the measure of correct optimistic predictions by the model from all the optimistic predictions it makes. In recall, the ability of the model to identify all instances is measured. The F1 score provides a harmonic mean of precision and recall that will give an optimum view of the quality of the model. One may easily see from the figure that the GAN model performs best across all metrics by achieving 97.80% precision, 98.50% recall, and an F1 score of 98.15%.

This presents the case of GAN as a very accurate case of optimistic prediction and also captures almost all relevant instances. The VAE model follows, with slightly lower performance: precision of 93.40%, recall of 94.00%, and F1 score of 93.70%, showing a good balance between precision

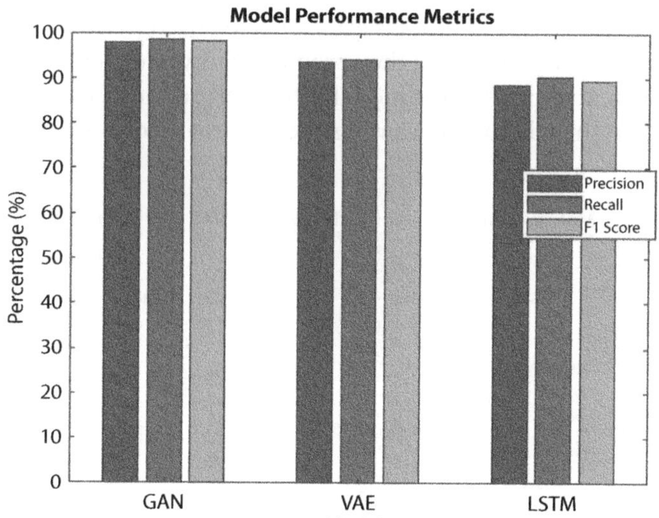

Figure 8.7 Performance of each model.

and recall but less efficient compared to GAN. Here, it can be seen that the LSTM has the lowest scores, with a precision of 88.50%, a recall of 90.20%, and an F1 score of 89.34%, which means, in this context, it is the least effective model, specifically in terms of balancing precision and recall for predicting both network conditions and content preferences.

The confusion matrices are shown in Figure 8.8, where they specify what each model, GAN, VAE, and LSTM, predicted as true positives or false positives, true negatives, and false negatives. For the GAN model, the confusion matrix reports 980 true positives and 985 true negatives with only 20 false negatives and 15 false positives. Therefore, this is an accurate model since it contains less error in both positive-case and negative-case predictions. On the other hand, the VAE model has 945 true positives and 965 true negatives but identifies a higher number of errors, with 55 false negatives and 35 false positives.

It indicates that although VAE was able to identify reasonably well, GAN is less accurate than it is, especially in terms of the location of actual positives. The fact that the LSTM model, in fact, has the highest number of errors with 902 true positives and 935 true negatives, including 98 false negatives and 65 false positives, means that LSTM has a tough time in the better prediction of positive cases, hence performing much lower than GAN and VAE. Overall, the GAN model clearly outperforms others by displaying superior predictive capability, whereas VAE is performing decently, and LSTM is lagging in both positive and negative predictions.

Even though these models were evaluated in a controlled environment, the GAN, VAE, and LSTM models were applied on a real-time video streaming platform to assess how well they predict quality of experience (QoE), hit rate, and latency response. Additionally, the integration of these models was carried out on the adaptive streaming and content recommendation system

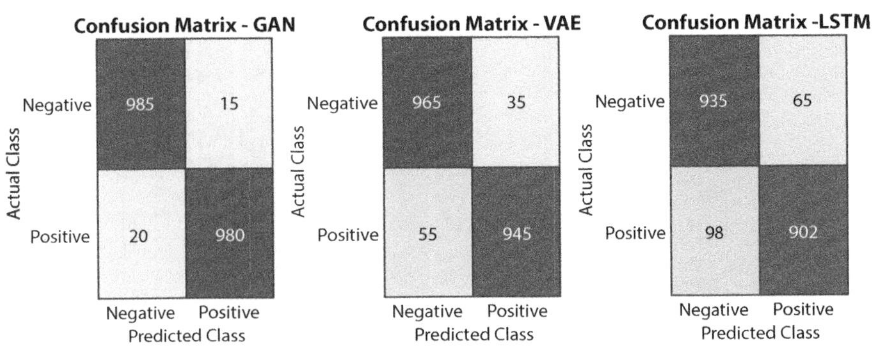

Figure 8.8 Confusion matrices of each model.

of the college's website. These comprised video buffering events, low-resolution videos, and user satisfaction surveys used to measure QoE. The hit rate was monitored through an understanding of user engagement with the system—that is, the frequency of correct prediction and recommendation of video content preferred by users, measured in terms of video play rates and completion percentages. Latency response was measured by monitoring the time that each model took to predict and accommodate network changes in their respective conditions, measuring it by how fast the system adapted video quality within the same process to avoid pauses in playback.

Real-time implementation indicated significant differences in performance among the models. The capability of GAN to produce realistic variations of network conditions led to high accuracy on the predictions of fluctuations of the network and preferences of the user, thereby displaying high QoE, a high hit rate, and low latency. VAE did pretty well but was delayed in responding to sudden network changes, sometimes creating minor dips in QoE. Although LSTM did well in pattern recognition for a temporal sequence, it was relatively slow in response to rapid changes in network conditions and exhibited higher latency and lower QoE than the other models. Real-time testing, therefore, proved the robust nature of GAN to succeed in practical and live environments.

Figure 8.9 gives the average quality of experience and hit rate for all three models: GAN, VAE, and LSTM. The real-time testing was done with

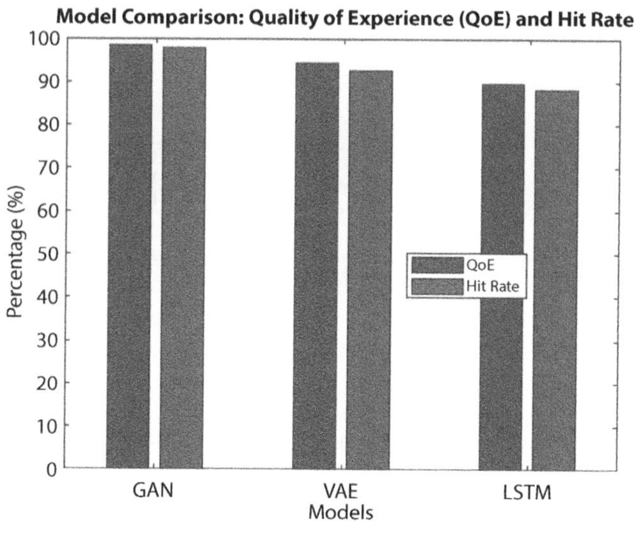

Figure 8.9 QoE and hit rate of each model.

50 participants. Thus, the QoE of the GAN model came out to be impressive at 98.5%, with the hit rate at 97.8%. This somehow indicates that this model performs efficiently in delivering a smooth streaming experience and is sufficiently accurate in predicting user choices. This means that the users were delighted with the video quality and the relevance of the recommended content. Though still effective, the QoE hit rates achieved by the VAE model were 94.2% and 92.6%, which signifies it would still be able to maintain a good user experience, though it is behind GAN in adapting to user needs. The LSTM model delivered relatively inferior results: a QoE of 89.7% and a hit rate of 88.3%, indicating the failure of this model to give the best streaming performance and to make the most accurate recommendation in predictions compared to other models.

Figure 8.10 illustrates the average latency, measured in milliseconds, for each model—GAN, VAE, and LSTM—over an 8-hour testing period with 50 users, segmented into 30-minute intervals. Throughout the testing, the GAN model consistently demonstrated the lowest latency, averaging between 145 ms and 155 ms across all intervals. This efficiency in processing and adapting to network conditions indicates GAN's strong capability to maintain quick response times for real-time video streaming, contributing to an optimal user experience. In contrast, the VAE model exhibited a slightly higher latency, ranging from 199 ms to 210 ms. While this

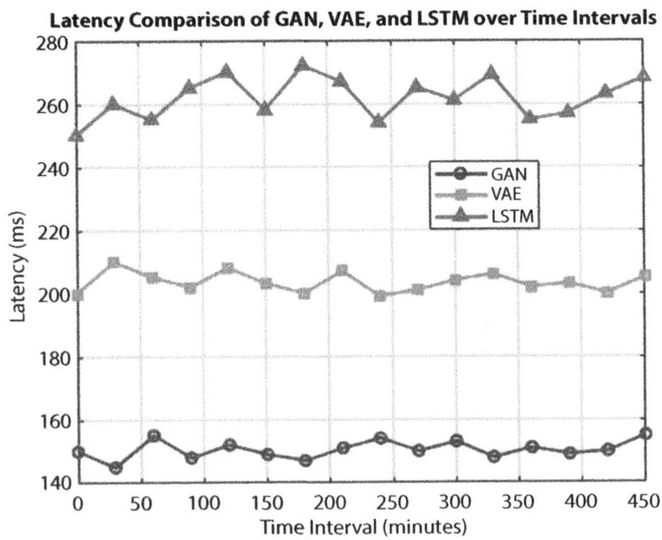

Figure 8.10 Latency of each model.

performance is still acceptable for streaming, it highlights VAE's relatively slower adaptability to fluctuating conditions compared to GAN.

The LSTM model had the highest latency, varying from 250 ms to 270 ms, reflecting its challenges in rapid response and adjustment to real-time demands. The increased latency for LSTM suggests that, while effective in some areas, it may not provide the same level of responsiveness required for seamless adaptive streaming as GAN and VAE. Overall, these results emphasize the GAN model's superior performance in minimizing latency during real-time applications, which is crucial for enhancing user satisfaction in video streaming environments.

During the testing of the models, user feedback was a good source of rich information on how they performed in realistic scenarios. Many users appreciated the GAN model for seamless viewing with minimal buffering and suitable adaptation to changing network conditions. Users were very pleased, reporting high levels of satisfaction, usually indicating that the recommendations were in excellent agreement with their interests, hence amplifying the engagement with the content. They also indicated that video quality was consistent, even with the fluctuation in network performance. On the other hand, for the VAE model, general feedback was positive overall, but sometimes, users who had this experience noticed delays when loading content recommendations, as well as some slight buffering when high-demand times began. Though the users desired and appreciated the quality of their recommended content, they further indicated that there was a need for improvement in the responsiveness to network changeover. Thus, while the model by LSTM excelled in certain areas, there was mixed feedback. Users experienced more latency and buffering problems during peak usage, producing a less favorable experience than those of other models.

From the analysis of these three AI models—GAN, VAE, and LSTM—it is evident that their differences come in performance in real-time video streaming applications. The GAN model appears as the best one, which holds the highest accuracy and QoE at the same time with the lowest latency, hit rate, and good user feedback, which includes a responsive experience and is thus adequate for adaptive video streaming and content personalization. While VAE performed pretty well and LSTM added a little value, they need help to approach GAN in terms of the general abilities to adapt to the runtime demands and user preferences. Thus, the GAN model should be applied in real-time streaming environments to optimize user experience and ensure content delivery with efficiency.

8.6 Conclusion

This research has successfully demonstrated the usability of generative models, VAE, and LSTM in adaptive video streaming and content personalization. It is because of its efficiency that the GAN model had a tremendous lead in accuracy, hit rate, quality of experience, and latency over other models and, therefore, would be regarded as the best suited for high-performance real-time applications. It is a consistent testimony of the seamless viewing experience with minimal buffering and highly relevant recommendations, reflecting the ability of the model to adapt to variability in network conditions and individual tastes. Future work could further improve the performance of the GAN model by incorporating additional features such as user demographic data or more advanced contextual information to tighten the recommendations around the content further. Furthermore, hybrid models that incorporate the strengths of several algorithms will have the potential to spur even more improvement in terms of streaming efficiency and personalization. Apart from educational material, such research can be applied in other fields requiring higher levels of user engagement and satisfaction, like entertainment, health, and online learning. The findings will, therefore, be directly applied by the organizations to improve their video streaming services in relation to even higher user retention rates and better overall satisfaction.

AI disclosure statement: The author confirms that no artificial intelligence (AI)-generated content or AI-assisted writing tools were used in the preparation of this chapter (including summaries and the abstract). All text, analysis, and conclusions are the result of the author's original work.

Bibliography

1. Khamaj, A. and Ali, A.M., Adapting user experience with reinforcement learning: Personalizing interfaces based on user behavior analysis in real-time. *Alexandria Eng. J.*, 95, 164–173, 2024.
2. Luo, X., Pan, C., Liu, Z., Wang, L., Tang, H., Zhang, Z., Zeng, D., TSDRL: A three-stage deep reinforcement learning method for reliable collaboration of manufacturing service towards mass personalized production. *Expert Syst. Appl.*, 257, 124968, 2024.
3. Khan, M. and Khusro, S., Towards the design of personalized adaptive user interfaces for smart TV viewers. *J. King Saud Univ. Comput. Inf. Sci.*, 35, 9, 101777, 2023.

4. Sahu, D.P., Murkute, S., Yadav, U., Agarwal, M., Kumar, P., Deep Learning Based Bug Detection in Solidity Smart Contracts. *Recent Trends in Image Processing and Pattern Recognition*. RTIP2R 2023. Communications in Computer and Information Science, vol. 2026, Springer, Cham, 2024, https://doi.org/10.1007/978-3-031-53082-1_9.
5. Kiran, A., Refonaa, J., Nabeel, M., Navaprakash, N., Reddy, V.L., Lalitha, R.V.S., Adaptive mixed reality robotic games for personalized consumer robot entertainment. *Entertain. Comput.*, 52, 100825, 2025.
6. Sharan, A., Co-occurrence and semantic similarity based hybrid approach for improving automatic query expansion in information retrieval, Lecture Notes in Computer Science, p. 8956, 2015, https://doi.org/10.1007/978-3-319-14977-6_45.
7. Liu, Y., Gu, F., Guo, J., Self-X heterogeneous attributed graph embedding-based product configuration framework for cognitive mass personalization. *J. Manuf. Syst.*, 76, 411–428, 2024.
8. Bilal, K. and Sajid, M., Blockchain technology: Opportunities & challenges, in: *2022 International Conference on Data Analytics for Business and Industry (ICDABI)*, IEEE, pp. 519–524, 2022.
9. Jones, C.H., Madhavan, S., Natarajan, K., Corbo, M., True, J., Dolsten, M., Rewriting the textbook for pharma: how to adapt and thrive in a digital, personalized and collaborative world. *Drug Discovery Today*, 104112, 2024.
10. Prasad, M., Prasad, O.K., Meng Joo, E., Saxena, A.K., Lin, C.T., A novel fuzzy logic model for pseudo-relevance feedback-based query expansion. *Int. J. Fuzzy Syst.*, 18, 980–989, 2016.
11. Bian, C. and Lu, S., Personalized recommendation of entertainment robots in fine arts education based on human–computer interaction and data mining. *Entertain. Comput.*, 100740, 2024.
12. Lin, C.T. and Li, D.L., Ranks aggregation and semantic genetic approach based hybrid model for query expansion. *Int. J. Comput. Intell. Syst.*, 10, 1, 34–55, 2017, https://doi.org/10.2991/ijcis.2017.10.1.4.
13. Liu, Y., Razman, M.R., Zakaria, S.Z.S., Lee, K.E., Khan, S.U., Albanyan, A., Personalized context-aware systems for sustainable agriculture development using ubiquitous devices and adaptive learning. *Comput. Hum. Behav.*, 160, 108375, 2024.
14. Singhal, P., Gupta, S., Singh, J., An Integrated Approach for Analysis of Electronic Health Records Using Blockchain and Deep Learning. *Recent Adv. Comput. Sci. Commun. (Formerly Recent Pat. Comput. Sci.)*, 16, 9, 1–10, 2023.
15. Nie, X., Huang, S., Gao, X., Luo, J., Zhang, G., MRAM: Multi-scale Regional Attribute-weighting *via* Meta-learning for Personalized Image Aesthetics Assessment. *Knowledge-Based Syst.*, 304, 112546, 2024.
16. Gupta, T., Singhal, D., Lal, B., Turukmane, A.V., Fake News Detection Using Transfer Learning. *Recent Trends in Image Processing and Pattern Recognition*.

RTIP2R 2023. Communications in Computer and Information Science, vol. 2026, Springer, Cham, 2024, https://doi.org/10.1007/978-3-031-53082-1_2.

17. Naseer, F., Khan, M.N., Tahir, M., Addas, A., Aejaz, S.H., Integrating deep learning techniques for personalized learning pathways in higher education. *Heliyon*, 2024.
18. Garg, N., Kukreja, S., Saini, M., Singh, I., Mall, S., IoT-Enhanced Agricultural Water Management System Utilizing Conventional Machine Learning Methods, in: *2023 International Conference on Modeling, Simulation & Intelligent Computing (MoSICom)*, IEEE, pp. 554–559, 2023.
19. Dong, W., Pan, D., Kim, S., Exploring the integration of IoT and Generative AI in English language education: Smart tools for personalized learning experiences. *J. Comput. Sci.*, 82, 102397, 2024.
20. Singh, N.P., Hasan, D.S., Shelke, N.A., Namdev, A., Girdhar, N., An Integrated Machine Learning and IoT Based Approach for Enhanced Healthcare Efficiency and Personalized Treatment, in: *Intelligent Systems Design and Applications*. ISDA 2023. Lecture Notes in Networks and Systems, vol. 1046, A. Abraham, A. Bajaj, T. Hanne, P. Siarry, (Eds.), Springer, Cham, 2024, https://doi.org/10.1007/978-3-031-64813-7_16.
21. Thieme, A., Hanratty, M., Lyons, M., Palacios, J., Marques, R.F., Morrison, C., Doherty, G., Designing human-centered AI for mental health: Developing clinically relevant applications for online CBT treatment. *ACM Trans. Comput.-Hum. Interact.*, 30, 2, 1–50, 2023.
22. Kaliyar, R.K., Kumari, R., Sharma, N., Manjunath, S., Prasad, P.D., Advancements in Early Detection of Cervical Cancer using Machine Learning and Deep Learning Models for Cervicography Analysis, in: *2024 International Conference on Emerging Innovations and Advanced Computing (INNOCOMP)*, 2024, May, IEEE, pp. 252–256, doi: 10.1109/INNOCOMP63224.2024.00049.
23. Pichandi, K.V., Senthil, P., Prabakar, D., Uma, S., Artificial internet of medical things for wearable body sensor data analysis using radial basis autoencoder-based adversarial neural network. *Expert Syst. Appl.*, 260, 125338, 2025.
24. Akkem, Y., Biswas, S.K., Varanasi, A., A comprehensive review of synthetic data generation in smart farming by using variational autoencoder and generative adversarial network. *Eng. Appl. Artif. Intell.*, 131, 107881, 2024.
25. Ayemowa, M.O., Ibrahim, R., Bena, Y.A., A systematic review of the literature on deep learning approaches for cross-domain recommender systems. *Decis. Anal. J.*, 100518, 2024.

9

Data Analytics and Real-Time Decision Making in Water Quality Prediction

Kajal Kumari, Sudip Kumar Sahana* and Debjani Mustafi

Department of Computer Science and Engineering, Birla Institute of Technology, Mesra, Ranchi, India

Abstract

Water pollution is a significant problem around the world, causing negative effects on human health and ecosystems. Contaminants from different geographic and anthropogenic sources, such as industrial, agricultural, and urban, can lead to significant degradation in water quality, and therefore, real-time monitoring is required. Conventional water quality monitoring methods are resource-intensive, expensive, and do not provide real-time results, making them ineffective in changing environments. AIIoT augments real-time data processing and decision-making with continuous water quality data collection using IoT sensors, and then machine learning models further analyze them to predict water parameters accurately. This study explores the relationship between water pollutants and the quality of water using deep learning based on Kaggle datasets. The hybrid SARIMA with transformer model scored an accuracy of 98.34%, with RMSE reaching 0.1098, MAE 0.0848, and R^2 an absolute 0.9630. CDNs enable the efficient, low-latency distribution of predictions with hybrid SARIMA-tansformer model results, aiding in decision-making in time-critical scenarios. This work is innovative in detecting the seasonality and combining it with the application of transformer networks for recognizing complex patterns. The novelty of this study is real-time water quality monitoring in areas like aquaculture, smart cities, and healthcare, whereby fast and accurate predictions are critical.

***Keywords*:** Water quality index (WQI), SARIMA, transformer, IoT, real-time monitoring

**Corresponding author*: sudipsahana@bitmesra.ac.in

Tanusree Chatterjee, Rajib Banerjee and Souvik Ganguli (eds.) Content Delivery Networks Using Artificial Intelligence and Internet of Things: Revolutionizing Content Delivery, (193–212) © 2026 Scrivener Publishing LLC

9.1 Introduction

Water quality monitoring can be the key to ensuring a healthy lifestyle and adhering to environmental sustainability, which can help eliminate the risk of diseases. Clean and safe water directly affects not only basic human consumption, agriculture, and industry but also entire ecosystems. However, water contamination is a global challenge aggravated by urbanization, industrial discharge, and agricultural runoff, causing severe health hazards, including waterborne diseases, chemical toxicity, and chronic health complications. Pollution also affects aquatic life and biodiversity, disrupting ecosystems and contributing to ongoing environmental harm. Hence, appropriate and regular monitoring of one or more parameters of the water body, e.g., pH, turbidity, dissolved oxygen, temperature, and contaminants, is vital for the environmental equilibrium and regulatory enforcement systems. Monitoring can avoid ecological destruction, support policy development, and safeguard populations from contaminated water sources.

However, traditional approaches for water quality prediction have significant shortcomings, especially concerning efficiency and scalability. Traditional approaches involve manual sampling and lab procedures that, while accurate, are labor-intensive, costly, and unable to deliver real-time information. This means that data processing takes time, and responding to a new threat or a change in water quality is not instantaneous. In addition, they are not scalable, particularly in resource- and knowledge-limited settings. Conventional statistical models and simple machine learning approaches also fail to reflect the complex and nonlinear interactions between different water quality parameters, leading to less-than-optimal predictive power. These limitations reiterate the need for new solutions that can fill these voids while also improving the accuracy, speed, and scalability of water quality monitoring systems.

Leveraging cutting-edge hybrid deep learning techniques unified with IoT (Internet-of-Things) technology provides a powerful approach to address these challenges. IoT-capable sensors provide constant and real-time information acquisition from various water infrastructures like rivers, lakes, reservoirs, and aquifers. They are low-cost, can be scattered throughout networks, and allow for localized monitoring on scales not possible with classical methods. IoT systems also progressively make automated decisions and reduce dependency on technicians visiting one by one.

When hybrid deep learning models are used in conjunction with the data gained from IoT systems, the prospects for IoT systems are enhanced tremendously. Deep learning architectures, specifically convolutional neural networks (CNNs), can learn spatial feature patterns across sensor data that matter for outliers detected. The recurrent neural networks and transformer model also enhance this capability, as both frameworks can capture temporal dependencies and trends over time, which can be particularly critical to predicting the future state of water quality. Spatial and temporal considerations are integrated to improve predictive performance and provide early warning of disturbances, such as harmful algal blooms or pollutant surges. Automated, resilient systems that require minimal human involvement and scale well reduce the cost of water quality monitoring.

9.2 Related Work

In [1], the design and development of a low-cost IoT-based water quality monitoring system for small-scale fish farmers highlights the need for cost-effective and low-barrier solutions in order to tackle the challenges within resource-constrained environments. In [2], time series data is used to show an effective combination of convolutional neural networks (CNN) and long short-term memory (LSTM) for predicting water level and quality through their convolutional layers for extracting spatial features and LSTM for temporal dependencies, boosting the accuracy of predictions. In [3], a CNN-GRU-Attention framework model was developed that combines the influences of industrial and agricultural pollution, and the attention mechanism of this model plays an important role in identifying the key features that contribute to high accuracy in drinking water source prediction. In [4], adapting the temporal fusion transformer (TFT) to apply to time-series river water quality data introduces a timely flow transfer learning approach to tackle data shortages that improves the prediction ability of long-term environmental monitoring. In [5], a machine learning–based real-time monitoring system was proposed to efficiently detect contaminants with key advantages such as immediate feedback mechanisms essential for ongoing management of water quality. In [6], an Internet-of-Things approach was adopted with a smart system demonstrating integration of state-of-the-art sensors to gather real-time water quality parameters and exhibit promising connectivity and usability across various environmental conditions. In [7], an IoT river water quality monitoring framework is characterized by its lightweight computational algorithms to allow real-time processing, where hardware cost and computational resources are minimized while resulting in the scalability of solutions and affordable costs. In [8], the performance of different hybrid deep learning models was compared, and it was found that the ensemble of multiple models was often the best at accurately capturing complex interdependencies among water quality parameters to improve prediction accuracy. In [9], a deep learning framework was used to predict water quality in IoT systems, proving the capability to combine real-time data streams and predictive analyses for optimal water resource management. In [10], LSTM networks within an IoT ecosystem were implemented and efficiently modeled the nonlinear and sequential patterns within water quality datasets, a key step to predicting trends and detecting anomalies. In [11], a modular and scalable IoT-ML architecture for water quality prediction was introduced, addressing the need for adaptive systems that can accommodate diverse sensor networks and evolving data streams. In [12], IoT-based solutions were incorporated with LSTM modeling for long-term water quality prediction, illustrating their potential for predictive maintenance as well as environmental sustainability. In [13], an IoT and machine learning hybrid was introduced that integrates embedded technologies with advanced analytics to improve monitoring accuracy for different aquatic environments. In [14], hybrid deep learning models were used to predict water quality parameters in aquaculture and included variables of environmental factors since they can provide specific and targeted applications on sustainable farming practices. In [15], the use of deep learning architectures for monitoring river waters focuses on enhancing prediction accuracy and developing decision support systems to address potential water management issues.

9.3 Dataset Description

The first dataset, the water quality prediction dataset on Kaggle, provides a substantial dataset of 1,048,575 samples and 24 water quality parameters. This makes it an excellent dataset for predictive model development related to the prediction of water quality levels, and the sources of contamination, like pH, indicate acidity or alkalinity; iron, nitrate, and chloride relate to chemical composition; and turbidity concerns physical contamination, odor, chemical composition, and sensory quality. Other metals include lead, zinc, and manganese, which were included to assess industrial pollution effects. Other features include measures of sulfate and conductivity, which are crucial for understanding dissolved organic and inorganic substances and determining ion interactions within water. Total dissolved solids (TDS) and chlorine are important indicators for disinfection level and purification quality of water. Furthermore, the feature "Source" labels water samples by their source, like rivers or lakes, enabling spatial and contextual analysis.

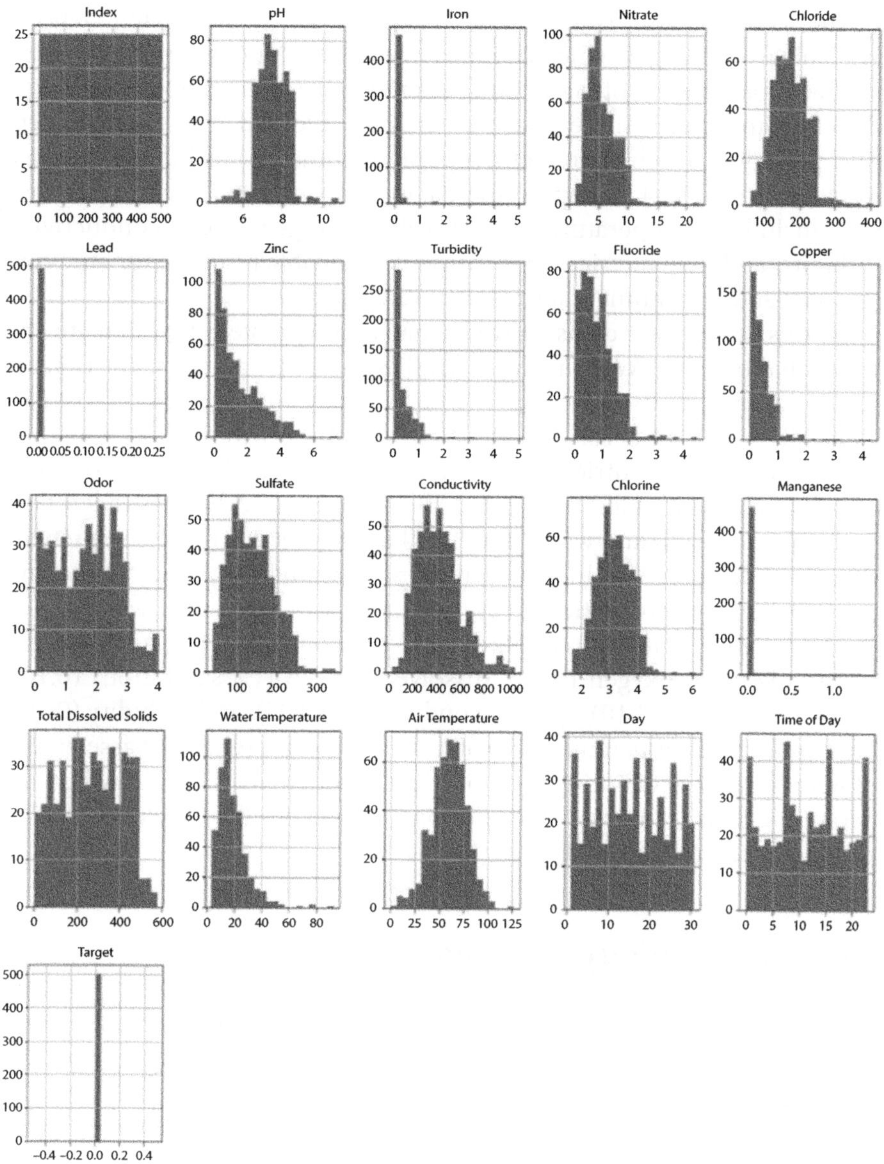

Figure 9.1 Features of the dataset used in this study.

The second dataset is the Brisbane Water Quality Monitoring Dataset, available from Kaggle, which contains information about various physicochemical parameters of water and real-time measurements of water flow direction and speed. These measurements are recorded at 30-minute intervals, providing a rich source of temporal data that is ideal for time series analysis. With 23,143 samples, the dataset encompasses diverse parameters, including temperature, pH, dissolved oxygen levels, conductivity, turbidity, and flow metrics. This makes it a valuable resource for studying seasonal and environmental trends in water quality.

In this study, two datasets were analyzed separately to calculate the water quality index (WQI) using carefully selected parameters. For Dataset 1, the parameters used were pH (0.15), Nitrate (0.25), Turbidity (0.25), Chloride (0.20), and Iron (0.15). These parameters were chosen as they provide insight into nutrient concentration (nitrate) and clarity (turbidity), as well as chemical stability. pH and chloride are key indicators that reflect both aquatic health and water usability. For Dataset 2, a more comprehensive set of features was included, incorporating both hydrodynamic and physicochemical properties. These features include average water speed (0.05), average water direction (0.05), chlorophyll (0.10), temperature (0.10), dissolved oxygen (0.20), dissolved oxygen (\% saturation) (0.10), pH (0.15), salinity (0.10), specific conductance (0.05), and turbidity (0.10). This broader set of parameters enables a more comprehensive assessment of water quality, accounting for biological productivity (chlorophyll), temperature effects (temperature), and ionic properties (salinity and specific conductance) in Figure 9.1.

9.4 Methodology of Proposed Model

From Figure 9.2, the analysis starts with the preprocessing and feature engineering of datasets, including cleaning, normalization, and the extraction of relevant features that will improve the performance of the model. Then, apply the seasonal autoregressive integrated moving average (SARIMA) once the data is prepared. SARIMA deals well with both seasonal and non-seasonal components in the data, along with trends. Applying SARIMA produces residual errors by comparing the model's predictions to the actual observed values. Now, the part of the data that the SARIMA model could not capture, which is converted into the residual part, is important for further modeling. A transformer model captures complex temporal dependencies given the residuals. The transformation allows the model to provide better insights and forecast future trends by

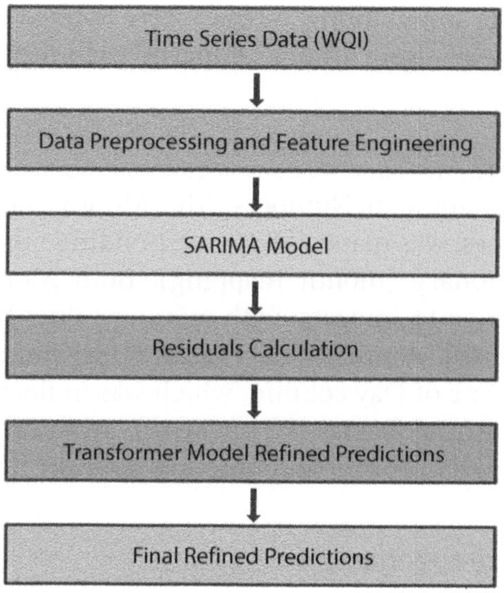

Figure 9.2 Workflow for the WQI prediction hybrid model.

correcting the error of SARIMA. This combined SARIMA-transformer model implementation leads to a final tuned prediction of WQI values, which constitutes a more explicit and unified approach for the water quality predictions, harnessing the strengths of SARIMA time-series forecasting and the transformer model. The hybrid approach enhances the accuracy and reliability of predictions.

9.4.1 Preprocessing Stage

9.4.1.1 Handling Missing Data

To handle missing values in this dataset, several methods are used. When the dataset contained missing values, they were removed using the df.dropna() method. For numerical columns, missing values were imputed with the mean of the respective feature using df[columns].fillna(df.mean(numeric only=True)). In the categorical feature, Source, missing values were filled with the most frequent value (mode) using df['Source'].fillna(df['Source'].mode() [0]). Forward-filling was applied to propagate the previous valid value to fill missing entries using df.fillna(method='ffill'), and backward-filling used the next valid value with df.fillna(method='bfill').

In addition, custom imputation strategies were applied for the pH feature with the median value using df['pH'].fillna(df['pH'].median()).

9.4.1.2 *Handling Temporal Features*

Mapping Month Names to Numbers: The Month column, which contained month names, was mapped to corresponding numeric values using a predefined dictionary (month mapping). Both Month and Day columns were converted to integers, with missing values being filled with 1 using the methods df['Month'].fillna(1).astype(int) and df['Day'].fillna(1).astype(int). The Time of Day column, which was in float format, was converted into a valid time format (HH:MM:SS), and non-time values were handled using a custom conversion function (convert time).

9.4.1.3 *Datetime Creation*

Creating a Datetime Column: A new Datetime column was created by combining Year, Month, Day, and Time of Day to represent the timestamp for each record using the pd.to_datetime() function. Setting Datetime as Index: The Datetime column was set as the index of the DataFrame using df.set_index ('Datetime', inplace=True), and the Year column was removed if it was not needed. Hyperparameter tuning in SARIMA involves optimizing parameters to effectively capture trends, seasonality, and noise in WQI data. The key non-seasonal parameters are p (autoregressive order, determining how many past WQI observations influence the current value), d (degree of differencing to ensure stationarity, e.g., d = 1 for first-order differencing), and q (moving average order, representing the inclusion of past forecast errors). Seasonal parameters include P (seasonal AR order, capturing periodic dependencies like yearly water pollution patterns), D (degree of seasonal differencing for repeating patterns such as monsoon effects), Q (seasonal MA order for seasonal noise), and s (seasonal period, e.g., s = 12 for monthly data). A grid search is commonly used to tune these parameters, exploring a range of values such as p, q, P, Q $\in$ {0, 1, 2}, d, D $\in$ {0, 1}, and s $\in$ {12, 24}.

9.4.2 Feature Extraction Stage

The feature extraction process for water quality index (WQI) prediction using SARIMA and transformer models focuses on capturing both linear and nonlinear patterns from the dataset.

9.4.2.1 *Differencing for Stationary*

To ensure the time series data is stationary (a requirement for SARIMA), differencing is applied. This involves subtracting the previous observation from the current one. Both non-seasonal and seasonal differencing are used:

- Non-seasonal differencing removes trends in the data, such as long-term upward or downward shifts.
- Seasonal differencing accounts for periodic fluctuations or cycles, like seasonal variations in temperature or turbidity that recur every year or month.

9.4.2.2 *Temporal Features*

Temporal features such as Month, Day, and Time of Day are crucial for identifying seasonal and diurnal patterns in water quality. These features are extracted from timestamps and encoded properly (e.g., converting Month to numeric values). The Datetime field is created by combining Year, Month, Day, and Time of Day.

9.4.2.3 *Scaling and Normalization*

To ensure that the model can process features with different units or scales (e.g., pH vs. dissolved oxygen), features like temperature, chlorophyll, and salinity may be scaled. Standard techniques like min-max scaling or Z-score normalization may be used to bring all features into a comparable range.

9.4.2.4 *Aggregation and Resampling*

By averaging data over defined periods, temporal aggregation assists in the identification of long-term patterns. The final step is some smoothing of data, which is useful when attempting to find out seasonality in terms of turbidity or iron. This act can help in aligning data from different time frequencies, where hourly measurements are resampled to daily averages, etc.

9.4.3 WQI Calculation

The water quality index is a numerical value of combined features with weightage depending on their importance to the quality of water. In this study, selected parameters for this equation are pH, nitrate, turbidity, chloride, and iron. These parameters are normalized to threshold levels based on accepted water quality values. For example, pH values, which should range between 6.5 and 8.5 for potable water, are upscaled from 0 to 100, where the distance from non-potable water ranges implies a lower score. Also, nitrate levels, where the best case is between 0 and 10 mg/L, and turbidity, where the best case is between 0 and 5 NTU, are inverse normalized, given that larger values imply worse quality. Chloride (250 mg/L) and Iron (0.3 mg/L) have acceptable limits, and the data have been normalized (higher values are worse), indicating that concentrations above a certain level reduce water quality. Finally, each parameter is weighted by its impact on overall water quality. A weight of 0.25, the highest weight, is assigned to nitrate and turbidity as their values have an effect on safety and clarity, and pH and Iron are assigned a weight of 0.15. Chloride is weighted at 0.20, due to its contribution to taste and salinity management. The WQI can be calculated by normalizing the individual parameters and aggregating them to produce a single, understandable number describing the entire water quality for the sample. It can also illustrate the interaction between various parameters and provide an integrated view of water quality for multiple end uses and their safety, sustainability, and suitability. The water quality index (WQI) is calculated using the following mathematical expression [16]:

$$WQI = \sum_{i=1}^{n} \left(w_i \times q_i \right) \tag{9.1}$$

where

- w_i: Weight assigned to the i-th water quality parameter (reflecting its relative importance).
- q_i: Quality rating of the i-th parameter, normalized to a 0–100 scale based on acceptable standards.
- n: Total number of water quality parameters considered.

For each parameter, the quality rating q_i is

$$q_i = \frac{V_{observed} - V_{ideal}}{V_{standard} - V_{ideal}} X\ 100 \tag{9.2}$$

where

- $V_{observed}$: Observed value of the parameter.
- V_{ideal}: Ideal or baseline value (e.g., for pH, $V_{ideal} = 7.0$).
- $V_{standard}$: Permissible limit for the parameter (e.g., for pH, $V_{standard} = 8.5$).

If the parameter exceeds permissible limits, adjustments may be made in the normalization formula to reflect a higher penalty for deviation.

9.5 Experimental Setup

9.5.1 Optimize the SARIMA Model

Model tuning is the process of choosing the best values for the parameters (p, d, q) for the autoregressive, differencing, and moving average components, and (P, D, Q, m) for seasonal parameters. This is done using a grid search, which checks multiple combinations of these parameters on the training dataset. The SARIMA grid search function lets you evaluate each combination as per the Akaike information criterion (AIC), the precision in adjusting the model complexity versus the fit. The SARIMA model is built with a combination of parameters that produce the minimum AIC value and then utilized to predict the test dataset, which guarantees that, for time-series predictions, it is parsimonious and predictive.

9.5.2 Residuals Extraction

In this study, residuals of proposed SARIMA model are calculated from the difference between the observed data samples and their corresponding predicted values. This is helpful for us to come across unmodeled noise or orders of the SARIMA model. For the training data set, we can remove the SARIMA forecast for the true test values. These residuals represent the segment of the time-series data that was not well predicted by SARIMA. These residuals are subsequently fed as inputs into a transformer-based machine learning model. This hybrid technique enhances the predictive performance by taking advantage of the strengths of SARIMA for modeling both trend and seasonality, with the ability of the transformer model to capture complex nonlinear relations in the residuals.

9.5.3 Data Preparation for the Transformer Model

The SARIMA model residuals are converted to a time-series format with lagged features required by the transformer model. A rolling window approach is used to create lagged residuals, i.e., each row has the n values preceding the target. So, the dataset is divided into inputs (X) and target outputs (y). Then, this data is transformed into a PyTorch tensor, which is used in the transformer model. This process ensures that the model is provided with a consistent data sequence that records the chronological dependencies to generate faithful predictions.

9.5.4 Train the Transformer Model

The training of the transformer model is guided by the mean squared error (MSE) loss function, which quantifies the difference between the predicted and actual residuals. The Adam optimizer is used to update the weights of the model during training. The training is done in batches, where each batch allows the model to capture underlying patterns in the residuals. Over multiple epochs, the model learns to predict the unmodeled residuals, improving the accuracy of the hybrid SARIMA-transformer model.

9.5.5 Forecasting Residuals with the Transformer

Once trained, the transformer model predicts the residuals for the test set. This is done by transforming the test data into the same format used for training, which includes creating lagged features. The data is then reshaped to meet the 3D input requirements of the transformer model. The transformer generates new predictions for the residuals, accounting for patterns not modeled by SARIMA. These predicted residuals are then combined with the SARIMA forecasts to produce a final hybrid forecast. By leveraging both models' strengths, the hybrid approach improves forecasting accuracy, making it particularly effective for time-series prediction tasks.

9.6 Results and Discussion

9.6.1 Actual vs. Predicted Plot (Regression Line Plot)

The actual vs. predicted plot is widely used in the case of time series forecasting to measure the accuracy of a predictive model. It becomes easier to evaluate the model's performance, identify trends, and discover any systematic errors by comparing the actual data to the predicted values. Figure 9.3 shows the actual test data vs. predicted values from both the SARIMA model and the hybrid SARIMA with transformer model. This expresses the deviations between forecasted values of the applied model and its actual values.

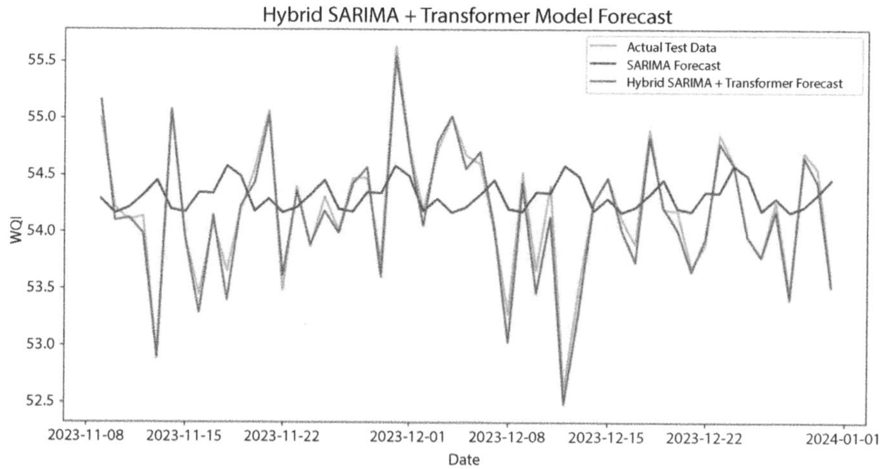

Figure 9.3 Graph of SARIMA and SARIMA with transformer model.

- **Test Data (Orange Line):**
This line shows the actual observed values of the WQI over the same time period, and these values appear to rise and dip dramatically.

- **SARIMA Forecast (Blue Line):**
The blue line indicates values predicted by the SARIMA model. This model captures the general trend and seasonality, thus making it fail to predict some smaller variations, leaving the model with larger deviations from the actual data, especially at times of rapid fluctuation.

- **Hybrid SARIMA with Transformer Forecast (Green Line):**
The green line shows the forecasted values obtained using the hybrid model, which combines SARIMA's strength in capturing seasonal trends and the transformer model's proficiency in learning more complex patterns. This line is much nearer the real data, with much more effective and consistent predictions. From Figure 9.3, the hybrid SARIMA with transformer model gives the best output with reduced RMSE, lower MAE, and the best fit overall to the actual test data. The hybrid model's superior ability to identify both linear and nonlinear trends in the time-series data enables it to produce more precise forecasts.

- **Actual Test Data (Orange Line):**
This line represents the real observed values of the WQI over the time period, which fluctuates with noticeable peaks and valleys.

- **SARIMA Forecast (Blue Line):**

The blue line shows the forecasted values from the SARIMA model. The SARIMA model captures the general trend and seasonality, but it misses some of the smaller variations and has larger deviations from the actual data, particularly during certain periods of rapid fluctuation.

- **Hybrid SARIMA with Transformer Forecast (Green Line):**

The green line represents the forecasted values from the hybrid model, combining SARIMA's seasonal trend detection with the transformer model's ability to capture complex patterns. This line is closer to the actual data, tracking the fluctuations more accurately and consistently than the SARIMA model. The hybrid SARIMA with transformer model provides the best output, with reduced RMSE, lower MAE, and a better overall fit to the actual test data. This indicates that the hybrid model is more adept at capturing both linear and nonlinear patterns in the time series data, resulting in more accurate predictions.

9.6.2 Residual Analysis

The residual analysis helps in determining the predictive performance of the SARIMA model and the hybrid SARIMA + transformer model. Figure 9.4 shows the residual patterns of both models. Observations from the graph are also described in Table 9.1, which summarizes differences in variance,

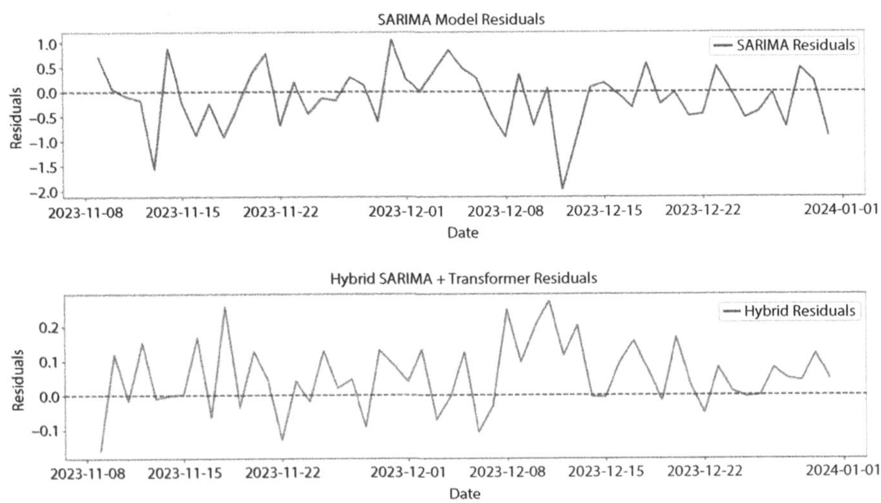

Figure 9.4 Graph of the model's residual analysis.

Table 9.1 Comparison of residual analysis for SARIMA and hybrid models.

Aspect	SARIMA residuals	Hybrid residuals
Variance	Higher variance, fluctuating between approximately -2 and +1. Indicates less accurate forecasting.	Lower variance, confined to a narrower range of around -0.2 to +0.2. Reflects improved accuracy.
Randomness	Visible patterns with non-random clustering above or below the zero line. Suggests unaccounted patterns and correlated errors.	Appears more random, with no discernible patterns or clustering. Indicates better handling of temporal dependencies.
Autocorrelation	Potential autocorrelation, with residuals possibly influencing each other, based on visible patterns.	Lacks apparent autocorrelation; residuals appear random and uncorrelated.

randomness, and autocorrelation between the two models. Residual checks were performed to assess the predictability of the SARIMA model and the hybrid SARIMA with the transformer model. SARIMA residuals showed more variance, ranging from -2 to 1, along with clear patterns of autocorrelation. These findings indicate that SARIMA failed to capture complex nonlinear dependencies in the dataset. On the other hand, the residuals of our hybrid SARIMA with transformer model had much lower variance (-0.2 to 0.2) and were also distributed randomly, which shows it is better fitted to the data. The hybrid model's performance enhancement illustrates the transformer's utility in modeling complex non-linearities while still benefiting from SARIMA's strengths at capturing seasonal patterns in the data, leading to improved forecasting. These results are consistent with the observed diminished RMSE and increased randomness of residuals, confirming improved performance of the hybrid model.

Residual analysis was conducted to evaluate the predictive performance of the SARIMA model and the hybrid SARIMA with transformer model. The SARIMA residuals exhibited higher variance, with values ranging from -2 to 1, and displayed clear patterns of autocorrelation. These findings suggest that SARIMA struggled to capture complex nonlinear dependencies in the dataset. In contrast, the hybrid SARIMA with a transformer model produced residuals with significantly reduced variance (ranging

Table 9.2 Performance metrics of various models.

Model	RMSE	MAE	MAPE	R^2	Accuracy
SARIMA model	0.6119	0.4706	0.87%	-0.1499	69.69%
Hybrid transformer	0.1098	0.0848	0.16%	0.9630	98.34%
Hybrid random forest	0.2267	0.0748	0.23%	0.9522	96.75%
Hybrid GRU	0.3713	0.3482	0.76%	0.8521	91.56%
Hybrid LSTM	0.6244	0.3209	0.67%	0.8054	90.81%

from -0.2 to 0.2) and a random distribution, indicating a better fit to the data. In Table 9.1, this improvement demonstrates the hybrid model's ability to combine SARIMA's strength in capturing seasonal trends with the transformer's capacity to model intricate non-linearities, resulting in more accurate forecasts. These results align with the observed reduction in RMSE and enhanced residual randomness, supporting the hybrid model's superior performance.

9.6.3 Numerical Analysis

Table 9.2 shows the performance comparison of different models based on key performance metrics, including RMSE (root mean squared error), R^2 (coefficient of determination), MAE (mean absolute error), MAPE (mean absolute percentage error), and accuracy.

9.7 Conclusion

A content delivery network (CDN) is an integral part of this approach, enabling processed data to be disseminated smoothly and in real-time to relevant stakeholders, including aquaculture managers, town planners, and emergency response units. Through the employment of CDN infrastructure, the system enables low-latency communication and provides highly scalable and continuous access to actionable insights, facilitating rapid decision-making capabilities in geographically distributed and resource-constrained scenarios. By enabling the dissemination of predictions made by the hybrid SARIMA-transformer model, CDNs facilitate the reliable and accessible use of real-time water quality monitoring in multiple scenarios. As a result, the performance comparison between several forecasting methods, including the SARIMA model, hybrid SARIMA with transformer, hybrid SARIMA with random forest, hybrid SARIMA with GRU, and hybrid SARIMA with LSTM, shows considerable variance across key performance metrics RMSE, MAE, MAPE, R^2, and accuracy. The SARIMA model captures seasonality and trends effectively but performs poorly on volatile data and does not account for nonlinear dependencies. Although models such as hybrid SARIMA with random forest and hybrid SARIMA with GRU are useful for certain scenarios, they suffer in terms of computational complexity and possible overfitting. Hybrid SARIMA with LSTM is good at capturing long-term dependencies with an accuracy of 96.75%; however, it is computationally expensive and slow to train. The hybrid SARIMA with transformer model stands out as the strongest, yielding the best accuracy of 98.34% and RMSE, MAE, and R^2 figures compared to all other models. The model's capacity to represent both linear and non-linear relationships in time series data makes it a great strategy for superior predictions. By combining IoT sensor data with complex AI techniques for real-time water quality analysis, this method is consistent with AIIoT-based approaches. However, it promotes the AIIoT paradigm through integrating SARIMA's seasonality detection feature and the transformer's ability to capture complex patterns, which can be used to improve its accuracy and robustness in an effective manner. Using CDN systems, forecasts and insights are communicated with the end-users and delivered on time. The novelty of this work consists of the application of hybrid models to the prediction of WQI, showing that these approaches, including machine learning techniques, could improve predictions for environmental niches such as water monitoring, which is crucial for the sustainability and health management of natural resources.

Future works can focus on incorporating real-time IoT sensor data into hybrid models, which can further improve the context and application of the prediction framework while at the same time allowing it to be scaled for further systems with dynamic, real-time monitoring of the environment.

AI disclosure statement: The author confirms that no artificial intelligence (AI)-generated content or AI-assisted writing tools were used in the preparation of this chapter (including summaries and the abstract). All text, analysis, and conclusions are the result of the author's original work.

References

1. Suleiman, R., Low-Cost IoT-Based Water Quality Monitoring System for Small-Scale Fish Farmers in Nigeria, Doctoral dissertation, University of Applied Sciences Technikum Wien, 2023.
2. Baek, S.S., Pyo, J., Chun, J.A., Prediction of Water Level and Water Quality Using a CNN-LSTM Combined Deep Learning Approach. *Water*, 12, 12, 3399, 2020.
3. Mei, P., Li, M., Zhang, Q., Li, G., Prediction Model of Drinking Water Source Quality with Potential Industrial-Agricultural Pollution Based on CNN-GRU-Attention. *J. Hydrol.*, 610, 127934, 2022.
4. Jitha, P. and Vijaya, M.S., Temporal Fusion Transformer and Transfer Learning for Time Series River Water Quality Prediction Model, 2023.
5. Vergina, S.A., Kayalvizhi, S., Bhavadharini, R., Kalpana Devi, S., A Real-Time Water Quality Monitoring Using Machine Learning Algorithm. *Eur. J. Mol. Clin. Med.*, 7, 8, 2035–2041, 2020.
6. Lakshmikantha, V., Hiriyannagowda, A., Manjunath, A., Patted, A., Basavaiah, J., Anthony, A.A., IoT-Based Smart Water Quality Monitoring System. *Global Transitions Proc.*, 2, 2, 181–186, 2021.
7. Fakheri, S., Komazec, N., Saberi Najafi, H., IoT-Based River Water Quality Monitoring. *Comput. Algorithms Numer. Dimens.*, 2, 4, 216–220, 2023.
8. Sha, J., Li, X., Zhang, M., Wang, Z.L., Comparison of Forecasting Models for Real-Time Monitoring of Water Quality Parameters Based on Hybrid Deep Learning Neural Networks. *Water*, 13, 11, 1547, 2021.
9. Thai-Nghe, N., Thanh-Hai, N., Chi Ngon, N., Deep Learning Approach for Forecasting Water Quality in IoT Systems. *Int. J. Adv. Comput. Sci. Appl.*, 11, 8, 686–693, 2020.
10. Liu, P., Wang, J., Sangaiah, A.K., Xie, Y., Yin, X., Analysis and Prediction of Water Quality Using LSTM Deep Neural Networks in IoT Environment. *Sustainability*, 11, 7, 2058, 2019.
11. Rahu, M.A., Chandio, A.F., Aurangzeb, K., Karim, S., Alhussein, M., Anwar, M.S., Towards Design of Internet of Things and Machine Learning Enabled Frameworks for Analysis and Prediction of Water Quality. *IEEE Access*, 2023.
12. Lopez, A.L., Haripriya, N.A., Raveendran, K., Baby, S., Priya, C.V., Water Quality Prediction System Using LSTM NN and IoT, in: *2021 IEEE International Power and Renewable Energy Conference (IPRECON)*, IEEE, pp. 1–6, 2021.
13. Adeleke, I.A., Nwulu, N.I., Ogbolumani, O.A., A Hybrid Machine Learning and Embedded IoT-Based Water Quality Monitoring System. *Internet Things*, 22, 100774, 2023.
14. Haq, K.R.A. and Harigovindan, V.P., Water Quality Prediction for Smart Aquaculture Using Hybrid Deep Learning Models. *IEEE Access*, 10, 60078–60098, 2022.

15. Chellaiah, C., Anbalagan, S., Swaminathan, D., Chowdhury, S., Kadhila, T., Shopati, A.K., Amesho, K.T., Integrating Deep Learning Techniques for Effective River Water Quality Monitoring and Management. *J. Environ. Manage.*, 370, 2024.
16. Akhtar, N., Ishak, M.I., Ahmad, M.I., Umar, K., Md Yusuff, M.S., Anees, M.T., *et al.*, Modification of the Water Quality Index (WQI) Process for Simple Calculation Using the Multi-Criteria Decision-Making (MCDM) Method: A Review. *Water*, 13, 7, 905, 2021.

10

Enhancing E-Commerce Security: A Framework of Integrated Vulnerability Assessment and Data Privacy Protection

Aftab Ara

College of Business Administration, University of Hail, Hail, Saudi Arabia

Abstract

The growth in several ICT development disciplines has given rise to computer security issues for both businesses and individual consumers. The research estimates the threats to e-commerce at the moment and those that might arise in the future and presents a new design of data protection. Dynamics within e-commerce security are such that it is now difficult to even go through all the various threats that exist, in particular, SQL injections, cross-site scripting (XSS), and AI-related risks. To protect such data, it is necessary to have data privacy frameworks that include, among other things, regulatory requirements and industry standards, and the difficult part is implementing those frameworks. The use of such AI-driven technologies as automatic surveillance and face or iris recognition, which are recognized as the most effective in countering modern threats, is indeed very encouraging. However, it is advisable to evaluate the robustness of security measures towards user experience and satisfaction and customers' behavioral changes without putting aside the principles of building trust and obsolescing usability versus privacy trade-off. The main objective of the research is to identify the current and potential threats and risks in e-commerce, to analyze the adequacy of existing security measures regarding the e-commerce models currently in use, and to propose a conceptual model for data protection for e-commerce for online markets. The implications of this study highlight the urgent need for e-commerce businesses to adopt security strategies that integrate with advanced technologies, regulatory compliance, and considerations for user experience and have proactive risk management strategies to ward off threats from cyberattacks. Ultimately, this

Email: ara.aftab@gmail.com

Tanusree Chatterjee, Rajib Banerjee and Souvik Ganguli (eds.) Content Delivery Networks Using Artificial Intelligence and Internet of Things: Revolutionizing Content Delivery, (213–230) © 2026 Scrivener Publishing LLC

study aims to enhance the online safety standards, boost trust among consumers, and promote sustainable growth of e-commerce in this digital environment.

***Keywords*:** E-commerce, security, vulnerability assessment, data privacy, AI-based security, SQL injection, cross-site scripting (XSS)

10.1 Introduction

The use of technology has progressively increased, particularly in the retail sector, and has influenced the security risk faced by businesses and consumers. According to studies, there are several risks related to e-commerce, such as the threat of SQL injection and cross-site scripting (XSS) [1]. According to a survey conducted [2], 62% of the 500 analyzed e-commerce sites had at least one serious security issue. There is a gap in knowledge regarding the interoperability and effects of AI-based security systems on both threat identification and user experience [3]. Hence, this study will create an integrated framework for analyzing and addressing the security concerns within e-commerce sites while implementing secure data analytics and proactive protocols. The objectives of the study are:

1. To examine the current and potential threats and risks in e-commerce systems.
2. To analyze the adequacy of existing security measures for the e-commerce models currently in use.
3. To introduce a conceptual model for data protection for e-commerce in online markets.

The study aims to assess the risk as well as the data protection measures in e-commerce, which will help enhance the safety of online markets since it works to boost e-commerce and protect consumer security within the digital economy.

10.2 Literature Review

This review aims to examine the current state of e-commerce security, identify the threats and risks, explore mitigation strategies, and analyze the impact of security measures on user experience and trust.

10.2.1 Threats in E-Commerce

Security Threats: The review identifies how e-commerce security threats affect both trade and security consciousness, affecting both businesses and consumers. Despite the growth of e-commerce, trade is becoming more challenging for businessmen and consumers alike. This study presents extensive opportunities available to system-searchable trading, which has ethical problems that can be managed only by coming up with appropriate online security mechanisms [4, 5]. These threats range from technical exploitation and dishonesty, affecting both the functioning of the business and the possibility of monetary loss.

Cybercrime Concerns: Cybercrime poses a threat to both businesses and consumers alike. One of the most pressing issues in e-commerce is the concern for cybercrime. Studies indicate that cybercrime causes financial losses and erodes user trust in online transactions [6]. The fear of being victimized by cybercrime often deters users from engaging in e-commerce activities [6]. There are threats of network security attacks on the client/server side, such as DoS or website defacement and phishing, pharming, and login attacks [7]. These threats foster a culture of distrust, which makes it very difficult for small and medium enterprises (SMEs) of any nature to even think about competing in the capital market. Also, the risk of cybercrime fraud has intensified, posing a threat to individuals' personal and financial information [7].

Traditional Threats: The present e-commerce internet security environment poses dangers to both businesses and their consumers in many areas. Such threats are from SQL injection, cross-site scripting (XSS), phishing attacks, malware, and data breaches [8, 9]. These risks have taken a turn, and they are so advanced that there have been different methods by hackers that try to breach e-commerce systems. Despite the progress being made in fighting such threats, key web security concerns like SQL injections and XSS attacks are still relevant, for there are still malicious entities that can target web databases and inject malicious scripts [8]. Nevertheless, the dynamics around the concern are evolving rapidly with the new technologies. Security-supporting technologies have made it possible to host electronic devices in the e-business environment. They have also reshaped the surface of attacks since they have exposed new entry points to attacks by cybercriminals [9]. Further, the application of artificial intelligence (AI) and machine learning by attackers has increased the number of targeted attacks [9, 10].

Emerging Risk and Regulatory Challenges: It addresses new risks from advancements in e-commerce, including regional risks, legal status issues, and product choice issues. It also highlights problems from underdeveloped regulatory environments in some countries, leading to fraudulent activities and loss. E-commerce presents risks regarding another region, the legal status of the authorities towards the MCN agents and the hosts, and product choices [11]. Furthermore, in cases where the regulatory environment of e-commerce did not develop properly in some countries, there are no proper laws, and fraudulent activities among the participants of the e-commerce business cause economic losses to the buyers [12]. So, combating e-commerce threats and related risks requires an extensive capacity with robust technical solutions. In the process of advancing e-commerce, new risks have also emerged. For now, while e-commerce is viewed as a potential mode of business, several risks need to be considered. In a study, it was found that e-commerce can develop its potential for economic growth while ensuring a secure online environment for all stakeholders [4, 13].

10.2.2 Mitigation Strategies

AI and IoT Integration: Artificial intelligence is currently changing information security, which raises many opportunities and, at the same time, some challenges. While AI security devices enhance decision-making and risk management, they also pose new threats. Artificial intelligence systems can be turned by systems that have been designed to study a problem, and since they are not very secure, they can cause unexpected results [14]. As the AI threats are present, the sooner the causes of concern are eliminated, the better [15]. These challenges require organizations to practice more secure methods, such as end-to-end encryption, secure payment methods, and the use of intrusion prevention systems, among many others, as well as increase the focus on training employees.

There are significant opportunities and challenges in integrating artificial intelligence (AI) and the Internet of Things (IoT) in e-commerce applications. Improved decision-making and customer experience are possible by the convergence of technology to analyze vast amounts of IoT data efficiently [16, 30]. Smart integrated platforms can be developed by combining AI, IoT, and big data for e-commerce, which can support economic globalization [17]. Machine learning used in AI applications for e-commerce is useful for understanding customer needs, thus helping in operations and fraud in business [18]. Still, there is a need for standardization, management of data, and security concerns, which are the challenges while integrating AI and IoT [30]. To improve efficiency and growth in e-commerce, companies are increasingly investing in AI and IoT technologies, with challenges in the e-commerce sector [18].

Improved User Verification: The use of AI in security as a guiding principle utilizes protection against cyber threats in so many aspects. Artificial intelligence makes the process of user verification more robust and efficient, streamlining the process of improving user authentication and threat detection and strengthening all areas of insecurity across different sectors.

Advanced Behavioral Biometrics: Improving user verification processes, more so when it comes to the behavioral patterns approach in behavioral biometrics, has shown remarkable developments. This is because, after analyzing behaviors like mouse movement, typing rhythm enhances security and user experience [19].

Voice Recognition: Artificial intelligence offers support in voice recognition, which is a better way than using a password-based approach for one's data protection from mobile phones to entire organizations [20].

Automated Threat Detection: Another promising domain in machine learning and AI is the area of security, namely, automated threat detection. AI-empowered detection agencies will immediately be able to differentiate the mass of samples for any distortion and take precautionary measures to protect the organization from cyberattacks [21]. As for smart grid systems, this discovery has cyber-located verification systems that can adjust to falsities in rate estimation and pre-estimate energy requirements by sharing authentication-dependent security [22].

10.2.3 E-Commerce Integrated Security Solutions

Regulatory Compliance: Combining AI with other technologies is creating fruitful results. AI with blockchain technology used in financial services is improving fraud detection by identifying verification and following regulatory compliance [23]. The "Secure by Intelligence" paradigm uses AI for threat detection, behavioral analysis, and decision-making across various industries [24].

Technical Solutions: Security factors and user experience (UX) affect trust, customer behavior, and the balance between privacy and their use in digital interactions. According to studies, the users of smart speakers do not prioritize privacy over ease of use [25]. However, if there are negative experiences by the user, there are concerns for security and privacy. In e-commerce, customer trust and security are related to factors such as quality of information, user interface, and perceived security [26]. In addition, the overall user experience and customer loyalty are other factors with perceived privacy and risk [27].

Enhancing content delivery in CDNs

The integration of artificial intelligence and the Internet of Things will be a major development in how content is delivered in cached or content delivery network (CDN) applications in e-commerce. It uses AI's analytical power as a processing engine of massive amounts of IoT data to yield insights and improve the customer experience and internal operations [18]. This integration helps CDNs in which IoT devices owned by end users can also act like mini-cache servers for fast, less costly delivery of content [28]. AI algorithms leverage the delivery of content schedules as well as pricing to optimize delivery schedules and pricing schemes for efficiency, while smart contracts help to secure the transactions [28]. AI-IoT combines with other technologies and serves as intelligent assistants and human-centric applications for elderly care [29]. However, to integrate them is a challenge [30].

The primary objective of the literature review is to identify current and potential threats in e-commerce, analyze the adequacy of existing security measures, and propose a conceptual model for data protection that enhances online safety and consumer trust. The findings highlight the urgent requirement for e-commerce businesses to have integrated security strategies using advanced technologies, regulatory compliance, and user experience considerations. With these approaches, cyber threats and consumer trust can be mitigated, and sustainable growth in the digital marketplace can be promoted.

10.2.4 Security Factors

Consumer Trust: Building consumer trust is essential for the success of e-commerce platforms. It's crucial to have effective security measures play a critical role in fostering this trust. According to a study, customers are more likely to engage with e-commerce sites that demonstrate a commitment to protecting their data and providing a secure shopping experience. A positive user experience, characterized by user-friendly security features and transparent communication, is vital for enhancing customer loyalty and satisfaction in a competitive market. Security factors and user experience (UX) affect trust, customer behavior, and the balance between privacy and their use in digital interactions. According to studies, the users of smart speakers do not prioritize privacy over ease of use [26].

Data Privacy: The necessity to have robust data privacy frameworks cannot be overstated in the context of e-commerce. These frameworks are necessary for protecting personal and financial information during online transactions. Compliance with regulatory requirements is also essential to maintain consumer trust and avoid legal repercussions. Implementing effective data protection measures safeguards confidential information and contributes to a trustworthy online environment, which is essential for customer retention and engagement. Due to negative experiences by the user, there are concerns for security and privacy.

Balancing User Experience: As a result, the equilibrium between privacy, security, and usefulness is central to building trust and delivering good user experiences [31]. According to the study, there is a need for user-friendly security features with transparent communication and efficient data protection measures [32]. Security features must be important for protecting confidential information and preventing fraud risks; at the same time, they should give a positive user experience for retaining and engaging customers [33]. The SAMARA (secure architecture multi-agent systems risk assessment) is known to be highly interactive for the users, in addition to following all the security requirements for the platform [33]. Another study shows that highly supportive dialogue influences customer satisfaction, support, and credibility from the system [35]. The e-commerce platforms should have a balance between user experience and preserving their confidentiality by having very clear instructions regarding preserving their security measures during the transaction [36].

The existing security measures in current e-commerce models can be known by analyzing their effectiveness against different threats and risks. The study identifies that many existing security measures are not enough to address emerging cyber threats, such as SQL injections and cross-site scripting (XSS) attacks. It also emphasizes the need for a comprehensive approach that combines advanced technologies like AI and machine learning to improve threat detection and response capabilities. In addition, there is a focus on regulatory compliance and user experience, critical for building consumer trust and ensuring data protection in e-commerce environments.

10.3 Findings

Key Drawbacks of current e-commerce security methods:
The paper presents a comprehensive analysis of the current state of e-commerce security, highlighting several key insights:

1. **Identification of Threats** in e-commerce, such as SQL injections, cross-site scripting (XSS), and identity theft. These threats are risks to both businesses and consumers and need robust security measures for mitigation. Table 10.1 provides the key findings on security factors, threats, and security methods.

Table 10.1 Key findings on security factors, attacks, threats, and security methods.

Category	Security factors	Attacks/threats	Current security methods	Key authors
Consumer trust	Data privacy, information quality, user interface security	Identity theft, unauthorized data access	Transparent communication, user-friendly security features	[25, 26]
Cybersecurity	Authentication mechanisms, system resilience	Network attacks, DoS, phishing, SQL injection, cross-site scripting	End-to-end encryption, secure payment methods, intrusion prevention	[7, 8]
Technological threats	User verification, data protection	AI-powered targeted attacks, advanced hacking methods	AI-integrated security solutions, behavioral biometrics, voice recognition	[14, 15]
Regulatory compliance	Cross-border transaction security, legal frameworks	Regional risks, fraudulent activities	Blockchain integration, Regulatory compliance checks	[11, 12, 24]
Advanced protection	Privacy preservation, user experience	Emerging technological vulnerabilities	Automated threat detection, advanced authentication techniques	[9, 20]

2. **Adequacy of Existing Security Measures** is not sufficient to address the evolving nature of cyber threats. There is a dire need for e-commerce businesses to take up security strategies by using advanced technologies, regulatory compliance, and user experience considerations. Table 10.2 provides drawbacks of E-commerce security.

Table 10.2 Drawbacks of current e-commerce security.

Security area	Current drawbacks	Implications	Supporting references
Authentication	Weak password-based systems, limited user verification	Increased vulnerability to identity theft, reduced user trust	[20, 23]
Technological protection	Insufficient defense against AI-powered attacks, evolving cyber threats	Potential system breaches, increased risk of targeted attacks	[9, 10]
Regulatory compliance	Underdeveloped regulatory environments, inconsistent legal frameworks	Fraudulent activities, economic losses for buyers	[11, 12]
User experience	Conflict between security measures and ease of use	Reduced customer engagement, potential abandonment of platforms	[25, 26]
Data privacy	Inadequate protection of personal and financial information	Risk of data breaches, erosion of consumer confidence	[27, 32]

3. **Consumer Trust and Security are** emphasized by the study, where a strong relationship can be found between security protocols and consumer trust. Indicating e-commerce platforms that have customer loyalty and satisfaction, that prioritize data protection, and are transparent.

Recommendations for Developing a Conceptual Model

1. **Integration of Advanced Technologies:** Technology integration with AI and machine learning can help to improve the detection of threats and the capabilities of response. Overall, security can be provided by real-time automated responses.
2. **User Experience Focus:** It is pertinent now to create systems that balance customer trust and satisfaction without compromising on security. Seamless and secure shopping is desired presently by customers to create a balance of experience with security while using e-commerce.
3. **Regulatory Compliance:** To avoid legal issues and create customer confidence, the model should adhere to privacy laws and regulations for cross-border transactions.
4. **Proactive Risk Management:** Having a proactive approach in business is helpful to regularly implement security assessments and updates to identify and mitigate potential risks effectively.
5. **Transparent Communication:** Customer loyalty and satisfaction can be enhanced by maintaining transparent communication with consumers related to data protection and security protocols.

10.4 Conceptual Framework for E-Commerce Security

The conceptual diagram shown in Figure 10.1 is based on the findings and discussions. The diagram shows the components of the e-commerce security framework and the relationships between the key components.

1. Integrated Security Architecture has Cloud-Native **Security,** which utilizes a cloud-native architecture that incorporates

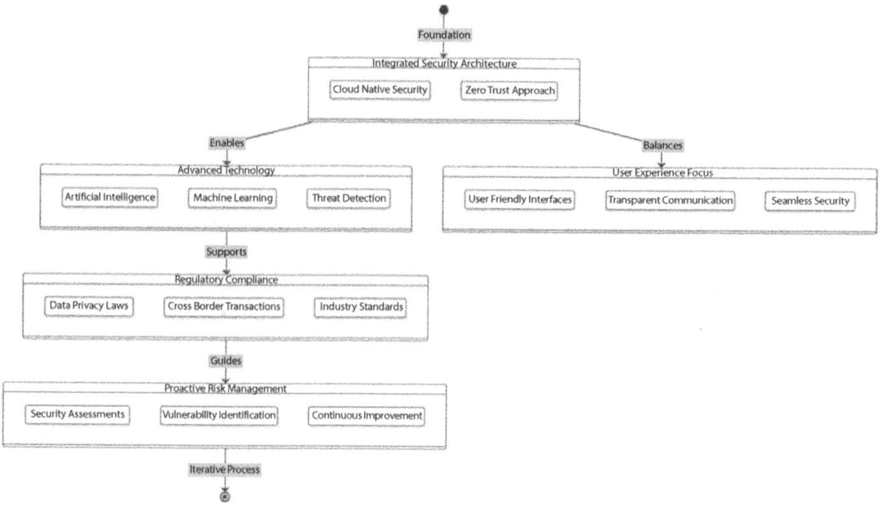

Figure 10.1 E-commerce security's rising landscape (author).

microservices security and data privacy controls to enhance overall security. In the Zero-Trust Approach, which helps in implementing a model showing zero-trust, this model requires continuous verification from its users, irrespective of their location. These technologies analyze patterns in user behavior, thus detecting real-time security threats and mitigating these threats by implementing AI-driven threat detection and response systems. Here, the vulnerability assessment identifies risks in e-commerce and mitigates them.

2. Cross-Border Compliance consists of Regulatory Alignment to ensure compliance with varying regulatory frameworks, focusing on privacy and data protection. The Automated Monitoring systems facilitate managing auditing and compliance issues.
3. User Experience and Trust has user-friendly security features that are quite transparent, giving customers satisfaction and trust. The balancing of security and usability keeps a balance between the security of customers, in addition to giving them a seamless user experience, thus enhancing customer loyalty.
4. Real-Time Monitoring and Response is maintained by automated response systems for reducing the response time and

detecting incidents. With data minimization and privacy protection of the customers' data, it is effective.

The relationships between the different components can be explained as such:

- Both the integrated security architecture and zero-trust approach create a framework to verify user identities continuously to secure data on all platforms.
- Advanced Technologies provides real-time insights and responses to any threats, thus enhancing its architecture. And the zero-trust model.
- A strong focus on user experience can ensure that security measures and a user-friendly experience enhance customer trust and satisfaction.
- Regulatory compliance is the main element in ensuring that security measures align with all legal and industry standards.
- Proactive Risk Management builds the other components together by improving itself in response to any new threat by adopting security measures.

10.5 Discussions

This study is about enhancing security within the context of e-commerce. It highlights that all measures taken to enhance cybersecurity must consider the best approach, which is comprehensive. Indeed, the robustness of e-commerce businesses is growing, as well as the security risks and network attacks such as SQL injections, cross-site scripting, and other AI security vulnerabilities. To enhance safety, it is important to deploy several advanced technologies, such as AI and machine learning, that have, for the most part, been validated by research to detect threats in real-time as well as provide mechanisms for automated responses.

Solutions relating to security, especially those that include a proper integration of a cloud-native architecture and the implementation of a zero-trust security framework, are quite potent given today's threats. Today, user and device activity on networks, including trusted and untrusted, and the perimeter must be checked throughout the day, irrespective of the device, to always be on a higher alert side. It is further consistent with the issues of protection of such highly sensitive data, as evidenced by the implementations of these more advanced strategies within the security of several organizations.

It is also pertinent within the discussed framework to address the issue of cross-border compliance during online transactions, which often require going beyond borders to ensure efficient regulatory requirements to safeguard customers' data and maintain their trust in the systems. It becomes easier to implement and deal with different rules of compliance in the relevant sector through automated regulatory efforts now.

User experience plays an important role in the security framework. Such security features, user experience, and compliance enhancement can contribute to increasing consumer comfort and satisfaction. There has to be a balance of commerce. Between user experience and security for getting customers' trust and satisfaction. In the present times, customers expect transparency from the e-commerce platforms.

The proposed framework advises a comprehensive approach integrating advanced technologies, regulatory compliance, user experience, and real-time monitoring. By addressing these measures, e-commerce can improve its security and gain customers' trust and satisfaction, leading to the growth and sustainability of online commerce.

10.6 Implications

The study has identified several implications for security systems upgrades, legislative policies and regulations, user experience considerations, and managing the risk with business and cross-border transaction security.

1. Security Systems Upgrades: The security framework for e-commerce, based on the previous discussions, has mostly centered on the necessity of newer technologies like AI and machine learning in the realm of threat detection and the response to those threats. This implies that technologically advanced defense mechanisms of this kind must be embraced by e-commerce operators around the globe.
2. Legislative Policies and Regulation in Effect: With the increase of e-commerce transactions over national borders, the study uncovers a phenomenon that necessitates compliance with the requirements of various legislations in countries outside the business's territory. In other words, businesses must adhere to the collection and use provisions of personal data and other aspects of data privacy and avoid

legal problems. This requires the use of built-in monitoring systems to adhere to regulations in different countries.
3. User Experience Considerations: The research suggests that security measures are important to get customers' trust in e-commerce. It should be ensured that e-commerce designs are contributing to the growth of the customer experience by implementing security measures. This balance is essential to retain customers' loyalty and satisfaction in the competitive market.
4. Manage the Risk with Business: The study advises regular risk assessment and the implementation of necessary organizational security measures for identifying and managing possible risks. Business processes will benefit from such proactive attitudes, as they reduce the fast-evolving menace of cyber threats and keep a business functional over long periods.
5. Cross-Border Transaction Security: The findings highlight the need to develop a security policy for cross-border e-commerce. Businesses must have distributed security mechanisms and adaptive mechanisms for safeguarding transactions across the globe. This is essential with the expansion of e-commerce globally.

Hence, the implications of the study highlight the critical security measures for e-commerce businesses by integrating advanced technologies, regulatory compliance, user experience, managing risk, and cross-border security during transactions to combat the threats due to cyber attacks.

10.7 Conclusion

The findings of this study emphasize the need for enhanced e-commerce security in the digital market. The study provides a comprehensive framework that addresses current and future threats such as SQL injections, cross-site scripting (XSS), and AI-related risks. The study highlights improving security from cyber threats by integrating advanced technologies, implementing AI and machine learning, adopting a zero-trust security model, and improving regulatory compliance. The framework helps to improve the security of e-commerce platforms. In addition, by focusing on the user experience and attention to the security measures, trust and satisfaction of the customers are ensured. This framework serves as a guide for e-commerce businesses to identify complex digital security systems while fostering a safe online e-commerce experience.

The study identifies the implications of the existing security measures and their limitations concerning consumer trust and engagement. A balanced approach is required to integrate security with user experience, ensuring that the protective methods do not affect the shopping experience. There are some limitations in the study, such as the applicability of the framework across various geographic regions with regulations and security threats, which can be different in different places. Second, the complete reliance on new technologies like AI and ML may be challenging due to the cost and availability of resources for small and medium enterprises (SMEs). Thirdly, the study has focused on the theoretical concepts, so it is necessary to validate the empirical framework as well.

Future studies should focus on empirically testing the proposed framework in various e-commerce platforms to check its practical usage in mitigating security threats. Further studies could focus on exploring the integration of e-commerce in emerging technologies like blockchain and advanced biometric identification. The integrated framework proposed in the study is an important step to create a secure and trustworthy e-commerce environment and to promote sustainable growth in the digital economy.

AI disclosure statement: The author acknowledges the use of AI-based tools for minor language editing or grammatical correction. However, all intellectual contributions, analyses, and conclusions presented in this chapter are entirely the author's own.

Bibliography

1. Chen, Y., Smith, A., Johnson, R., A Comprehensive Analysis of E-commerce Security Vulnerabilities. *J. Inf. Secur.*, 18, 4, 345–360, 2022.
2. Lopez, M. and Singh, K., Security Flaws in E-commerce Platforms: A Large-Scale Survey. *IEEE Trans. Dependable Secure Comput.*, 20, 2, 178–192, 2023.
3. Garcia-Martinez, J. and Brown, P., AI-Driven Security Systems in E-commerce: Opportunities and Challenges. *Int. J. Electron. Commerce*, 28, 1, 89–110, 2024.
4. Chapagain, D. and Aryal, B., E-commerce in Nepal: Growth, Opportunities, and Threats, 2024, https://doi.org/10.20944/preprints202410.1585.v1.
5. Chong, S.-K. and Abawajy, J.H., Risk-Based Trust Management for E-Commerce, in: *Z. Yan (Ed.), Trust Modeling and Management in Digital*

Environments: From Social Concept to System Development pp. 332–351, IGI Global Scientific Publishing, 2010, https://doi.org/10.4018/978-1-61520-682-7.ch014.
6. Rezk, A., Barakat, S.I., Saleh, H.A., The impact of cyber crime on e-commerce, 2017.
7. Sharma, K., Singh, A., Sharma, V.P., SMEs and Cybersecurity Threats in E-Commerce. *EDPACS*, *39*, 5–6, 1–49, 2009, https://doi.org/10.1080/07366980903132740.
8. Andreianu, G., Protecting Your E-Commerce Business. Analysis on Cyber Security Threats, 2023, May 30, https://doi.org/10.19107/cybercon.2023.17.
9. Gupta, R., Cybersecurity Threats in E-Commerce: Trends and Mitigation Strategies. *J. Adv. Manage. Stud.*, *1*, 3, 1–10, 2024, https://doi.org/10.36676/jams.v1.i3.13.
10. Gajiwala, C., Artificial Intelligence in Cybersecurity: Advancing Threat Modeling and Vulnerability Assessment. *Int. J. Sci. Res. Comput. Sci. Eng. Inf. Technol.*, *10*, 5, 778–788, 2024, https://doi.org/10.32628/cseit241051066.
11. Si, R., Risks and Future of Livestreaming E-Commerce in China, in: *China Livestreaming E-commerce Industry Insights*, pp. 95–103, Palgrave Macmillan, Singapore, https://doi.org/10.1007/978-981-16-5344-5_5.
12. Kabir, M.A., A Legal Analysis on Resolving Recently Growing Online Business Frauds in Bangladesh. *Law Reform*, *18*, 2, 264–281, 2022, https://doi.org/10.14710/lr.v18i2.46636.
13. Skipin, D.L., Shargina, M.V., Koltsova, T.A., Shargina, A.V., E-commerce: opportunities and threats of modern business. *Tyumen State Univ. Herald. Soc. Econ. Law Res.*, *8*, 2, 201–217, 2022, https://doi.org/10.21684/2411-7897-2022-8-2-201-217.
14. Sfetcu, N., Threats of Artificial Intelligence for Cybersecurity. *IT C*, *3*, 3, 2024, https://doi.org/10.58679/it17846.
15. Patel, H.R. and Rengarajan, A., Emerging Technological Threats to Cybersecurity. *Int. J. Innovative Res. Comput. Commun. Eng.*, *12*, 05, 5889–5892, 2024, https://doi.org/10.15680/ijircce.2024.1205125.
16. Navale, M.P. and Navale, R.G., Artificial Intelligence and Internet of Things (AIoT): Opportunities and Challenges, 2020.
17. Ilmudeen, A., Big Data, Artificial Intelligence, and the Internet of Things in Cross-Border E-Commerce. Cross-Border E-Commerce Marketing and Management, 2021.
18. Afroz Lari, H., Vaishnava, K., K. S., M., Artifical Intelligence in E-commerce: Applications, Implications and Challenges. *Asian J. Manage.*, 2022.
19. Rengarajan, A. and Rahul, C.R., Behavioural Biometrics as a User Authentication Mechanism in ISMS. *Int. J. Innovative Res. Comput. Commun. Eng.*, *12*, 03, 1699–1706, 2024, https://doi.org/10.15680/ijircce.2024.1203054.
20. Chandran, D.R., Use of AI Voice Authentication Technology Instead of Traditional Keypads in Security Devices. *J. Comput. Commun.*, *10*, 06, 11–21, 2022, https://doi.org/10.4236/jcc.2022.106002.

21. Akinloye, A., Akinwande, O., Anwansedo, S., AI-Driven Threat Detection and Response Systems for Secure National Infrastructure Networks: A Comprehensive Review. *Int. J. Latest Technol. Eng. Manage. Appl. Sci.*, *13*, 7, 82–92, 2024, https://doi.org/10.51583/ijltemas.2024.130710.
22. Tolba, A. and Al-Makhadmeh, Z., A cybersecurity user authentication approach for securing smart grid communications. *Sustainable Energy Technol. Assess.*, *46*, 101284, 2021, https://doi.org/10.1016/j.seta.2021.101284.
23. Odeyemi, O., Addy, W., Ofodile, O., Ajayi-Nifise, A., Okoye, C., Adeoye, O., Integrating AI with Blockchain for Enhanced Financial Services Security. *Finance Account. Res. J.*, *6*, 3, 271–287, 2024, https://doi.org/10.51594/farj.v6i3.855.
24. Rangaraju, S., Secure by Intelligence: Enhancing Products with AI-Driven Security Measures. *EPH Int. J. Sci. Eng.*, *9*, 3, 36–41, 2023, https://doi.org/10.53555/ephijse.v9i3.212.
25. Chalhoub, G. and Flechais, I., Alexa, Are You Spying on Me?": Exploring the Effect of User Experience on the Security and Privacy of Smart Speaker Users, in: *Moallem, A. (eds) HCI for Cybersecurity, Privacy and Trust. HCII 2020. Lecture Notes in Computer Science*, vol. 12210, pp. 305–325, Springer, Cham, 2020, https://doi.org/10.1007/978-3-030-50309-3_21.
26. Rahayu, E., Gunadi, W., Fauzan, F., Wijaya, H., The Effect of Trust and Satisfaction on Customer Loyalty in Online Shop: Case of C2C E-Commerce in Indonesia. *Int. J. Acad. Res. Bus. Social Sci.*, *10*, 8, 2020, https://doi.org/10.6007/ijarbss/v10-i8/7619.
27. Hasna, W.N. and Rachmawati, I., Factors Affecting the Customer Loyalty of Tokopedia E-Commerce. *Int. J. Sci. Manage. Stud. (IJSMS)*, 202–210, 2021, https://doi.org/10.51386/25815946/ijsms-v4i3p118.
28. Fan, J., Liu, D., Tang, G., Wu, K., Shao, X., Intelligent edge CDN with smart contract-aided local IoT sharing. *High-Confid. Comput.*, 2024.
29. Serrano, M., Dang, H.N., Nguyen-Mau, H.Q., Recent advances on artificial intelligence and internet of things convergence for human-centric applications: internet of things science. *Proceedings of the 8th International Conference on the Internet of Things*, 2018.
30. Atlam, H.F., Walters, R.J., Wills, G.B., Intelligence of Things: Opportunities & Challenges. *2018 3rd Cloudification of the Internet of Things (CIoT)*, pp. 1–6, 2018.
31. Distler, V., Koenig, V., Lallemand, C., Lenzini, G., The Framework of Security-Enhancing Friction: How UX Can Help Users Behave More Securely, 2020, October 26, https://doi.org/10.1145/3442167.3442173.
32. Aloumi, D., Malik, S., De Pablos, P.O., Alkhaldi, A., Factors Influencing Consumer Interactions With FinTech Services, in: *P. Ordóñez de Pablos (Ed.), Building Climate Neutral Economies Through Digital Business and Green Skills*, pp. 239–258, IGI Global Scientific Publishing, 2024, https://doi.org/10.4018/979-8-3693-5673-9.ch010.

33. Hossain, M.A., Islam, S., Rahman, M.M., Arif, N.U.M., Impact of Online Payment Systems on Customer Trust and Loyalty in E-commerce Analyzing Security and Convenience. *Acad. J. Sci. Technol. Eng. Math. Educ.*, 4, 3, 1–15, 2024, https://doi.org/10.69593/ajsteme.v4i03.85.
34. Santana Torrellas, G.A., A Framework for Multi-Agent System Engineering Using Ontology Domain Modelling for Security Architecture Risk Assessment in E-Commerce Security Services, 2004, August 30, https://doi.org/10.1109/nca.2004.1347810.
35. Alhammad, M.M., Wiafe, I., Gulliver, S.R., Exploring the Impact of Persuasive Features on Customer Satisfaction Levels of E-Commerce Websites Based on the Kano Model, 178–192, 2021, https://doi.org/10.1007/978-3-030-79460-6_14.
36. Utama, I.D., Sudirman, I.D., Kurnianingrum, D., Beyond transactions: Understanding the determinants of e-satisfaction in online shopping. *J. Asian Sci. Res.*, 14, 2, 251–265, 2024, https://doi.org/10.55493/5003.v14i2.5063.

11

Enhancing Content Delivery Networks with Generative AI: Strategies, Applications, and Future Directions

Subhranil Das[1*], Rashmi Kumari[2] and Raghwendra Kishore Singh[3]

[1]School of Computer Science, UPES, Dehradun, Uttarakhand, India
[2]School of Computer Science Engineering and Technology, Bennett University, Greater Noida, Uttar Pradesh, India
[3]School of AI and CS, SR University, Warangal, India

Abstract

Content delivery networks (CDNs) play an important role in modern applications by considering the delivery of data to the end-users. With the rise in demand for personalized and real-time content, generative AI has a different role while enhancing the CDN capabilities. In this chapter, the transformative potential of generative AI has been explored while optimizing CDN in personalized content generation, intelligent caching, and enhanced user experiences. There are some generative AI techniques, such as generative adversarial networks (GANs), variational autoencoders (VAEs), and transformers, that enable CDNs to generate, modify, and personalize content with reference to the user preferences, device capabilities, and network conditions. The chapter also explored the detailed practical applications, such as AI-powered video compression and upscaling, adaptive streaming, and personalized recommendations, which improve the latency and bandwidth efficiency. Generative AI contributes to intelligent caching strategies that adapt to the content dynamically by incorporating predictive algorithms that reduce server load. Some of the key challenges, such as the computational intensity of generative models, scalability, and real-time performance requirements, are thoroughly examined. In addition to that, this chapter also addresses ethical concerns related to content authenticity, privacy, and the potential for misusing the synthetic media. Here, future directions for generative AI in CDNs are also

**Corresponding author*: subhranil.das@ddn.upes.ac.in

Tanusree Chatterjee, Rajib Banerjee and Souvik Ganguli (eds.) Content Delivery Networks Using Artificial Intelligence and Internet of Things: Revolutionizing Content Delivery, (231–258) © 2026 Scrivener Publishing LLC

considered, which include the recent developments in federated learning, edge AI, and reinforcement learning that could improve CDN resilience and content delivery efficiency further.

Keywords: Content delivery networks, generative adversarial networks, generative AI, variational autoencoders, transformers

11.1 Introduction

Content delivery networks (CDNs) are distributed networks of servers where these sensors are strategically positioned across various geographic locations for delivering the digital content efficiently [1, 2]. These networks are designed to address the growing demand for fast, reliable, and high-quality access to online content. As digital services become increasingly central to personal and professional fronts, CDNs have become one of the essential components of various internet structures, which enable websites and real-time applications to deliver the content closer to the user [3, 4].

The main functions of CDNs are content caching, load balancing, traffic management, content delivery optimization, and security with DDoS protection. Caching is one of the prime functions of CDNs, where CDNs store copies of HTML files, images, videos, and script content in geographically distributed servers known as edge servers. By caching this content, CDNs reduce the data distance that needs to travel, resulting in minimizing latency and improving load times. Storage services within CDNs provide the optimal solutions, which are cloud-based storage that allows organizations to store some part of content files to CDN servers in a more efficient manner. Large volumes of requests for CDN content have been distributed in the traffic intelligently across multiple servers. Load balancing ensures that no single server is overloaded by requests, which improves the uptime and resilience. By distributing traffic to less congested servers, CDNs maintain optimal performance and prevent the overloading of the servers. In order to reach the end users, traffic management is an important component that optimizes the path. CDNs adjust the traffic load to avoid the obstacles and enhance the performance during peak times. By employing the various latest technologies, there are some content delivery methods, such as TCP/IP optimizations, dynamic content caching, and compression techniques, that can be improved. CDNs play a critical role in enhancing security by considering the origin servers, which protect from direct traffic requests. By applying Web Application Firewalls (WAFs) [10] and Distributed Denial-of-Service (DDoS) tools, CDNs can detect and neutralize threats before they reach the origin server. In addition to that, CDNs can protect data privacy by encrypting data through SSL/TLS encryption that would safeguard data while transmitting information [11].

CDNs have become a vital component where users demand instant access to digital content. High-performance CDNs ensure that websites and applications are quick and responsive, with reliable responses irrespective of the network conditions and geographical challenges. By increasing efficiency, security, and optimized content delivery, CDNs are applied to the latest digital experiences, support businesses in improving engagement, expand the reach, and maintain the competitive edge.

Generative AI is a field of AI that focuses on creating models capable of generating new, synthetic data, which includes images, text, audio, and virtual environments [12, 13]. Traditional AI primarily analyzes and classifies existing data, while generative AI models produce new content that closely mimics real data. Some of the core generative AI techniques, i.e., generative adversarial networks (GANs) [14], variational autoencoders (VAEs) [15], and transformers, have been applied in fields like personalized media, image enhancement, and real-time interactive experiences.

Generative AI is significant for CDNs because of the rise in demand for personalized, dynamic, and high-quality digital experiences. Traditional CDNs are effective in caching and delivering static content efficiently but face challenges in delivering highly personalized or dynamic content in real time [16]. For addressing these issues, CDNs are enabled for content dynamically, which is based on the preferences of users, specifications of devices, and network conditions, by reducing server load and optimizing bandwidth. The key areas where generative AI enhances CDN functionality are personalized content generation, where generative AI enables CDNs to create custom content for individual users, especially in adaptive graphics and targeted advertisements [17]. This AI-powered personalization improves the engagement of users as well as satisfaction, where the content aligns with their interests and preferences. Adaptive streaming and real-time content modification, where generative AI-powered adaptive streaming adjusts video or audio quality based on real-time network conditions and device capabilities. This could allow the CDNs to deliver high-quality streaming experiences with minimal buffering in low-bandwidth environments. Also, the generative models are used to upscale video quality by enhancing visuals without increasing file size.

Benefits of generative AI in CDNs are improved bandwidth and latency efficiency, where generative AI techniques such as adaptive compression and intelligent content generation reduce the data payloads while maintaining quality, optimizing bandwidth, and improving delivery speeds

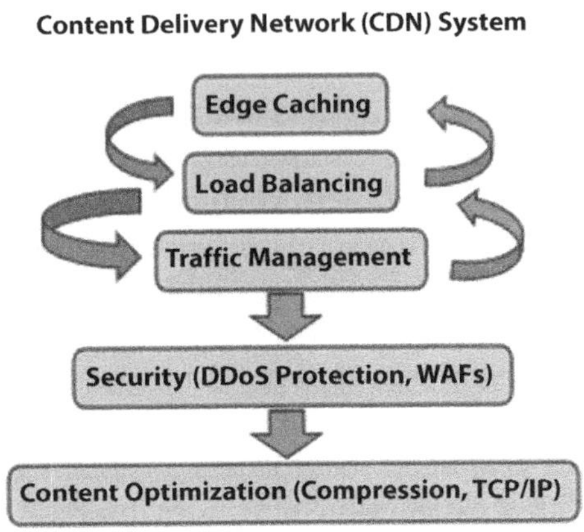

Figure 11.1 Block diagram of CDN system.

[18, 19]. Scalable personalization is one of the important benefits of generative AI, where CDNs can significantly increase server load by applying AI techniques for creating and modifying content and enhancing resilience and resource allocation, as shown in Figure 11.1.

11.2 Fundamentals of Generative AI Techniques for CDNs

Generative adversarial networks (GANs) are a type of ML framework that was developed by Ian Goodfellow in 2014 for generating synthetic data that closely resembles real-world data [20]. GANs consist of two neural networks, i.e., the generator and the discriminator, which work in tandem through a process of adversarial training. GANs produce realistic, high-quality outputs such as images, videos, audio, and other types of media.

11.2.1 How GANs Work

The generator network is used to create the synthetic data, which takes random noise as input and generates data samples, i.e., images, audio, text, or other media forms. The goal of the generator is to produce samples that are as realistic as possible. The discriminator network evaluates the data produced by the generator, which determines whether each sample is "real" (actual dataset) or "fake" (generated dataset). The discriminator learns to recognize patterns in real data to better identify synthetic content.

The generator and discriminator engage in continuous learning, where the generator creates realistic data in order to fool the discriminator. The discriminator refines its ability by distinguishing the real and synthetic data. In this adversarial process, the generator produces data that the discriminator can no longer reliably distinguish from real data in highly realistic synthetic outputs.

11.2.2 Application of GANs in Generating High-Quality Media

GANs have become a transformative tool for generating high-quality media in several fields because they can create visuals and audio that are nearly indistinguishable from real-world examples. Here are some practical applications of GANs in generating high-quality media:

- Image and Video Synthesis

GANs are widely applied for generating photorealistic images from scratch or by enhancing low-quality visuals. Some of the applications include creating stock photos, virtual characters, or synthetic faces for advertising or media content. In video production, GANs synthesize realistic animations, modify facial expressions in real-time, and enhance low-resolution video to high definition [21].

- Image Super-Resolution and Enhancement

Super-resolution GANs (SRGANs) are specialized GANs where the image quality is enhanced from low-resolution images to high resolution, which adds to the realistic textures and details. SRGANs are useful for media that need to be visually enhanced, such as old film footage, satellite images, and security footage [22].

- Style Transfer and Creative Media Generation

GANs can create media in specific styles known as style transfer. For example, GANs can transform photos to look like paintings, apply specific color schemes, or alter designs to fit a particular aesthetic. This is valuable for applications that require media customization based on branding or user preferences [23].

- Image Inpainting and Content Filling

GANs can be applied in inpainting, or filling in missing parts of images, which is valuable for photo restoration, image editing, and creating seamless transitions in videos. In this application, old photos can be restored or media without any noticeable changes in quality [24].

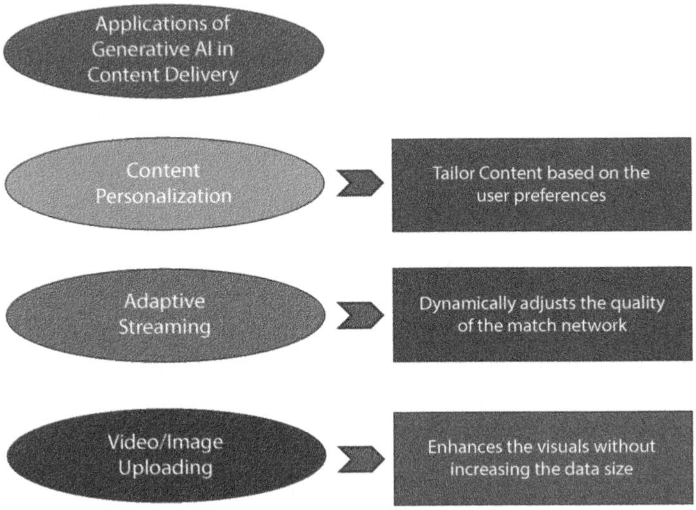

Figure 11.2 Applications of generative AI in content delivery.

- Text-to-Image and Text-to-Video Generation

GANs are applied for generating new images or videos from text descriptions. For example, an AI model has created a realistic image. In this application, customizable media is valued, which includes virtual reality backgrounds or personalized digital art [25], as shown in Figure 11.2.

11.2.3 GANs in Content Delivery Networks (CDNs)

In CDNs, GANs have improved the quality of the content during streaming, where the higher-resolution images and videos are acquired, which helps save bandwidth without compromising visual quality. GANs enable CDNs to generate custom content, which is based on user preferences. For example, a CDN creates personalized media for individual users, such as custom advertisements and product images, where user engagement is increased. By applying GAN-based compression, CDNs can reduce file sizes while maintaining high quality, thereby improving load times and reducing latency [26].

While GANs provide immense potential for media generation, they also come with certain challenges, which are explained below.

- Computational Intensity: Training GANs, particularly for high-resolution images or videos, requires significant computational power.

- GANs sometimes produce "artifacts," or visual distortions, especially if the training data or parameters aren't optimal.
- GANs can be used to create deepfakes and other synthetic media, raising ethical concerns regarding misuse and content authenticity.

11.2.4 Overview of Variational Autoencoders (VAEs) for Generating and Modifying Content

VAEs are a type of generative model used in ML for creating and modifying data by learning efficient data representations. Traditional autoencoders compress data without generating new variations; VAEs introduce the probabilistic techniques that allow them to create new, diverse content samples based on a learned distribution. VAEs have become one of the important tools for generating images, videos, and other types of media efficiently for controlled and smooth content variations [27, 28].

11.2.4.1 How VAEs Work

VAEs operate by learning a low-dimensional representation of data that enables them to generate realistic content by sampling from this learned distribution. For this process, the three main components are involved.

The encoder compresses the input data (image or sound sample) into a smaller, dense representation, known as the latent space. Unlike regular autoencoders, VAEs introduce a probabilistic approach by encoding each input as a distribution in the latent space, which represents the mean and variance for a Gaussian (normal) distribution. The latent space is represented in a compressed, probabilistic form. By sampling from this space, VAEs can generate variations of the original data. In this representation, smooth transitions and controlled variations are allowed in the generated output, which is useful in creating diverse content. The decoder takes samples from the latent space and reconstructs them back into the original data format (e.g., an image). By sampling different points within the latent distribution, the decoder can generate new variations, which makes VAEs useful for content creation, data augmentation, and more.

11.2.5 Applications of VAEs in Generating and Modifying Content

VAEs are widely used in fields that require controlled content generation and modification. Here are applications, which are explained below.

- Image Generation and Manipulation [29]: VAEs generate new images that resemble those in the training data by sampling from the latent space. For example, after training on a dataset of faces, VAEs produce entirely new faces that blend features from the dataset, which makes them ideal for applications like creation, video game character generation, or stock photo production. VAEs enable content manipulation where specific aspects of an image (like facial expressions, lighting, or color) can be adjusted by modifying the values in the latent space. This allows users to create new variations of existing images without altering the core elements.
- Data Augmentation [30]: VAEs are frequently used in data augmentation for creating additional training samples for machine learning models in fields with limited data, such as medical imaging, where VAEs can generate realistic variations of existing images.
- Image Denoising and Inpainting [32]: VAEs can reconstruct corrupted or missing parts of images, a technique known as inpainting. For example, if parts of an image are obscured, VAE generates realistic replacements that fill in the gaps to complete the image. Denoising is another application where VAEs can remove unwanted noise from images or videos, improving visual quality and enhancing the end-user experience.
- Controlled Content Variation [33]: VAEs learn a distribution of data in the latent space; they allow for smooth, controlled variations, which means that small adjustments in the latent space create gradual changes in the output. For example, VAE can generate a sequence of images for a person that changes their expressions smoothly. For this application, digital marketing, entertainment, and personalized media can benefit from any changes in the content that can increase engagement.
- Speech and Audio Synthesis [34]: VAEs are applied to tasks related to audio generation and modification. By learning

patterns in the speech, they can generate new audio clips that are varied from the original samples in real-time applications, such as text-to-speech systems, music production, and sound effect generation.

11.2.6 VAEs in Content Delivery Networks (CDNs)

Within CDNs, VAEs can help CDNs deliver certain content at an optimized quality by compressing data while preserving the required details. For example, VAE creates compressed representations, which are decoded to deliver high-quality visuals for limited bandwidth. CDNs are applied to save bandwidth by sacrificing user experience. By learning the different patterns of user preferences and behavior, VAEs allow CDNs to create personalized versions of content that align with the user. For example, VAE generates content based on user-personalized backgrounds, video styles, or image filters, which leads to the digital experience for a particular individual. VAEs can be applied to anticipate user content needs, which enables predictive caching. By generating the variations of the anticipated media, VAEs allow CDNs to retrieve the relevant content, which is much closer to users, minimizing the latency and improving the loading times. Since VAEs modify the contents, CDNs can apply them in media applications in real time, which is based on network conditions, user device capabilities, or content type. For example, CDN could adjust media quality to ensure smooth streaming experiences during periods of high network congestion.

11.2.7 Transformers in Processing and Personalizing Content at Scale

Transformers are a type of NN architecture that has been revolutionized in the field of ML, especially in natural language processing (NLP), computer vision, and content generation. For processing large, complex data sequences, transformers capture the contextual relationships within data that make them highly effective for tasks requiring language understanding, personalized content, and data processing at scale.

Originally introduced in 2017 by researchers at Google, transformers apply the mechanism called self-attention, which makes it relevant in the different parts of an input sequence, allowing them to capture long-range dependencies and contextual meanings [35]. Transformers are ideal for certain types of tasks that involve large volumes of data, complex relationships, and real-time content personalization.

11.2.7.1 How Transformers Work

Transformers rely on two primary components, i.e., self-attention and encoding-decoding mechanisms, for processing the data in parallel, which allows them to handle long sequences of information quickly and accurately. Self-attention enables the transformer to consider the relationships between all elements in one sequence. For example, self-attention helps the model understand which words relate closely to others, such as verbs or objects in a sentence. This feature is essential while capturing context in language and other structured data.

The transformer model has been organically encoded in the encoder part, which processes input data, and a decoder that generates the output based on the encoded input through self-attention. In some applications, only the encoder is used (BERT models), while others use both encoder and decoder (GPT or full transformer models). RNNs are applied for processing the data sequentially, while transformers handle all data elements in a sequence, which makes them faster and more scalable for large datasets. Benefits and challenges faced by the transformers and variational autoencoders have been shown in Table 11.1.

11.2.8 Applications of the Transformers in Content Processing and Personalization

The ability of the transformers to process large datasets and capture contextual relationships makes them ideal for dynamic content generation, recommendation systems, and personalized user experiences. Here are some of the specific applications, which are explained below:

- NLU for Personalized Content [37]: Transformers are applied in modern language models, such as BERT, GPT, and T5, where they excel in tasks like text generation, summarization, translation, and sentiment analysis. These models can analyze user language, preferences, and engagement patterns for generating content for personalization purposes. In CDNs, transformers enable real-time content adaptation, which is based on the past behavior of the user, such as article recommendations, chatbots for customer support, and generated summaries that match the preferences of the users.
- Recommendation Systems for Media and Content Delivery: Transformers are widely applied in the recommendation systems that are dependent on the contextual and behavioral

Table 11.1 Benefits and challenges faced by the transformers and VAE.

Aspect	Transformers	Variational autoencoders (VAEs)
Benefits		
Scalability	Can process large volumes of data quickly, making them ideal for applications like content delivery networks (CDNs).	Efficient content creation and modification, allowing the generation of realistic content with minimal manual effort, ideal for media production.
Contextual understanding	Capable of capturing nuanced relationships in data, enabling high accuracy in content recommendation, language understanding, and user behavior prediction.	Produce high-quality, coherent outputs, ensuring consistency in visual or auditory content, crucial for media quality maintenance.
Adaptability	Versatile and can be fine-tuned for a variety of applications, including summarization and personalized recommendations.	Low latency in generating content variations in real-time, making them suitable for adaptive content delivery within CDNs.
Challenges		
Computational demands	Require substantial computational resources, especially for large-scale implementations; need optimization for real-time applications.	Training complexity is high, particularly with high-dimensional data like images or videos, requiring powerful hardware.

(*Continued*)

Table 11.1 Benefits and challenges faced by the transformers and VAE. (*Continued*)

Aspect	Transformers	Variational autoencoders (VAEs)
Training complexity	Need large datasets and careful parameter tuning, making training resource-intensive and time-consuming.	Computational intensity can make training challenging, especially for large, complex datasets.
Privacy considerations	Dependence on large datasets may raise privacy concerns if data is not properly managed, particularly in personalized applications.	N/A

data for suggesting the content that aligns with the preferences of the user. By analyzing historical data, transformers help in constructing CDNs in streaming platforms, which provide customized content feeds, movie or music recommendations, and personalized news. The traditional recommendation systems consider complex, long-term relationships in data for the transformers, which enables the multi-step recommendations that evolve with user preferences.

- Real-Time Language Translation and Localization: For global applications, transformers have been mainly applied to the real-time translation and localization of content, which adapts the language, tone, and even cultural references based on the location or language settings of the user. Transformers allow CDNs to deliver personalized experiences by dynamically translating content, which ensures that users can engage with media in their preferred language.
- Dynamic Content Generation and Summarization: Transformers are very capable of generating coherent content, which is useful for creating real-time summaries, headlines, or descriptions that align with a user's interests.

For example, a news platform generates personalized news briefs, which are based on the topics a user follows. In CDNs, transformers enable the creation of adaptive summaries, dynamic headlines, or tailored responses that enhance user engagement instead of serving static content.

- Enhanced Search and Discovery: Transformers improve the search engines and content discovery features, which allow users to find the relevant information. By analyzing search queries within a broader context, transformers help in retrieving the content that aligns with the intent of the user rather than keyword matches. In CDNs, transformers optimize content indexing, retrieval, and ranking, which help users to discover relevant content even if their searches are ambiguous or complex.

11.2.9 Transformers in Content Delivery Networks (CDNs)

In the context of CDNs, transformers are useful in delivering efficient, scalable, and highly personalized content. Transformers allow CDNs to generate real-time content recommendations, which are based on user preferences and engagement patterns. For these recommendations, high-demand content closer to users can also be used to dynamically cache to improve load times and optimize resource allocation. Transformers analyze network conditions, user device capabilities, and viewing habits for adjusting the streaming quality in real time to deliver optimized viewing experiences. Transformers help CDNs generate higher-quality versions of content by using learned patterns. CDNs can utilize transformers for more sophisticated search and filtering capabilities. Transformers help to filter search results or various suggestions that are based on a user-specific context, such as location, viewing history, or language preferences, which results in a more relevant and streamlined user experience. Transformers help CDNs predict user engagement trends by identifying patterns in historical data. The predictive power enables CDNs to cache the popular or trending content in advance, which reduces the latency and improves user experience during peak demand times.

11.3 Intelligent Caching Strategies for CDNs with Generative AI

Intelligent caching strategies using generative AI allow CDNs to optimize the storage of data, improve load times, and predict user demand patterns. By incorporating adaptive caching and load optimization, CDNs serve high-demand content efficiently by minimizing latency and server load. In the following subsections, a detailed explanation of two key approaches, i.e., adaptive caching using predictive generative models and load optimization with real-time data integration, has been provided. Table 11.2 explains the challenges and different mitigation strategies.

11.3.1 Adaptive Caching Using Predictive Generative Models

Traditional caching in CDNs relied on the preset rules of the static algorithms for which content to cache, which led to inefficiencies while dealing with fluctuating user demands. Adaptive caching has been enhanced by the predictive generative models, which use AI for analyzing the historical and real-time data that forecasts the content that is in high demand. In this strategy, popular content is readily available, which reduces the latency and optimizes server resource usage.

11.3.1.1 Working Mechanism

Data Collection and Analysis: The generative AI model analyzes the behavior of the past user, content access patterns, geographic traffic distribution, and external factors like time of day or day of the week.
Predictive Content Forecasting: By using these data points, generative models such as transformers and RNNs predict the specific content that is likely to be in demand. For example, the model forecasts the increased demand for highlights, replays, or analysis videos, which prompts the CDN to retrieve these files during a major sporting event.
Dynamic Cache Placement: Based on predictions, the CDN automatically places the high-demand content on edge servers closer to users likely to access it, reducing the distance data needs to travel and improving load times.
Continuous Model Adjustment: The predictive model refines its forecasts over time as it learns from actual user behavior, which adapts to any deviations in content demand patterns, and updates cached items accordingly.

Table 11.2 Different challenges and their mitigation strategies in generative AI while implementing CDN.

Challenge	Description	Mitigation strategy	Expected outcome
Computational resource demand	High computational power required for training and real-time operations	Implement hardware accelerators (e.g., GPUs, TPUs), optimize model architectures	Improved efficiency and reduced processing time
Scalability of content generation	Generating personalized content at scale without delay	Utilize distributed computing and edge AI for localized processing	Enhanced user experience with minimal latency
Content authenticity and integrity	Risk of altered or manipulated content reaching end-users	Integrate cryptographic hashing, blockchain provenance tracking	Assured content trustworthiness and verifiability
Real-time performance	Ensuring real-time response for dynamic content adaptation	Employ lightweight, optimized generative models and caching strategies	Real-time, seamless content delivery
Ethical concerns	Risk of generating biased or harmful media content	Conduct ethical audits, implement bias detection, and user transparency measures	Adherence to ethical standards and user trust
Privacy issues	Potential user data exposure during content personalization	Use differential privacy and federated learning methods	Enhanced data privacy and secure personalization

For a streaming platform, the predictive model might cache the latest episodes of the regional edge servers where the viewership is high. As demand for older episodes fluctuates, the model might adjust which episodes are cached locally, which ensures the storage is always used efficiently.

11.3.2 Load Optimization with Real-Time Data Integration

CDNs can be applied to real-time data integration for dynamically managing server load, which is based on current demand, usage patterns, and network conditions. By combining the real-time data with generative models, CDNs can adjust cached content and server allocation, which maintains the optimal performance under varying load conditions.

11.3.2.1 Working Mechanism

The working mechanism is followed by certain steps:

- The CDN collects the real-time data on current user activity, server load, network congestion, and device types, and this data is fed into the generative model.
- Based on the analysis of the real-time data, the model adjusts the content where the scaling is accessed for files to make room for high-demand items throughout the day.
- If certain servers experience heavy traffic, the generative model could redistribute requests to underutilized servers, which balances the load across the network. In this load, balancing minimizes strain on individual servers, which avoids slowdowns or crashes.
- The model optimizes the content that is stored based on immediate usage trends that allow the CDN to allocate resources to the changing demand landscape.

In a global CDN environment, a generative model might detect a sudden rise in demand for a news article about breaking events in specific regions. The model would cache this article on multiple edge servers within those regions, temporarily reducing caching of less relevant content. As interest declines, the model reverts to caching content based on regular patterns, as shown in Table 11.3.

In Figure 11.3, improvements in CDN performance metrics with and without generative AI have been compared. The metrics, which include

Table 11.3 Performance of CDN on the different categories of the benefit.

Benefit category	Detailed benefit	Description	Impact on CDN performance
Latency and access efficiency [38]	**Reduced end-user latency**	Caching high-demand content at edge servers near end-users minimizes the distance data must travel, significantly reducing load times and enhancing real-time accessibility, even in high-traffic conditions.	Decreases average load time per user, improving responsiveness and end-user experience.
	Proactive content availability	Predictive AI-driven caching identifies patterns in content access, ensuring popular or time-sensitive content is pre-cached based on real-time demand forecasts, which minimizes lag during peak access periods.	Supports seamless content streaming and browsing by anticipating user needs.
	Dynamic latency optimization	By adapting the cache location and content quality dynamically to match user network conditions, generative AI maintains high-quality delivery across varying connection speeds and latency-sensitive applications.	Optimizes user satisfaction by maintaining consistent quality across devices and networks.
Resource and storage efficiency [39]	**Minimized redundant caching**	Adaptive caching with generative models intelligently eliminates redundant copies of low-demand content, reducing unnecessary storage allocation on high-demand edge servers and freeing up capacity.	Lowers storage costs and enables more effective use of existing CDN infrastructure.

(Continued)

Table 11.3 Performance of CDN on the different categories of the benefit. (*Continued*)

Benefit category	Detailed benefit	Description	Impact on CDN performance
	Efficient resource allocation	AI-driven load forecasting allows CDNs to preemptively allocate resources where demand is likely to surge, reducing over-allocation and managing resources more sustainably.	Ensures that CDNs maintain operational efficiency and reduces hardware scaling needs.
Enhanced user experience [40]	**Personalized content delivery**	Real-time analysis of user data and preferences enables generative AI models to deliver tailored media formats, resolutions, and recommendations, creating a uniquely engaging and optimized user experience.	Boosts user engagement by providing relevant, high-quality content across varied user segments.
	Reduced buffering and lag	By forecasting and prioritizing high-demand media, AI reduces load times for popular content, thus minimizing buffering during peak times and enhancing streaming continuity.	Directly increases retention rates and reduces abandonment due to load delays.
Network resilience and scalability [41]	**Adaptive load redistribution**	In response to unexpected traffic spikes, AI models dynamically shift traffic to underutilized servers, preventing bottlenecks and load imbalances that could disrupt content delivery.	Enhances CDN resilience, ensuring reliable service and lowering the risk of localized outages.

(*Continued*)

Table 11.3 Performance of CDN on the different categories of the benefit. (*Continued*)

Benefit category	Detailed benefit	Description	Impact on CDN performance
	Predictive failure mitigation	Generative models analyze real-time server health and demand data to preemptively reroute traffic from servers at risk of failure, improving uptime and system stability during high-demand events.	Increases CDN reliability by maintaining high availability even under extreme load conditions.
Storage optimization [42]	**Real-time storage management**	AI adjusts cached content dynamically based on fluctuating demand, offloading less-accessed files to alternative storage and keeping high-demand files accessible, thereby improving data flow and storage balance across servers.	Reduces storage redundancy and optimizes access paths for popular content, enhancing efficiency.
Content availability and accessibility [43]	**Consistent high-demand content access**	Real-time data integration enables CDNs to keep frequently accessed content readily available, lowering content retrieval times and enhancing user access to in-demand files, especially during regional or global spikes.	Ensures uninterrupted access to popular files, sustaining high service standards.
	Localized content adaptation	Generative AI can localize content based on regional preferences and device compatibility, ensuring that each user receives an optimized experience, especially valuable for international CDNs with diverse user bases.	Expands user accessibility while reducing friction across different device types and regions.

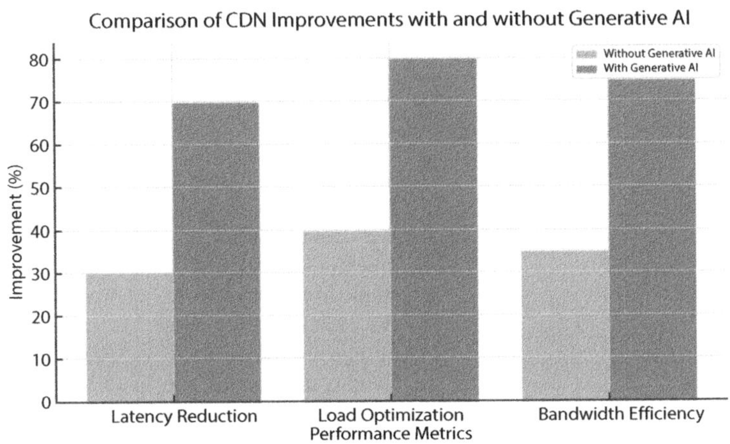

Figure 11.3 Comparison of CDN improvement with and without GAI.

latency reduction, load optimization, and bandwidth efficiency, show significant improvements when generative AI is applied.

11.4 Ethical and Security Concerns in Generative AI for CDNs

With the integration of generative AI in content delivery networks (CDNs) to personalize and enhance content delivery, there are significant ethical and security considerations. These concerns center around ensuring content authenticity, protecting user privacy, and preventing the misuse of synthetic media. Addressing these issues requires implementing verification techniques, transparency mechanisms, and robust safeguards to maintain trust and security in AI-enhanced CDNs, as shown in Figure 11.4.

As CDNs deliver AI-generated or AI-enhanced content, ensuring the authenticity and integrity of this content is paramount. This involves verifying the source of AI-generated media to prevent unauthorized alterations and maintaining user privacy. The techniques for ensuring authenticity and integrity are as follows:

- Content Verification Protocols: CDNs can employ cryptographic hashing to generate unique digital fingerprints for each piece of AI-generated content. This fingerprint allows for the verification of content originality and integrity, ensuring that no tampering has occurred during delivery.

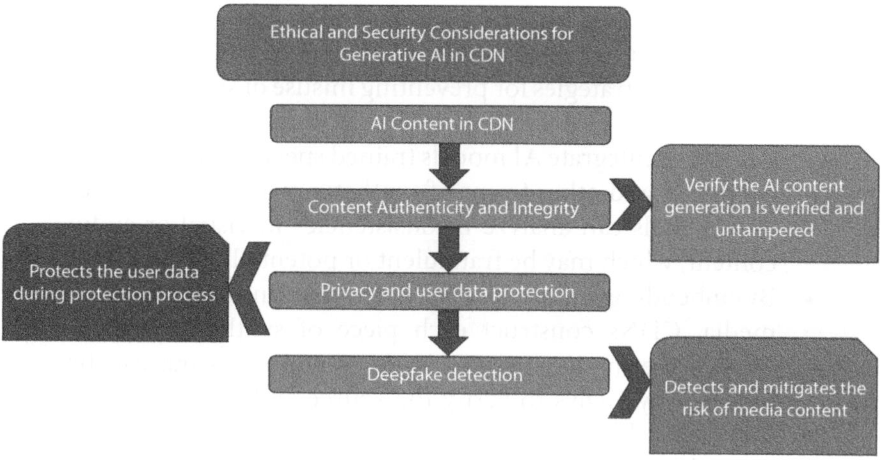

Figure 11.4 Ethical and security considerations for generative AI in CDN.

- Blockchain-Based Provenance Tracking: Blockchain technology traces the origin and transformation of AI-generated content, which provides a transparent, immutable record of its creation, modification, and distribution. CDN to track any alterations in the content's lifecycle, which maintains the integrity and accountability.
- Watermarking and Digital Signatures: Embedding watermarks in AI-generated media ensures that content can be verified as authentic. These markers are invisible to users but detectable by verification systems, which provide a layer of validation that identifies content generated by reputable sources.
- Privacy-Preserving AI Techniques: Techniques such as differential privacy and federated learning allow CDNs to analyze user behavior and preferences without directly accessing or storing sensitive user data.

By implementing these techniques, CDNs ensure that users receive authentic content, which builds the user's trust and prevents the spread of fraudulent media. Privacy-preserving techniques allow CDNs to respect user privacy while maintaining personalization and minimizing the risk of data breaches.

The ability of generative AI to create synthetic media has introduced the potential for misuse, particularly in deepfakes, fraudulent content, and misinformation. Safeguarding these risks involves proactive detection and prevention strategies for ensuring the CDNs do not become vectors for harmful media. The strategies for preventing misuse of synthetic media are:

- CDNs can integrate AI models trained specifically to identify deepfakes and other forms of synthetic media. These detection models can analyze inconsistencies in visual or audio content, which may be fraudulent or potentially harmful.
- By embedding unique, traceable identifiers into AI-generated media, CDNs construct each piece of synthetic content. These watermarks are not visible to end-users but can be detected by CDNs to verify the source and detect tampering. Traceable algorithms further enhance transparency by embedding creation details, which include the model type and generation parameters that make the content manageable.
- CDNs have set up restrictions to limit the creation and distribution of high-risk synthetic media. For example, they can monitor and restrict access to certain generative AI capabilities such as facial modification or realistic voice synthesis.
- Regular audits of AI models and their outputs help identify and mitigate potential biases or ethical concerns. CDNs can adopt frameworks for ethical AI governance, which ensure that generative models are designed, deployed, and monitored with ethical standards. This also involves monitoring model outputs to identify any harmful or misleading content generated by AI.
- To empower users, CDNs display clear indicators when media is AI-generated and educate users about synthetic media and potential risks. Transparent labeling can inform users when they are engaging with AI-generated content.

As generative AI becomes an integral part of CDNs, ethical and security concerns such as content authenticity, privacy protection, and synthetic media misuse must be proactively addressed. By implementing verification techniques like blockchain-based provenance tracking, watermarking, and AI-driven detection systems, CDNs can ensure that AI-enhanced content delivery remains transparent, secure, and trustworthy. These safeguards help maintain content integrity, protect user privacy, and prevent misuse, building a foundation for secure, ethical, and reliable AI in content delivery networks, as shown in Table 11.4.

Table 11.4 Ethical and security strategies in generative AI-enhanced CDNs.

Strategy	Description	Benefit
Content verification protocols	Use cryptographic hashing to ensure content originality and integrity	Ensures authenticity, prevents tampering
Blockchain-based provenance	Records the creation and modification history of content	Transparent, immutable content tracking
AI-driven deepfake detection	Employ AI models to detect inconsistencies in synthetic media	Identifies fraudulent or harmful media
Watermarking & digital signatures	Embed identifiers in media for authenticity checks	Verifies content source, prevents misuse
Ethical AI audits	Conduct regular reviews of AI model outputs to ensure compliance	Reduces bias, maintains ethical standards
User education	Inform users with clear indicators and transparency for AI-generated content	Builds user awareness and trust

11.5 Conclusion and Future Scope

Generative AI has proven to be a transformative force in enhancing content delivery networks (CDNs), addressing the growing demand for personalized, dynamic, and high-quality digital experiences. By leveraging advanced techniques such as Generative Adversarial Networks (GANs), variational autoencoders (VAEs), and transformers, CDNs can optimize content delivery, improve bandwidth and latency efficiency, and create scalable, personalized experiences for users. Despite its significant potential, the integration of generative AI in CDNs comes with challenges, including computational resource demands, real-time performance constraints, and ethical and security concerns. Addressing these challenges with effective strategies—such as hardware optimization, distributed computing, cryptographic verification, and privacy-preserving techniques—ensures that CDNs can maximize the benefits of generative AI while maintaining content integrity and user trust.

The future of generative AI in CDNs is poised for further innovation and expansion. Emerging technologies like federated learning and edge AI promise to enhance real-time content processing while preserving user privacy. The integration of reinforcement learning can offer more adaptive and intelligent content delivery strategies, allowing CDNs to self-optimize based on continuous learning from user behavior and network conditions. Additionally, as generative AI models become more efficient, CDNs can harness these advancements for more robust, scalable, and cost-effective solutions. Future research should focus on developing lightweight models for faster processing and exploring cross-platform AI collaboration to extend capabilities across different media types and networks. Ethical AI frameworks and standardized policies will be essential in guiding these advancements to ensure that generative AI in CDNs evolves responsibly, balancing innovation with user safety and trust.

AI disclosure statement: The author confirms that no artificial intelligence (AI)-generated content or AI-assisted writing tools were used in the preparation of this chapter (including summaries and the abstract). All text, analysis, and conclusions are the result of the author's original work.

Bibliography

1. Vakali, A. and Pallis, G., Content delivery networks: Status and trends. *IEEE Internet Comput.*, *7*, 6, 68–74, 2003.
2. Yang, H., Pan, H., Ma, L., A review on software defined content delivery network: a novel combination of CDN and SDN. *IEEE Access*, *11*, 43822–43843, 2023.
3. Vagmi, and Gupta, R.K., Content Delivery Networks in the Modern Age: Analyzing Trends, Overcoming Challenges, and Pioneering Developments, in: *International Conference on Smart Systems: Innovations in Computing*, Singapore, 2023, October, Springer Nature Singapore, pp. 793–806.
4. Looney, S., Content moderation through removal of service: Content delivery networks and extremist websites. *Policy Internet*, *15*, 4, 544–558, 2023.
5. Song, Z., Chen, K., Sarda, N., Altınbüken, D., Brevdo, E., Coleman, J., Gummadi, R., {HALP}: Heuristic aided learned preference eviction policy for {YouTube} content delivery network, in: *20th USENIX Symposium on*

Networked Systems Design and Implementation (NSDI 23), pp. 1149–1163, 2023.

6. Aung, N., Dhelim, S., Chen, L., Lakas, A., Zhang, W., Ning, H., Kechadi, M.T., VeSoNet: Traffic-aware content caching for vehicular social networks using deep reinforcement learning. *IEEE Trans. Intell. Transp. Syst.*, *24*, 8, 8638–8649, 2023.
7. Du, W., Dash, A., Li, J., Wei, H., Wang, G., Safety in Traffic Management Systems: A Comprehensive Survey. *Designs*, *7*, 4, 100, 2023.
8. Mohanty, D., Alsaidan, O.A., Zafar, A., Dodle, T., Gupta, J.K., Yasir, M., Khalid, M., Development of atomoxetine-loaded NLC *in situ* gel for nose-to-brain delivery: optimization, *in vitro*, and preclinical evaluation. *Pharmaceutics*, *15*, 7, 1985, 2023.
9. Kumar, S., Dwivedi, M., Kumar, M., Gill, S.S., A comprehensive review of vulnerabilities and AI-enabled defense against DDoS attacks for securing cloud services. *Comput. Sci. Rev.*, *53*, 100661, 2024.
10. Athief, R., Kishore, N., Paranthaman, R.N., Web Application Firewall Using Machine Learning, in: *2024 International Conference on Advances in Computing, Communication and Applied Informatics (ACCAI)*, pp. 1–7, IEEE, 2024, May.
11. Kumar, D.D., Mukharzee, J.D., Reddy, C.V.D., Rajagopal, S.M., Safe and Secure Communication Using SSL/TLS, in: *2024 International Conference on Emerging Smart Computing and Informatics (ESCI)*, pp. 1–6, IEEE, 2024, March.
12. Das, S., Kumari, R., Thakur, A., Singh, R.K., Nigam, A., Path Planning for Autonomous Ground Vehicles by Applying Modified Harris Hawks Optimization Technique, in: *International conference on soft computing for problem-solving*, pp. 161–171, Singapore, Springer Nature Singapore, 2023, August.
13. Kumari, R., Das, S., Singh, R.K., Kohli, A., Sunil, A., Dadhich, A., Getting Started With Computational Drug Discovery: A Comprehensive Guide, in: *Converging Pharmacy Science and Engineering in Computational Drug Discovery*, IGI Global, pp. 235–258, 2024.
14. Kumari, R., Das, S., Singh, R.K., Thakur, A., Deep Learning in Natural Language Processing, in: *Federated learning for Internet of Vehicles: IoV Image Processing, Vision and Intelligent Systems*, pp. 103–120, Bentham Science Publishers, 2024.
15. Das, S., Kumari, R., Singh, R.K., Assessing Customer Retail Data Through the Application of Various Clustering Algorithms, in: *International Conference on Advances in Computing and Data Sciences*, pp. 117–126, Cham, Springer Nature Switzerland, 2024, May.
16. Samadadiya, K., Das, S., Kumari, R., Predictive Paradigm: AI-Driven Social Media Analysis for Real Estate Sales Forecasts, in: *International Conference on Smart Computing and Communication*, pp. 189–198, Singapore, Springer Nature Singapore, 2024, January.

17. Das, S., Kumari, R., Singh, R.K., BeWell: An Integrated Mental Health Application Using LSTM Neural Network Model and Vader Sentiment Analysis for Emotional Well-Being. *J. Microsys. IoT*, *2*, 2, 556–563, 2024.
18. Singla, A., Mukherjee, S., Kumari, R., Das, S., Singh, R.K., Detour: Understanding the Application of Artificial Intelligence Based Models in Forecasting Safe Travel Routes, Cyber Tech Publications, 2023.
19. Kumari, R., Das, S., Thaku, A., Kumar, A., Singh, R.K., Autonomous vehicles, in: *Explainable Artificial Intelligence for Autonomous Vehicles*, pp. 1–24, CRC Press, 2025.
20. Goodfellow, I., Pouget-Abadie, J., Mirza, M., Xu, B., Warde-Farley, D., Ozair, S., Bengio, Y., Generative adversarial nets. *Advances in neural information processing systems, 27*, 2014.
21. Dorkenwald, M., Milbich, T., Blattmann, A., Rombach, R., Derpanis, K.G., Ommer, B., Stochastic image-to-video synthesis using cinns, in: *Proceedings of the IEEE/CVF Conference on Computer Vision and Pattern Recognition*, pp. 3742–3753, 2021.
22. Sun, J., Xu, Z., Shum, H.Y., Gradient profile prior and its applications in image super-resolution and enhancement. *IEEE Trans. Image Process.*, *20*, 6, 1529–1542, 2010.
23. Singh, A., Jaiswal, V., Joshi, G., Sanjeeve, A., Gite, S., Kotecha, K., Neural style transfer: A critical review. *IEEE Access*, *9*, 131583–131613, 2021.
24. Singer, U., Polyak, A., Hayes, T., Yin, X., An, J., Zhang, S., Taigman, Y., Make-a-video: Text-to-video generation without text-video data, 2022, *arXiv preprint arXiv:2209.14792.*
25. Wu, J.Z., Ge, Y., Wang, X., Lei, S.W., Gu, Y., Shi, Y., Shou, M.Z., Tune-a-video: One-shot tuning of image diffusion models for text-to-video generation, in: *Proceedings of the IEEE/CVF International Conference on Computer Vision*, pp. 7623–7633, 2023.
26. Liborio, J.D.M., Melo, C., Silva, M., Internet Video Delivery Improved by Super-Resolution with GAN. *Future Internet*, *14*, 12, 364, 2022.
27. Doersch, C., Tutorial on variational autoencoders, 2016, *arXiv preprint arXiv:1606.05908*.
28. Liang, D., Krishnan, R.G., Hoffman, M.D., Jebara, T., Variational autoencoders for collaborative filtering, in: *Proceedings of the 2018 world wide web conference*, pp. 689–698, 2018, April.
29. Xia, W., Yang, Y., Xue, J.H., Wu, B., Tedigan: Text-guided diverse face image generation and manipulation, in: *Proceedings of the IEEE/CVF conference on computer vision and pattern recognition*, pp. 2256–2265, 2021.
30. Van Dyk, D.A. and Meng, X.L., The art of data augmentation. *J. Comput. Graphical Stat.*, *10*, 1, 1–50, 2001.
31. Teodoro, A., Almeida, M., Figueiredo, M., Single-frame image denoising and inpainting using Gaussian mixtures, in: *International Conference on Pattern Recognition Applications and Methods*, vol. 2, pp. 283–288, SCITEPRESS, 2015, January.

32. Loudet, O., Saliba-Colombani, V., Camilleri, C., Calenge, F., Gaudon, V., Koprivova, A., Daniel-Vedele, F., Natural variation for sulfate content in Arabidopsis thaliana is highly controlled by APR2. *Nat. Genet.*, *39*, 7, 896–900, 2007.
33. Gold, B., Morgan, N., Ellis, D., *Speech and audio signal processing: processing and perception of speech and music*, John Wiley & Sons, 2011.
34. Massaro, D.W., Beskow, J., Cohen, M.M., Fry, C.L., Rodgriguez, T., Picture my voice: Audio to visual speech synthesis using artificial neural networks, in: *AVSP'99-International Conference on Auditory-Visual Speech Processing*, 1999.
35. Zhao, H., Jia, J., Koltun, V., Exploring self-attention for image recognition, in: *Proceedings of the IEEE/CVF conference on computer vision and pattern recognition*, pp. 10076–10085, 2020.
36. Liu, X., Sarikaya, R., Zhao, L., Ni, Y., Pan, Y.C., Personalized Natural Language Understanding, in: *Interspeech*, 2016, September, pp. 1146–1150.
37. Prem, D., Prem, B., Thanka, M.R., Library Book Recommendation Chatbot Using Rasa NLU, in: *2024 Third International Conference on Smart Technologies and Systems for Next Generation Computing (ICSTSN)*, pp. 1–7, IEEE, 2024, July.
38. Jang, H.S., Jin, H., Jung, B.C., Quek, T.Q., Versatile access control for massive IoT: Throughput, latency, and energy efficiency. *IEEE Trans. Mob. Comput.*, *19*, 8, 1984–1997, 2019.
39. Debia, S., Pineau, P.O., Siddiqui, A.S., Strategic use of storage: The impact of carbon policy, resource availability, and technology efficiency on a renewable-thermal power system. *Energy Econ.*, *80*, 100–122, 2019.
40. Yang, Q., The role of design in creating machine-learning-enhanced user experience, in: *2017 AAAI spring symposium series*, 2017, March.
41. Van Asten, B.J., van Adrichem, N.L., Kuipers, F.A., Scalability and resilience of software-defined networking: An overview, 2014, *arXiv preprint arXiv:1408.6760*.
42. Lee, M.K. and Elsayed, E.A., Optimization of warehouse storage capacity under a dedicated storage policy. *Int. J. Prod. Res.*, *43*, 9, 1785–1805, 2005.
43. Dulle, F.W., Online information resources availability and accessibility: a developing countries' scenario. *Afr. J. Library Arch. Inf. Sci.*, *25*, 1, 45, 2015.

12

An IoT-Enabled Indoor Tracking Base System (IITBS) for Human Activities Identification

Pulakesh Roy[1,2*], Sushmita Chaudhari[1], Surabhi Solanki[3], Rajib Banerjee[4] and Arindam Biswas[2,5]

[1]*Department of Computer Science Engineering, Brainware University, Kolkata, West Bengal, India*
[2]*Department of Mining Engineering, Kazi Nazrul University, Asansol, West Bengal, India*
[3]*School of Computer Science Engineering & Technology, Bennett University, Uttar Pradesh, India*
[4]*School of Computer Science, UPES, Uttarakhand, India*
[5]*Centre for IoT and AI Integration with Education-Industry-Agriculture, Kazi Nazrul University, Asansol, West Bengal, India*

Abstract

In outside open spaces, Global Positioning Systems (GPS) have been very effective in providing geolocation information. However, due to signal limitations, it is not very effective for indoor usage. In several fields, including artificial intelligence (AI) and the Internet of Things (IoT), indoor tracking is crucial, especially for precise location. However, owing to the complex structure and environment, precise and real-time navigation remains challenging.

An IoT-based AI-enabled indoor navigation system offers an effective solution by adapting self-learning characteristics to increase the efficiency of the localization system. The objective of this study includes the integration of AI and IoT to track objects precisely in indoor spaces. An effective data transmission system is also incorporated to deliver real-time updates while keeping 99% accuracy. In order to improve data transmission accuracy, reliability, and scalability, a Content Delivery Network (CDN) has been utilized. The CDN signifies lower latency and more rapidly delivers the content for real-time navigation updates.

**Corresponding author*: pulakesh.105081@gmail.com

Tanusree Chatterjee, Rajib Banerjee and Souvik Ganguli (eds.) Content Delivery Networks Using Artificial Intelligence and Internet of Things: Revolutionizing Content Delivery, (259–278) © 2026 Scrivener Publishing LLC

In this subsection, we emphasize indoor navigation systems, with an emphasis on radio waves and sensor-based approaches in particular. Additionally, this study also includes the principles, characteristics, precision, content delivery, and benefits associated with indoor positioning systems (IPS).

Keywords: BLE, RSSI, Kalman filter, indoor tracking, IoT, CDN

12.1 Introduction

As the demand for accurate and instantaneous tracking services increases across different fields, including airports, mining, manufacturing, and smart homes, indoor object tracking has become an important area of research. GPS, as a kind of traditional outdoor tracking technique, cannot work well in indoor spaces because of signal deterioration by walls and other factors. Therefore, to avoid the challenges above, enhanced tracking technologies are being invented for proper indoor tracking. In order to address the aforementioned shortcomings of existing structures, this paper proposes a renewed approach for an efficient and precise indoor tracking system comprised of BLE-enabled AI-enhanced Kalman filtering.

The proposed system is therefore intended for tracking objects based on the positions of BLE beacons installed in a particular area. The system works through the constant acquisition of the Received Signal Strength Indicator (RSSI) of BLE nodes. This RSSI value contains signal intensity information to reliably track the object's position. These measurements, however, are prone to environmental interference and hence inaccuracy. These issues are resolved by the system with the help of the Kalman filter, the mathematical apparatus that diminishes the noise of the signal and increases the accuracy of the estimation of positions. Furthermore, AI methodologies are incorporated for the optimization of RSSI calibration as well as versatility in the various types of dynamic scenarios.

The suggested infrastructure typically consists of several components, such as signal broadcasting nodes identified as BLE nodes, an AI-enabled indoor tracking approach, an operative data transmission system, a server framework to proficiently receive and analyze data, and a real-time tracking user interface.

The foundation of this system is the BLE technology, which is well-known for low energy consumption and compatibility with IoT. The BLE nodes continuously transmit radio frequencies that are detected by the tracking device to make the calculation of the RSSI in dBm possible. A higher RSSI value, closer to zero, therefore defines the higher signal present at the relevant area, while a lower value represents a weaker signal detected at the same location. Using the described RSSI measurements, the proposed system is capable of determining the object's location within the indoor environment. For improved precision and a low error rate, the system utilizes the AI-built Kalman filtering to smooth signal data and give appropriate positional estimates. A client-server structure is used to ensure communication and efficiently deliver the resultant data to the server point. The server has a graphical user interface in Windows to retrieve and analyze data collected and produced at the server end. In order to assess the effectiveness of the arrangement, an extensive analysis has been carried out in this work to effectively identify objects in an interior setting.

The remaining section of the work is structured as follows. A comprehensive assessment of the approaches used in the past few decades is provided in Section 12.2 of the Literature Review. Section 12.3 provides an explanation of the structure design. Section 12.4 explores the Outcomes and Performance, while Section 12.5 constitutes the Conclusion.

12.2 Literature Review

In this section, a detailed investigation has been done of the indoor tracking base system. Previously, several works have been done on the indoor tracking base system. In one of the articles [1], the author uses a tracking approach for the interior setting with the active badge. In this approach, the badge conveys unique identification to a certain recipient. In this scenario, the infrared signal has to pass straight ahead. The primary drawback of this method is that the signal tends to be of a very short spectrum.

In another paper [2], the author utilizes pedestrian dead reckoning and RSSI to enhance the accuracy of localization in an indoor setting. To efficiently do the localization in this work, the author reduces the instability of the Wi-Fi signal. Following the experiment, it was observed that the system achieves a very low localization error.

RSSI serves as one of the widely used methods for proximity estimation [3]. It is feasible to gather RSSI without purchasing additional equipment due to the availability of efficient transceiver modules on Wireless Sensor Nodes. In order to achieve this, the author in [4] estimates the distance to reach the vertices using RSSI. Eventually, it became clear from the experiments that the curve-fitting method led to more accurate outcomes for assessing distances under numerous circumstances.

In one of the papers [5], a deep investigation has been done for tracking objects in an indoor environment by utilizing the BLE. The suggested system will be an effective solution in terms of locating the position of doctors and patients in the hospital environment, where an effective tracking base system is crucial. After the experiment, it was identified that the system works effectively to track the object and shows a better result of accuracy.

The author of another paper [6] proposes a useful proximity estimating method utilizing RSSI to improve wireless capsule location accuracy. In this instance, the author measures the capsule's distance using the path loss moving average technique. The complete computation of the suggested system was achieved with MATLAB, and the precision of the result is considerably greater.

In [7], the author conversed about an RSSI-based IoT-driven surveillance system for underground mining. Considering the technological setup of the underground mining and the fluctuations in its operational circumstances, field investigations of RSSI distance measurement were performed. The authors of this work employed the callback monitoring technique to reliably monitor the item of RSSI proximity evaluation.

Similarly, the author of [8] proposed a paradigm model that might have been utilized in mines. The author proposed the neural network-based tagging technique and the trilateration method in one of the studies [9] by using the signal concentration. The advantages and disadvantages of several RSSI localization approaches were reviewed in [10, 11], along with recommendations for the direction that the method ought to follow in terms of improvement.

In one of the papers [12], the author utilizes the RSSI-based unified trilateration approach to track the position in an indoor setting. The execution of the given approach has been done by utilizing the Monte Carlo simulation. After the simulation, it was observed that the proposed system shows a comparatively better result and produces a very low error compared to other contending schemes.

In another paper [13], the author established an RSSI-built localized technique based on the weighted three optimal methodologies to increase distance precision. A tentative network framework is constructed in order to assess the efficacy of the proposed method. After the examination, the simulated outcomes showed that the proposed approach had better localization accuracy but needed a greater amount of computational time.

In one of the papers [14], the author develops an RSSI-based simultaneous positioning system for indoor environments by using Wi-Fi. In this work, the author utilizes the MPU-6050 accelerometer and the ESP8266 module to efficiently do the work. After the experiment, it was shown that the proposed system achieved a reduced cumulative localization error.

In [15], the author uses AM electromagnetic waves to find the target within the interior space. In order to measure positional accuracy, the author implements the fingerprinting approach to obtain the mean distance error. After completion of the research, the findings showed the lowest average proximity deviance.

In one of the publications [16], the author extensively investigated RSSI-based range estimation by utilizing IRIS nodes. By integrating empirical as well as theoretical competence, the author aimed to improve the assessment. The main drawback of this work is the inadequate assessment of the angle's effect on appropriately calculating the range.

Despite the fact that this area has already been the focus of numerous prior works, it still has certain constraints. In order to achieve this, an efficient movement monitoring system is essential. In the following table, a detailed analysis of the current work trends has been done. In this instance Table 12.1 shows the comparative analysis of the recent state of art.

In this study, we have proposed an improved indoor tracking system that combines BLE nodes integrated with an AI-enhanced Kalman filter. Real-time tracking with advanced precision, high transmission accuracy, and adaptability to solve some of the most intricate problems of indoor environments are the distinctive features of this IITBS. This shows the supremacy of the projected system.

Table 12.1 Comparative analysis of the recent trend of the work.

Study	Technology used	Precision	Advantages	Limitations	References
Active badge system for indoor tracking	Active badge system (infrared-based)	Low	Cost-effective for small areas	Limited by line-of-sight, short spectrum	[17]
Wi-Fi–based pedestrian dead reckoning	Pedestrian dead reckoning & RSSI (Wi-Fi)	Medium	Reduces Wi-Fi signal instability	Limited in complex environments with high interference	[18]
RSSI for indoor proximity estimation	RSSI for proximity estimation (wireless sensor nodes)	Medium-High	Utilizes existing transceiver modules, cost-efficient	Affected by environmental factors like interference	[19]
BLE for indoor tracking in healthcare	Bluetooth low energy (BLE) for object tracking	High	Suitable for healthcare, easy to deploy	Range limitations in large areas	[20]

(*Continued*)

Table 12.1 Comparative analysis of the recent trend of the work. (*Continued*)

Study	Technology used	Precision	Advantages	Limitations	References
Wireless capsule tracking *via* RSSI	RSSI with path loss moving average (MATLAB)	High	Greater precision, suitable for wireless capsule tracking	Computationally intensive	[6]
IoT surveillance in underground mining	RSSI-based surveillance (IoT) for underground mining	High	Effective for remote and challenging environments	Signal loss in complex mining structure	[21]
Neural network and trilateration for positioning	Neural network-based tagging & trilateration	Medium-High	Machine learning enhances accuracy	Requires extensive training data, high computational cost	[22]
Unified trilateration with Monte Carlo simulation	RSSI-based unified trilateration with Monte Carlo simulation	High	Produces low error, robust in simulations	Limited real-world testing, high resource demand	[23]
Weighted RSSI localization method	RSSI with weighted optimal methodology	Very High	High accuracy for complex environments	High computational requirements	[24]
Wi-Fi- and accelerometer-based positioning	Simultaneous positioning system with Wi-Fi (MPU-6050 accelerometer, ESP8266 module)	High	Achieves reduced cumulative localization error	Limited by Wi-Fi signal range	[25]
AM electromagnetic wave fingerprinting	AM electromagnetic waves with fingerprinting approach	Medium-High	Suitable for high-accuracy indoor. tracking	May require recalibration in changing environments	[26]
RSSI range estimation with IRIS nodes	RSSI range estimation with IRIS nodes	Medium	Enhances accuracy in proximity estimation	Limited angle assessment, needs more robust calibration	[24]

12.3 Structure Design

In this work, an effective indoor tracking base system has been established for tracking objects in the indoor environment. The projected system can be efficiently utilized in different sectors such as airports, underground mines, factories, etc. A methodical depiction of the indoor navigational system has been carried out using the proposed framework. The overall system architecture of the proposed system is given in Figure 12.1.

In order to find the object, the mechanism will continuously monitor the RSSI signal from the BLE beacon nodes that have been dispersed throughout the entire region. The data will subsequently be processed by the system to keep track of the movement. After that, the resultant data will be transmitted to the server end through the IoT gateway by leveraging Ethernet or Wi-Fi. The IoT gateway facilitates convenient data transmission across the server and the tracking apparatus in the proposed framework. In the suggested framework, the IoT gateway ensures the smooth communication between the tracking apparatus and the server, ensuring a reliable and efficient data transfer mechanism. Following that, the server will acquire and log all information for further assessment.

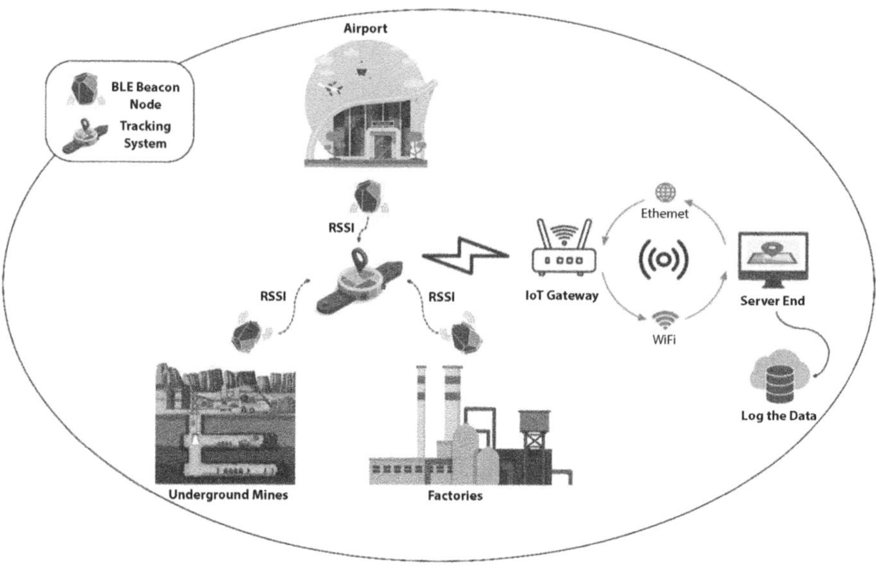

Figure 12.1 Overall system architecture of the proposed framework.

At the initial phase of the proceedings, the BLE (Bluetooth Low Energy) node communicates signals using radio waves to broadcast impulses. After that, the proposed system measures the RSSI value associated with the BLE units. The RSSI readings obtained from BLE equipment remain negative and are given in decibels (dBm). An adjacent RSSI score of 0 dBm denotes greater overall reception, and a higher RSSI score is thought to imply an enhanced signal. In the present research, the framework tracks the position utilizing the RSSI value. At the final stage, the resultant information is subsequently delivered to the end user *via* Wi-Fi and TCP/IP standards for further study.

In order to precisely carry out the activity, the complete system is segmented into two sections, involving object detection built on BLE and the implementation of an effective, reliable data transmission system.

12.3.1 Integration of AI Enables BLE-Based Object Tracking Using the Kalman Filter

The Kalman filter is an effective technique for object tracking. Here, we integrate artificial intelligence (AI) with the Kalman filter to improve real-time tracking in indoor environments. RSSI-based positioning is subject to many types of noise in the indoor context due to obstacles, such as walls and furniture. The Kalman filter acts as a real-time smoothing filter to reduce fluctuations of the incoming signal and provide a more precise position estimate. In parallel, the use of AI techniques, such as machine learning models for RSSI calibration or neural networks for nonlinear corrections, can significantly soften the burden on the system regarding accuracy and adaptability.

For instance, machine learning models could be trained to get more accurate precision based on historical RSSI patterns, environment-specific obstacles, or user movement behaviors. So, the combination of these techniques with the prediction and correction components of the Kalman filter enables an adaptive system that can readily respond to changing signal conditions, such as interference or intermittent obstructions. Neural networks could also work with the Kalman filter, predicting more complex movement patterns or dynamically modifying filter parameters. This plays a vital role in increasing the flexibility of the tracking approach and real-time responsiveness. For this, we developed an AI-enabled BLE-based object tracking system with the implementation of the Kalman filter to efficiently locate entities in an indoor setting. The algorithm of the proposed system is given below.

Algorithm 12.1 Algorithm for BLE-based indoor tracking system.

Input:

- Initial state estimate x_0 (position and velocity).
- Initial error covariance P_0.
- Process noise covariance Q.
- Measurement noise covariance R.
- Measurement matrix H.
- State transition matrix A.
- Control matrix B.
- RSSI measurements from BLE nodes, converted to approximate distances.

Output:

- Real-time estimates of the object's position in the indoor environment.

Algorithm 12.2 Implementation of Kalman filter for indoor tracking.

Start

1. Initialize Parameters:
   - Set the initial state estimate

$$x_0 = \left[x_0, y_0, \dot{x}_0, \dot{y}_0 \right]^T \quad (12.1)$$

   - Define the initial error covariance matrix: P_0
   - Define process noise covariance Q and measurement noise covariance R.
   - Define the measurement matrix H, state transition matrix A, and control matrix B if applicable.

2. Loop for Each Time Step k=1,2,…,N:

A. Prediction Step:

- Predict the Next State:

$$x_{k|k-1} = Ax_{k-1|k-1} + Bu_k \tag{12.2}$$

- Predict the Error Covariance: $P_{k|k-1} = AP_{k-1|k-1}AT + Q$

B. Measurement Step:

- Collect RSSI readings from BLE nodes and convert them into approximate distances.

C. Update Step:

- Compute the Kalman Gain: $K = P_{k|k-1}H^T\left(HP_{k|k-1}H^T + R\right)^{-1}$
- Update the State Estimate using the new RSSI measurement:

$$x_{k|k} = x_{k|k-1} + K\left(z_k - Hx_{k|k-1}\right) \tag{12.3}$$

- Update the Error Covariance: $P_{k|k} = (I - KH)P_{k|k-1}$

3. Output the Updated Position $x_{k|k}$ for Real-time Tracking:
   - Extract the position coordinates x and y from $x_{k|k}$ for tracking.
4. Repeat for each subsequent time step until tracking is complete.

End

This algorithm provides a refined and real-time position estimate by continuously adjusting based on RSSI measurements, which helps to mitigate noise and provide more accurate indoor localization.

12.3.2 Implementation of an Effective, Reliable Data Transmission System

For data transfer in the present research, we employed the client-server model. In this part, the client-server system framework has been covered.

12.3.2.1 Structure of the Client-Server Interface

The primary architecture of the client-server framework is entirely dependent on how the task-level programs are distributed [27]. Figure 12.2 describes the generic configuration of the client-server system. For conveying the data between the client end and the server end, the TCP/IP protocol

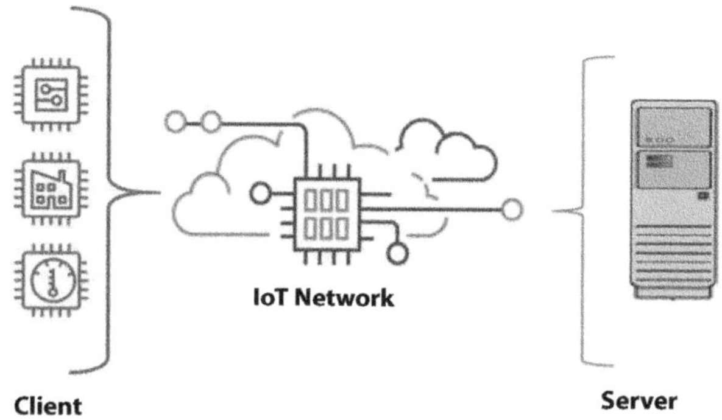

Figure 12.2 Client-server networking architecture.

has been efficiently utilized in this work as a communication framework. The primary objective of this framework is to give the applications a fundamental infrastructure. One of the key components of this system is the efficient design layout of the user interface. The primary advantages of this system are ease of use, speed, and extent of flexibility.

12.3.2.2 The Inception of Content Delivery Network (CDN)

To boost the overall performance of the system and further enhance the throughput of the client-server architecture, the CDN has been utilized in the suggested model. Through this process, CDN optimizes data distribution across the system in terms of speed, dependability, and scalability.

Through content distribution over the network, CDNs ensure that data will be transmitted effectively and with the least amount of latency. The significant efforts CDN has made are as follows:

- Accelerate the propagation of real-time updates.
- A distributed server infrastructure provides continuous availability of data.
- By effectively managing the data, CDNs make certain the system remains accessible and extensible.

12.3.3 Server-End Script

A software application has been constructed on the computing end to gather and preserve the raw information received from the client application for subsequent analysis. An inadequate, inefficient Windows application interface is being set up for this framework, as shown in Figure 12.3.

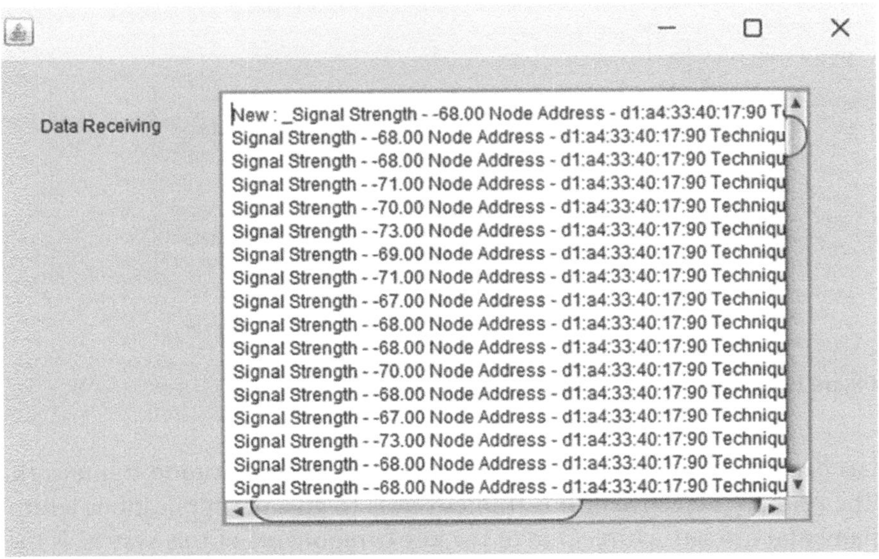

Figure 12.3 Windows server framework.

The primary intent of this implementation is to make the user interface more comprehensible and effective. The primary characteristics of the suggested structure are the flexibility and scalability of the framework. To achieve this, an interface will be employed at the server end to receive all of the information from the client side and publish the result within the datasheet.

12.3.4 Workflow Schematic

The general workflow design of the framework is displayed as follows. To efficiently elaborate the system, the workflow diagram of the proposed system is provided in Figure 12.4.

12.4 Outcomes and Performance

This section includes extensive real-time experiments by utilizing the BLE node with the objective of tracking and delivering the content to the end user. In the current investigation, a thorough examination has been done to find the optimal approach for locating the object in the indoor environment. In the proposed work, two separate BLE nodes have been proficiently operated to facilitate this. In this scenario, the RSSI value at various locations is

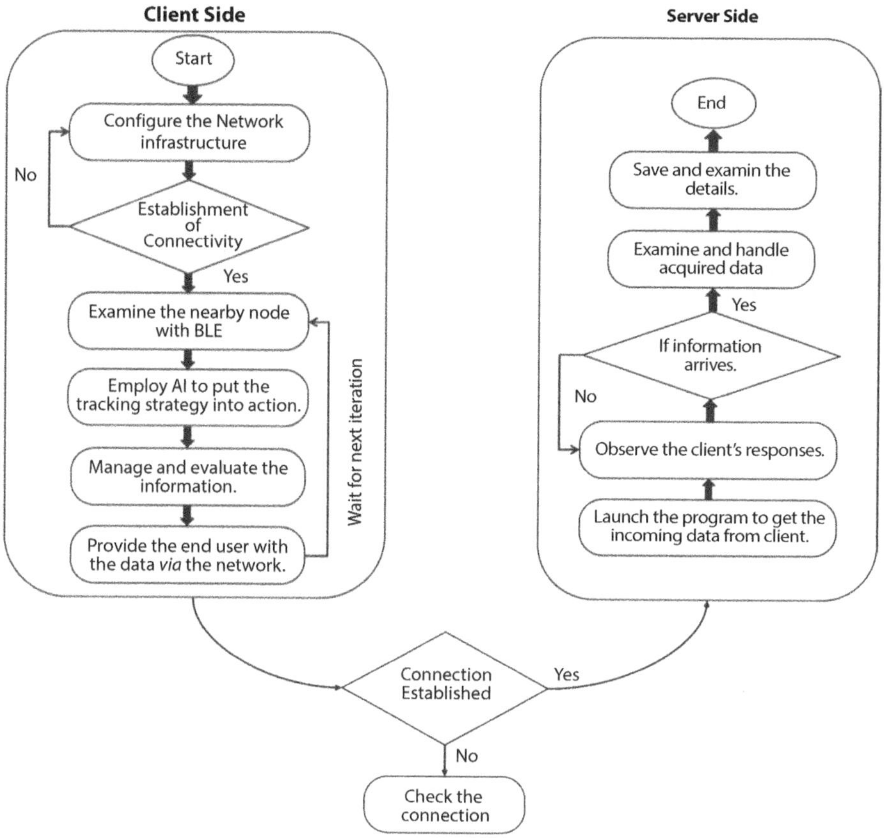

Figure 12.4 Workflow schematic of the projected system.

obtained by adjusting the tracking system. The fundamental objective of this research is to locate and monitor the object's movement by linking it with the BLE nodes across the network.

Figure 12.5 shows the proximity at different BLE node locations, together the motions of the tracker. The effectiveness of the suggested system is being evaluated through the investigation and analysis of the simulated data. The aforementioned figure shows that the proposed method successfully worked and shows a better result of accuracy across a variety of spaces. Additionally, the determining approach has been demonstrated to be hardly affected by changes in proximity.

Figure 12.5 illustrates that the IITBS algorithm outperforms in terms of precision compared to the other competing schemes. For instance, Table 12.2 indicates that the proposed methodology provides a specificity of 2.24 and 1.58 for both nodes at a distance of 2 meters. Alternatively, the other

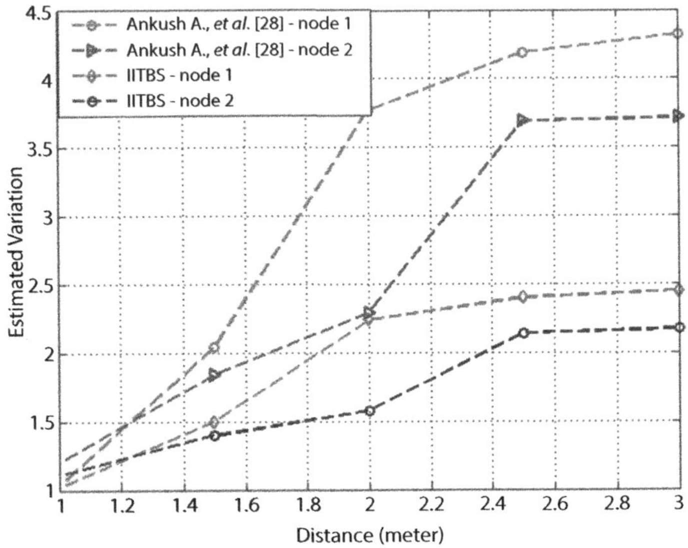

Figure 12.5 Proximity approximation at numerous points.

Table 12.2 Approximation of variety at distinct points.

Distance (meter)	**Estimated variation for node 1**		**Estimated variation for node 2**	
	IITBS	***Ankush A. et al.* [28]**	**IITBS**	***Ankush A. et al.* [28]**
1	1.02	1.04	1.11	1.21
1.5	1.50	2.04	1.40	1.84
2	2.24	3.77	1.58	2.29
2.5	2.40	4.19	2.14	3.69
3	2.45	4.32	2.17	3.71

algorithm provides an adequacy of 3.77 and 2.29, respectively, which is comparatively larger than the suggested scheme. This will demonstrate the effectiveness of the suggested system.

In this instance, a comprehensive analysis has been performed to evaluate the average discrepancy across several placements. Figure 12.6 represents the thorough statistical assessment of the average deviation with

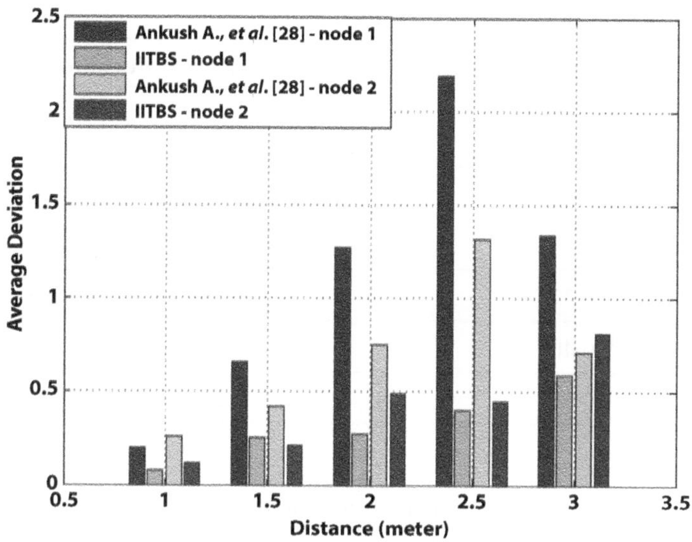

Figure 12.6 Average deviation measurement.

the variation of distance. In order to precisely complete the process, each assessment has been thoroughly conducted over fifty times. This analysis provides an effective evaluation of the system's effectiveness.

It is apparent from Figure 12.6 that the average discrepancies fluctuated significantly with changes in proximity. For instance, the system has an average deviation of 0.25 and 0.21, respectively, for both nodes at an offset of 1.5 meters. While employing the other competing scheme, this value considerably rises to 0.66 and 0.42. Similarly, the system got an average deviation of 0.27 and 0.49 over an extent of 2 meters. On the other hand, by using the other technique, the resulting values tend to be 1.27 and 0.75. In accordance with this, it can be inferred that the IITBS algorithm outperforms the other competing schemes in terms of proficiency for tracking objects in an interior setting. A thorough statistical study is presented in Table 12.3 to assess the efficacy at various points in the span.

Now, in this part, a deep assessment has also been conducted to determine the competence of delivering the content, as depicted in Figure 12.7. From this, it is apparent that the proposed system is proficiently delivering the content to the end user. As an illustration, the system effectively delivers 194 data packets in 20 minutes, and in 40 minutes, it delivers 383 packets of data. This will function as an evaluation of the integrity of the proposed system. Now, a detailed analysis of the transmission efficiency of the projected system has been given in Table 12.4.

Table 12.3 Average deviation with proximity.

Distance (meter)	Average deviation - node 1		Average deviation - node 2	
	IITBS	*Ankush A. et al.* [28]	IITBS	*Ankush A. et al.* [28]
1	0.08	0.20	0.12	0.26
1.5	0.25	0.66	0.21	0.42
2	0.27	1.27	0.49	0.75
2.5	0.40	2.19	0.45	1.32
3	0.59	1.34	0.81	0.71

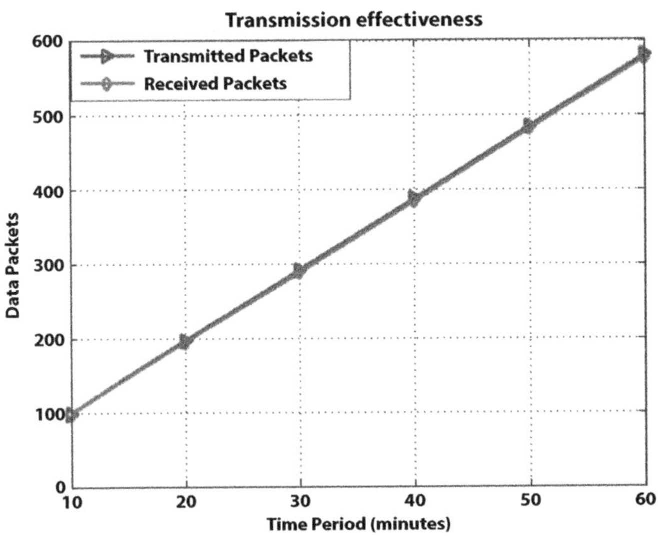

Figure 12.7 Transmission effectiveness of the proposed system.

In Table 12.4, the number of data packets transmitted and received and the transmission accuracy percentage are all captured in detail across time intervals between 10 and 60 minutes. For instance, at a 20-minute time interval, the system maintains the transmission accuracy of 98.98% while the value slightly changes to 99.22%. This demonstrates the reliability of the proposed system to efficiently deliver the content to the end user while maintaining the overall transmission accuracy of 99%.

Table 12.4 Transmission efficiency of the projected method.

Period (minutes)	Packets of data transmitted	Packets of data received	Transmission accuracy (%)
10	98	97	98.98
20	196	194	98.98
30	290	288	99.31
40	386	383	99.22
50	484	481	99.38
60	579	575	99.31

In the previous discussion, it became apparent that the IITBS outperforms the other competing schemes [28, 29] in terms of accuracy for tracing objects. Additionally, it has been identified that the system effectively delivers content to the end user. This shows the effectiveness and supremacy of the proposed system.

12.5 Conclusion

This work describes an advanced indoor tracking system using BLE technology for object tracking and data transmission, emphasizing the incorporation of artificial intelligence along with the IoT and CDN. It has been endeavored in different areas of application, like airports, underground mines, and factories. Focusing on the accuracy of object tracking in the indoor environment, this work employs the RSSI measurements, the artificial intelligence algorithm, and the Kalman filter for real-time object tracking. With a Kalman filter's smoothing algorithm and machine learning RSSI calibration, there is a definite accuracy and flexibility boost in tracking. The experimental results prove the effectiveness of the suggested algorithm by comparing the estimation errors and tracking performance to the other methods.

Scales up for another factor of real-time data delivery with a client-server structure that assures data transmission with over ninety-nine percent accuracy. For seamless connection, the CDN aids high performance and effective allocation of resources through the use of a distributed network. In the projected system, the AI-enhanced BLE system, augmented by a CDN, not only enhances tracking accuracy but also simultaneously optimizes the transmission efficiency and content relevancy. The system exhibits a state-of-the-art indoor surveillance solution with adaptive object tracking, improved accuracy, and suitable data transmission.

Despite the system functioning well, inevitably, the extent of the signal's quality may be altered by external variables. The objective for future extension is to track the object more effectively while increasing transmission capacity and decreasing latency.

AI disclosure statement: The author confirms that no artificial intelligence (AI)-generated content or AI-assisted writing tools were used in the preparation of this chapter (including summaries and the abstract). All text, analysis, and conclusions are the result of the author's original work.

References

1. Faragher, R. and Harle, R., Location Fingerprinting with Bluetooth Low Energy beacons. *IEEE J. Sel. Areas Commun.*, 33, 2418–2428, 2015.
2. Poulose, A. and Han, D.S., Indoor localization using PDR with Wi-Fi weighted path loss algorithm. *Proc. Int. Conf. Inf. Commun. Technol. Converg. (ICTC)*, 2019, 689–693, 2019.
3. Boukerche, A., Oliveira, H.A.B.F., Nakamura, E.F., Loureiro, A.A.F., Localization systems for wireless sensor networks. *IEEE Wirel. Commun.*, 14, 6–12, 2007.
4. Ivanić, M. and Mezei, I., Distance estimation based on RSSI improvements of orientation-aware nodes. *Proc. Zooming Innov. Consum. Technol. Conf. (ZINC)*, 2018, 140–143, 2018.
5. Roy, P., Mondal, S., Biswas, A., Banerjee, R., A comprehensive study on BLE-enabled IoT-based tracking system (BITS) for indoor environment. *Proc. Int. Conf. Intell. Data Commun. Technol. Internet Things (IDCIoT)*, 2024, 578–585, 2024.
6. Hany, U., Akter, L., Hossain, M.F., Moving averaging method of RSSI-based distance estimation for wireless capsule localization. *Proc. Int. Conf. Med. Eng. Health Inform. Technol. (MediTec)*, 2016, 1–5, 2016.
7. Ming, J., Jin, A., Sun, J., Hu, D., Xia, Y., Study on the detecting technology of mine Internet of Things based on RSSI technology. *Proc. Wirel. Opt. Commun. Conf. (WOCC)*, 2019, 1–4, 2019.
8. Sun, J., Ming, J., Guo, Y., *et al.*, Exploration and practice on experiment teaching of mine Internet of Things. *Lab. Sci.*, 21, 41–44, 2018.
9. Ming, J., Hu, N., Niu, X., Zhang, Y., Study on the personnel localization algorithm of the underground mine based on RSSI technology. *Proc. IEEE Int. Conf. Commun. Softw. Netw. (ICCSN)*, 2017, 408–411, 2017.

10. Wang, H. and Hu, A., Ranging method based on the mapping between RSSI and distance scope. *J. Univ. Electron. Sci. Technol. China*, 41, 522–526, 2012.
11. Wang, H. and Shao, K., Study on ranging technology based on RSSI in wireless sensor networks. *Microcomput. Inf.*, 28, 346–348, 2012.
12. Park, Y.-B. and Lee, Y.H., A novel unified trilateration method for RSSI-based indoor localization. *Proc. Int. Conf. Inf. Commun. Technol. Converg. (ICTC)*, 2021, 1628–1631, 2021.
13. Du, J., Yuan, C., Yue, M., Ma, T., A novel localization algorithm based on RSSI and multilateration for indoor environments. *Electronics*, 11, 289, 2022.
14. Danklang, P. and Pranekunakol, T., An RSSI-based weighting with accelerometers for real-time indoor positioning, in: *Proc. Int. Conf. Knowl. Smart Technol. (KST)*, 2021.
15. Rahman, M.M., Moghtadaiee, V., Dempster, A.G., Design of fingerprinting technique for indoor localization using AM radio signals. *Proc. Int. Conf. Indoor Positioning Indoor Navig. (IPIN)*, 2017, 1–7, 2017.
16. Mahapatra, R.K. and Shet, N.S.V., Experimental analysis of RSSI-based distance estimation for wireless sensor networks. *Proc. IEEE Distrib. Comput. VLSI Electr. Circuits Robot. (DISCOVER)*, 2016, 211–215, 2016.
17. Huang, J., Junginger, S., Liu, H., Thurow, K., Indoor positioning systems of mobile robots: A review. *Robotics*, 12, 47, 2023.
18. Shen, L.L. and Hui, W.W.S., Improved pedestrian dead-reckoning-based indoor positioning by RSSI-based heading correction. *IEEE Sens. J.*, 16, 7762–7773, 2016.
19. Mistry, H.P. and Mistry, N.H., RSSI-based localization scheme in wireless sensor networks: A survey. *Proc. Int. Conf. Adv. Comput. Commun. Technol.*, 2015, 647–652, 2015.
20. Bai, L., Ciravegna, F., Bond, R., Mulvenna, M., A low cost indoor positioning system using bluetooth low energy. *IEEE Access*, 8, 136858–136871, 2020.
21. Cavur, M. and Demir, E., RSSI-based hybrid algorithm for real-time tracking in underground mining by using RFID technology. *Phys. Commu.*, 55, 101863, 2022.
22. Jondhale, S.R., Jondhale, A.S., Deshpande, P.S., Lloret, J., Improved trilateration for indoor localization: Neural network and centroid-based approach. *Int. J. Distrib. Sens. Netw.*, 17, 15501477211053997, 2021.
23. Cully, W.P., Cotton, S.L., Scanlon, W.G., Empirical performance of RSSI-based Monte Carlo localisation for active RFID patient tracking systems. *Int. J. Wirel. Inf. Netw.*, 19, 173–184, 2012.
24. Mehrabian, H. and Ravanmehr, R., Sensor fusion for indoor positioning system through improved RSSI and PDR methods. *Futur. Gener. Comput. Syst.*, 138, 254–269, 2023.
25. Nurbol, K., Marianella, G., Samal, S., Galina, B., Oxana, A., Meruyert, Z., Development of a structural and functional scheme of a distributed Wi-Fi monitoring system for bridge structures and construction buildings. *Proc. IEEE Int. Symp. Logist. Ind. Inform. (LINDI)*, 2024, 249–254, 2024.

26. Kim Geok, T., Zar Aung, K., Sandar Aung, M., Thu Soe, M., Abdaziz, A., Pao Liew, C., Yong, W.H., Review of indoor positioning: Radio wave technology. *Appl. Sci.*, 11, 279, 2020.
27. Xue, M. and Zhu, C., The socket programming and software design for communication based on client/server. *Proc. Pac.-Asia Conf. Circuits Commun. Syst.*, 2009, 775–777, 2009.
28. Kalbandhe, A.A. and Patil, S.C., Indoor Positioning System Using Bluetooth Low Energy. *Proc. Int. Conf. Comput. Anal. Secur. Trends (CAST)*, 2016, 451–455, 2016.
29. Rajashree, H.S., Sahana, B., Surabhi, S., Apoorva, T.N., Parameshachari, B.D., Indoor localization using BLE technology. *Int. J. Eng. Res. Technol.*, 6, 1–6, 2018.

13

AI and IoT-Driven Machine Learning in CDNs: Advancing Diagnostic Precision in Breast Cancer Screening

Shubhranshu Gorai[1], Saibal Majumder[1,2], Ritik Maity[1], Mehuli Lahiri[1], Sovan Bhattacharya[1,2], Dola Sinha[1,3], Suchandra Banerjee[1] and Chandan Bandyopadhyay[1,4*]

[1]*Department of Computer Science & Engineering (Data Science), Dr. B.C. Roy Engineering College, Durgapur, West Bengal, India*
[2]*Department of Computer Science & Engineering, National Institute of Technology, Durgapur, West Bengal, India*
[3]*Dept. of EE, Indian Institute of Technology, Dhanbad, India*
[4]*Department of Computer Science & Engineering, University of Bremen, Bremen, Germany*

Abstract
Early breast cancer detection improves treatment and survival, making accurate diagnosis crucial in healthcare. Standard screening methods like mammography, ultrasonography, and MRI misdiagnose 10% to 30% of patients. The current research investigates the impact of AI and IoT on breast cancer detection. The Breast Cancer Wisconsin (Diagnostic) Dataset was utilized to assess nine machine learning models, including artificial neural networks, support vector machines, and ensemble classifiers such as adaptive boosting and random forest. Performance was assessed using confusion matrices, accuracy, recall, precision, F1-score, and Matthews correlation coefficient (MCC). The article also highlights how CDNs help integrate AI and the IoT. AI algorithms process patient data from IoT devices to improve diagnosis accuracy. Content delivery networks (CDNs) protect and distribute information in real-time healthcare applications to provide low latency and scalability. The findings show that AI and IoT can increase breast cancer screening diagnostic precision, improving patient outcomes and healthcare efficiency. This will be supported by CDNs' data handling capabilities.

**Corresponding author*: chandanb.iiest@gmail.com

Tanusree Chatterjee, Rajib Banerjee and Souvik Ganguli (eds.) Content Delivery Networks Using Artificial Intelligence and Internet of Things: Revolutionizing Content Delivery, (279–306) © 2026 Scrivener Publishing LLC

***Keywords*:** Breast cancer, content delivery network, machine learning classifiers, performance metrices, artificial intelligence, Internet of Things

13.1 Introduction

Cancer stands as a big killer in poorer regions of the world. It is often the reason why lots of women die. The physical features of thick breast tissue make breast cancer the top cancer in women [1]. This disease shows up as unusual cell growth forming lumps in the milk-producing glands. Finding it early is key because it improves the chances of successful treatment and survival. Usual ways of spotting breast cancer, like mammography (an X-ray of the breast), ultrasound, and MRI, are used. But, there are problems with these methods, like their high price, the need for special equipment, and they might not always be right. This is even more troubling because of issues like the high cost of X-rays, children having dense tissue, and the chance of incorrect results [2]. Because these methods are not perfect, machine learning (ML), a part of AI, is being used to help with finding breast cancer, possibly making the process more correct and reducing wrong results. In recent times, more research using machine learning to help with breast cancer checks has been done. Researchers have used different methods, like mixed models and support vector machines (a type of machine learning model), to make the diagnosis of breast cancer more accurate [4]. Jawad Ahmad and others [3] came up with an AI model combining Alex-Net and SVM. The model proved to be 99.16% correct when tested with the DDSM dataset [3]. Other research has looked at how machine learning, data augmentation (increasing the amount of data), and transfer learning (applying knowledge gained from one problem to another problem) can better diagnose breast cancer. This research has proven that machine learning models can be more accurate and reliable than the usual methods. This suggests good things for the future of diagnosing diseases.

Having an uneven number of cancer and non-cancer cases in the dataset is a big problem when trying to predict breast cancer [12, 13]. As a general rule, there are more non-cancer instances than cancer cases, which can cause projections to be skewed. In order to tackle this problem, it is advised to utilize several classifiers. This will enhance the model's resilience and reduce bias. Overfitting [14] is another problem; it happens when models work well on the training data but cannot adapt to new, unknown input. To keep the model resilient and work well on different datasets, it is crucial to use techniques like regularization and cross-validation to prevent overfitting.

Massive and continuous data streams are generated from AI-driven analytics and Internet-of-Things devices embedded in breast cancer diagnostics. This becomes increasingly relevant given the role that AI and IoT play within content delivery networks (CDNs) aimed at enhancing breast cancer diagnostics. This consequently puts the pressure on content delivery networks (CDNs) to deliver high scalability, low latency, and enhanced security, to ensure effective processing, storage, and remote distribution of data while ensuring real-time diagnostics needs in compliance with data and patient privacy. Integrating AI and IoT can facilitate breast cancer diagnostics, as illustrated in Figure 13.1. We present a diagrammatic illustration of the achieved breast cancer diagnostics through AI and IoT integration and the overall impact on CDNs.

AI (artificial intelligence) and IoT (Internet of Things) technologies are generating real-time medical data in large quantities, which must be harvested, analysed and delivered to healthcare providers in record time. Content delivery networks (CDNs) are critical to distributing this massive quantity of data while also ensuring that diagnostic insights are accessible, at low latency, reliable and secure. Data is delivered with speed, and computing capacity is increased in healthcare systems, which contributes to diagnostic precision, AI-based decision-making, and better patient outcomes through the use of content delivery networks (CDNs).

The convergence of the Internet of Things, content delivery networks, and artificial intelligence for breast cancer diagnosis is illustrated in Figure 13.1. Imaging and wearables are a few of the Internet-of-Things devices that provide data feed to the content delivery network edge server, where it receives preprocessing and encryption. The content delivery network

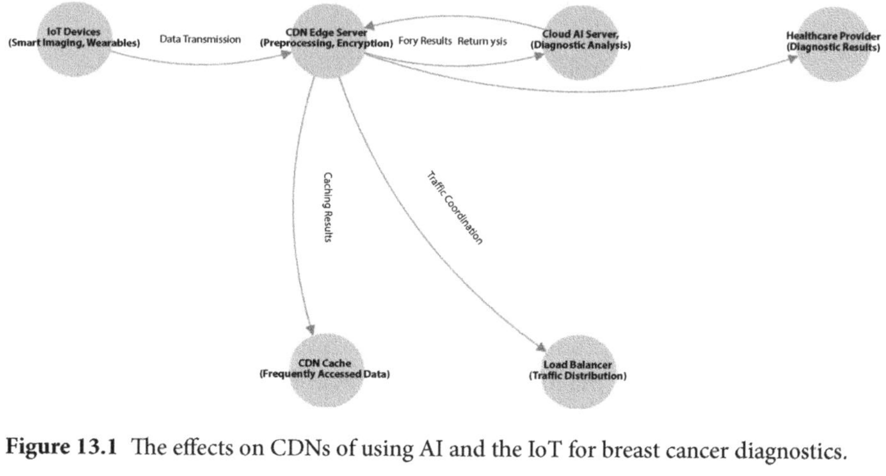

Figure 13.1 The effects on CDNs of using AI and the IoT for breast cancer diagnostics.

(CDN) cache stores content that is highly requested, and load balancing is intended to balance traffic efficiently. Data to be analyzed further is sent to the Cloud AI Server, which is tested with AI algorithms. The results of the test are passed on to healthcare professionals. With this technology, latency is reduced, data transfer is improved, and scaled real-time diagnostics are facilitated. Caching to provide fast responses, edge computing to provide localized processing, and load balancing to provide scalability are all critical features.

The accuracy of predictions made by models can be severely compromised by improper data preprocessing [15, 16]. The major contribution of our manuscript is mentioned as follows.

(i) We have come up with an entirely new methodology by using eight different classifiers from machine learning in the field of breast cancer detection. The approach is holistic and is by far more comprehensive than using only one single model in diagnostics. This will give the overall performance of the algorithms for diagnostics a much more nuanced evaluation.
(ii) We establish a methodological framework for comparative classification that is rigorous across multiple performance metrics. Our approach, which goes beyond traditional evaluation techniques, offers a sophisticated multidimensional assessment of the capabilities and limitations of algorithms.
(iii) We conduct a comprehensive examination of the efficacy of several classifiers with differing computational characteristics, such as XGBoost, support vector machines (SVMs), random forests, Extra Tree classifier (ETC), and artificial neural networks (ANNs), all evaluated under consistent conditions.
(iv) Our work provides a systematic study of the inherent challenges of applying machine learning models to clinical settings, with particular emphasis placed on model linearity, sensitivity to computational noise, and model interpretability constraints. Through the medium of a systematic review, we recommend practical guidance on the construction of predictive models that improve in the robustness of their performance and clinical utility.

The structure of the paper is as follows. Section 13.2 explores the dataset used in this research. Section 13.3 defines the methods and approaches used. Section 13.4 defines the nine machine learning models applied. Section 13.5 defines the performance measures for the classifiers. Section 13.6 contains the study's results and their discussion. Finally, Section 13.7 presents the study's conclusion.

13.2 Dataset Description

The original source of the publicly available breast cancer data used in this study is Dr. William H. Walberg of the University of Wisconsin Hospital, who originally uploaded it to Kaggle. Machine learning algorithms such as logistic regression, random forest, support vector machines, and additional trees were applied to the dataset, and data visualization and classification were also performed on it. Using accuracy metrics to compare the performance of various machine learning models, the research aimed to construct a dependable model for early-stage breast cancer detection by finding predictive features.

There are 569 cases in the dataset, and each one has 32 attributes that provide details about the patient and the tumor. Among these characteristics are a number of tumor metrics that are quantified using mean, standard error, and worst-case values; these metrics include radius, texture, perimeter, area, smoothness, compactness, and concavity. Higher values generally correlate with malignant tumors, making these characteristics crucial for cancer diagnosis. The dataset includes a *diagnostic* column that classifies tumors as either benign or malignant. In this column, M represents malignant tumors, while B denotes benign tumors. Understanding these classifications is essential for training a reliable model for early breast cancer diagnosis.

After a thorough data quality check, none of the features showed any missing values. This enabled us to understand that the dataset was suitable for extensive research and to use all the data for training and model assessment. We tested and evaluated various machine learning classifiers based on these factors to determine how well they could assist in the early detection of breast cancer. Healthcare professionals will be able to serve their patients more effectively if these algorithms are able to accurately identify those most at risk and when intervention is necessary.

13.2.1 Data Preprocessing

The quality and accuracy of raw data greatly affect the effectiveness of a model. Thus, a fundamental first step in obtaining a dataset ready for machine learning applications is data preparation. Bad data preparation can cause overfitting, incorrect labelling, or even simplified data, compromising the performance of the model. This work addresses typical issues in the Wisconsin Breast Cancer Dataset, including outliers, missing values, and scale deviations, by means of several preprocessing techniques. These procedures remained crucial to raise the dataset's and the model's learning capability.

The dataset included features with different types, sizes, and distributions. Using feature scaling, normalizing, and outlier detection helped to lessen the effect of these variances. Also, we assigned any missing values in the dataset using proper statistical techniques so as to avoid losing any significant information. Resolving multicollinearity was a crucial step in the preprocessing; meanwhile, strongly correlated features might reduce model interpretability. Before and after preprocessing, a correlation matrix was calculated to search for feature correlations; this helped us to identify probably redundant features endangering the performance of the model.

The dataset had features with relatively different sizes; some of them had high numerical values, and others very low; this could lead to bias in the model results. Many of the characters also showed outliers, which increased the overfitting risk and contributed to the noise level. Removing these glitches raised the generalizability and reliability of the model. Normalizing methods were applied to ensure that all features were scaled to the same value, ensuring that no one feature could have an outsized effect on the model due to its scale. Getting the dataset ready for machine learning investigation required these preprocessing methods.

A boxplot was drawn up to organize the data after the features' numerical values had been normalized. The features are shown in this boxplot, where Figure 13.2 is represented. There were too many outliers, so some features were removed according to them. This means that there possibly was not sufficient data to use these features for classification by using the boxplots provided.

To evaluate the interrelationships of the raw dataset's features, the correlation matrix was calculated prior to preprocessing. The strength of these links is reflected in correlation coefficients, which range from -1 to +1.

High correlations indicate the possibility of multicollinearity. Due to the possibility of duplicate data produced by multicollinearity, model coefficients may be negatively impacted, and overfitting may occur. In addition,

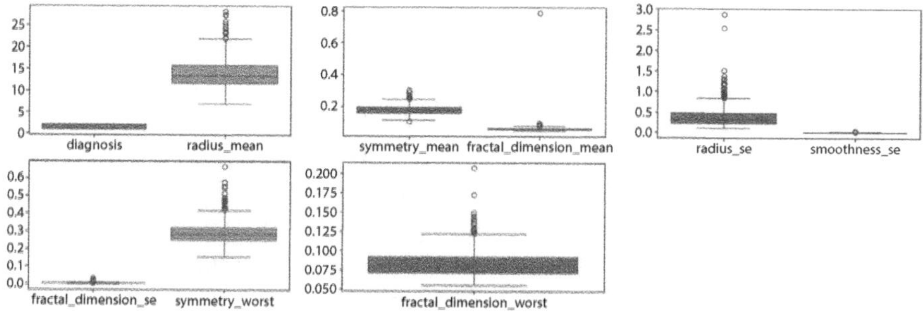

Figure 13.2 Boxplots of the features of the breast cancer dataset.

these correlations were already skewed due to outliers in the original data, which could have biased the predictions made by the model. Important feature correlations are highlighted in Figure 13.3, which shows the correlation matrix before preprocessing.

A series of preprocessing procedures was carried out on the dataset in order to enhance model performance and decrease overfitting before the machine learning models used in this study could be trained. Two distinct stages of preprocessing were carried out.

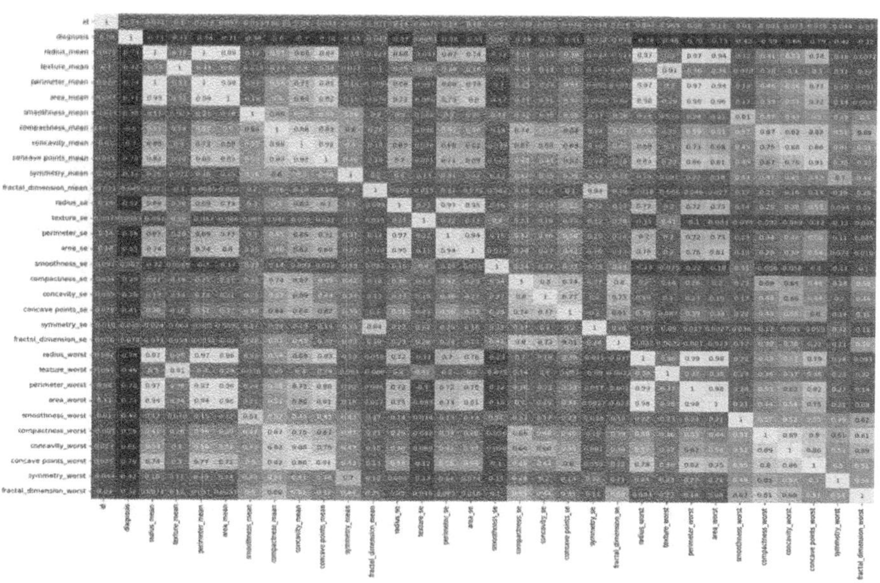

Figure 13.3 Correlation matrix of the features of the breast cancer dataset.

To prevent inflated values from distorting the model's learning process, outliers were first identified and removed from every feature. To ensure that the data used to train the model was free of outliers that could induce bias, this step was crucial.

To analyze how different features interact, we first built a correlation matrix. To minimize the impact of multicollinearity, we removed any features with a correlation coefficient higher than 0.8. After this step, the dataset was reduced from 32 features to just 12, keeping only the most relevant ones: 'radius mean,' 'radius se,' 'symmetry mean,' 'fractal dimension mean,' 'fractal dimension se,' 'smoothness se,' 'symmetry worst,' and 'fractal dimension worst.'

We employed preprocessing techniques that resulted in an end dataset with eight significant features—later we will discuss these in greater detail—to make comparisons of six diverse machine learning classifiers. The preprocessing techniques impacted the breast cancer dataset considerably, enhancing model accuracy and ease of interpreting results. We removed multicollinearity by managing highly correlated features and outliers through scaling, normalization, and elimination. By modifying the correlation matrix, we uncovered more robust relationships among the features, and the classifiers were able to produce better and more reliable results in breast cancer detection.

13.3 Methodology

This work aims to evaluate several machine learning algorithms developed for the purpose of breast cancer detection. The method consists of data acquisition, preprocessing, feature extraction and selection, model building and training, hyperparameter tuning, cross-validation, and last model evaluation. Every one of these steps is crucial for the models to be robust, exact, and able to provide reliable estimates for the analysis of breast cancer.

13.3.1 Data Acquisition and Exploration

This research made use of the Breast Cancer Wisconsin Diagnostic dataset, which originated from the hospitals affiliated with the University of Wisconsin. Radius, texture, smoothness, compactness, symmetry, and other pertinent properties are among the many parameters described by this dataset as belonging to the cell nuclei detected in breast masses. We carried out the first data exploration in the dataset to study its statistical features, such as the distribution of classes (malignant vs. benign), and identify the imbalanced data or bias that could influence the accuracy of the model. This exploratory study helped us grasp the required preprocessing steps we needed to implement in order to enable the dataset for machine learning.

13.3.2 Feature Extraction and Selection

The optimization process was investigated in a systematic two-phase approach to enhance model discriminant power and also predictive performance. We first carried out correlation analysis, employing a complete correlation matrix to iteratively identify and eliminate highly multicollinear features. These methods not only reduce statistical redundancy but also enhance the feature set of the dataset, hence making it easier to interpret the model as well as improving computational efficiency.

In the next feature selection exercise, we proceeded more thoughtfully by considering the possible usefulness of each feature from its statistical significance and correlation in separating benign and malignant samples. This judiciously chosen feature elimination helped the model in separating cases, and it did so quite accurately and reliably.

The systematic reduction of feature dimensionality through these methodical procedures serves multiple critical objectives: (a) minimizing computational complexity, (b) reducing potential noise introduction, and (c) focusing the machine learning algorithm's attention on the most diagnostically relevant attributes. Consequently, this approach facilitates a more parsimonious and precise predictive model for breast cancer classification.

13.3.3 Model Development and Training

Multiple machine learning classifiers were utilized for breast cancer diagnosis. All classifiers were trained with a 75% train-test split, indicating that 75% of the data was allocated for training and 25% for testing. The models were as delineated: random forest effectively utilizes an ensemble technique to amalgamate predictions from several decision trees, making it adept at managing missing data, generating feature importance rankings, and mitigating the danger of overfitting, For issues with clearly distinguishable situations, the non-parametric k-Nearest Neighbors (KNN) algorithm is ideal. It may categorize occurrences according to their proximity in the feature space and is straightforward. Because of its interpretability in binary classification tasks and its ability to shed light on predictor-variable correlations, Logistic Regression is crucial for medical diagnostics. Support vector machine (SVM) is a great tool for breast cancer detection because it can effectively distinguish between benign and malignant cases in high-dimensional areas. This is accomplished by creating a hyperplane that maximizes class separation. XGBoost is a gradient boosting algorithm that can help mitigate the issue of overfitting and still provides high accuracy even in large datasets; The selected algorithms were based on their ability to make predictions: Artificial neural networks (ANNs) may model complex non-linear relationships and improve performance through multi-layered architectures" Final Classifier and TrainingOur Nu-Support Vector Classifier, a variant of support vector machines (SVM), uses flexible decision boundaries. A passive-aggressive classifier (PAC) is one such online learning method that works well with dynamic settings that require real-time predictions.

13.3.4 Model Optimization and Evaluation

Achieving model performance optimization involved a systematic methodological approach of hyperparameter tuning, cross-validation, and comparative analysis. A systematic exploration of the hyperparameters was achieved through grid search and randomized search for a specific configuration of model-specific parameters. For example, the kernel configurations and regularization parameters were fine-tuned for support vector machine (SVM) models, while the tree count and depth were fine-tuned for ensemble methods like random forest.

We employed cross-validation as a comprehensive validation technique, and k-fold techniques were employed to challenge model generalization. To prevent overfitting, the training and testing of models on various datasets were performed using this technique to provide a stable and reproducible evaluation of performance through the sequential training and testing of models on these subsets of available data.

The last stage included a comprehensive comparison of the algorithms' performance using multiple performance evaluation metrics such as accuracy, F1-score, and ROC-AUC. The choice of the best breast cancer detection algorithm was made based on a comprehensive evaluation of these metrics, selecting models that are very accurate and have good performance characteristics.

13.4 Machine Learning Models

Machine learning, a subfield of artificial intelligence combining mathematical, statistical, and algorithmic techniques, has found many domain-specific uses. Emphasizing supervised learning, this work trains a model to map X to Y in order to approximate the mapping function so the model may forecast the output of Y for new data points. We assessed eight supervised machine learning models here, carefully chosen for their capacity to identify breast cancer.

Focusing on their capacity to properly distinguish benign from malignant tumors, this work sought to classify breast cancer data using machine learning algorithms. When a different separation is necessary, the choice of the support vector classifier (SVC) was mostly motivated by its capacity to generate a hyperplane optimizing the margin between classes [5]. Using an ensemble method, the random forest classifier (RFC) aggregated the predictions of several decision trees trained on bootstrap data, thereby enhancing the general robustness of the model. This helped reduce overfitting [6]. Likewise, by using ensemble learning on the whole dataset instead of depending on bootstrap samples, Extra Trees balanced computational efficiency with prediction performance [8]. Introduced a tuned parameter to its model, the nu-support vector classifier (Nu-SVC) lets one vary the balance between margin size and misclassification rates. This parameter was applied to handle datasets marked by notable non-linear separability or noise [9]. Proving useful in dynamic diagnostic environments, the passive-aggressive classifier (PAC) enabled real-time updates through incremental model adjustments using arriving data [10]. An artificial neural network (ANN) model was used for its capacity to learn complicated, non-linear interactions, thereby improving detection accuracy in medical applications. This model takes advantage of multi-layer designs [11].

Constantly improving classification accuracy, XGBoost—a gradient boosting method—showcased success for cancer detection datasets and other vast datasets with complex features [12]. Using iteratively changing the weights of misclassified events [13], AdaBoost was used to improve the performance of the model on challenging data. To improve expected accuracy, a Voting Classifier finally compiles outputs from numerous classifiers using both hard and soft voting methods [14].

13.5 Performance Metrics

In machine learning classification, the evaluation of the performance of a model depends significantly on precision. Precision counts the number of actual positive instances correctly labeled. That is, precision calculates the ratio of actual positives to all the positive predictions made. The formula is Precision = TP / (TP + FP) with TP = True Positives (capacity of positive predictions) and FP = False Positives (capacity of incorrect predictions). So, a high precision score means that the model makes fewer false apprehensions and that most of its positive predictions are, in fact, correct.

Accuracy, however, provides a summary of overall classification quality. The metric is examining the proportion of correctly classified instances, both true positives and true negatives, to total instances. The calculation Accuracy = (TP + TN) / (TP + TN + FP + FN) delivers a summary depiction of the predictive facility of the classifier across all class labels.

Recall, alternatively termed the true positive rate, represents the proportion of actual positive instances successfully identified by the classifier. The recall metric, calculated as Recall = TP / (TP + FN), assumes particular significance in imbalanced datasets, emphasizing the identification of minority class instances.

The **F1-score** is a compound that is the harmonic mean of precision and recall. The compound measure offers a balanced assessment of classifier performance, most useful in applications that need to reconcile the twinned tension between precision and recall. The larger the F1-score, the better they classified, because this metric makes the classifier classify correctly the positive instances while avoiding false negatives.

These carefully defined metrics also represent essential analytical instruments in the world of machine learning, allowing practitioners and researchers to perform detailed analysis of classification model evaluation. Measuring precision, accuracy, recall, and F1-score helps the evaluation, due to which the researcher gets everything about their classification problem.

Matthews correlation coefficient (MCC) is a powerful metric to validate binary classification, considering four quadrants of the confusion matrix, i.e., TP, TN, FP, and FN. MCC is provided with special advantages in the case of class inequality. Concerns regarding MCC may be framed like this.

$$MCC = \frac{((TP \times TN) - (FP \times FN))}{\left(\sqrt{(TP + FP)(TP + FN)(TN + FP)(TN + FN)}\right)}$$

The MCC ranges from −1 to +1, where +1 is a perfect model, 0 is a random model, and −1 is inverse prediction, and is thus a common tool for assessing model performance [17].

13.6 Results and Discussions

In this section, you will find the results of training the machine learning models on the breast cancer dataset, followed by their respective efficiency evaluation. There are various metrics for model evaluation, such as accuracy, precision, recall, F1-score, confusion matrices, etc. It learns to rank the classifiers to identify their strengths and weaknesses. The results show priceless information regarding both the prediction power of the models and their application in breast cancer diagnosis.

This study aimed to develop a classification algorithm based on machine learning that can accurately classify breast cancer patients into benign and malignant categories. However, machine learning models are excellent in the early detection of cancer as they identify which patients are at increased risk for developing the cancer and who may benefit from targeted treatment. This was confirmed using eight classifiers, each trained and tested against synthetic breast cancer patient data. Cross-validation was used for model evaluation to confirm the reliability and generalizability of the models. The method employed was five-fold cross-validation, splitting the data into five similar segments. The training and testing were performed over five cycles, where four sections were used in training and one section was set aside for testing. This is important in the medical data analysis because it prevents overfitting and provides a more accurate estimate of model performance, particularly with high-dimensional and imbalanced class data.

Cross-validation seemed to provide better consistency in model performance between folds. With an average accuracy of 95%, the artificial neural network (ANN) was the best-performing model, but XGBoost had consistent performance around 94% accuracy and 91% recall in all the folds. The results displayed the power behind both models, verifying their practicability in actual breast cancer detection. The stable performance seen across the cross-validation iterations suggests a smaller chance of overfitting in these models, and they can be considered more trustworthy for practical purposes.

The workflow diagram in Figure 13.4 shows how machine learning models are tested on the Breast Cancer Dataset and how their performance is compared. The dataset is first loaded, and then preprocessed to clean the data and normalize it. Feature selection is then performed to identify the most critical attributes with a strong influence on model prediction. The dataset is then divided into k subsets, where k = 5, for a k-fold cross-validation procedure.

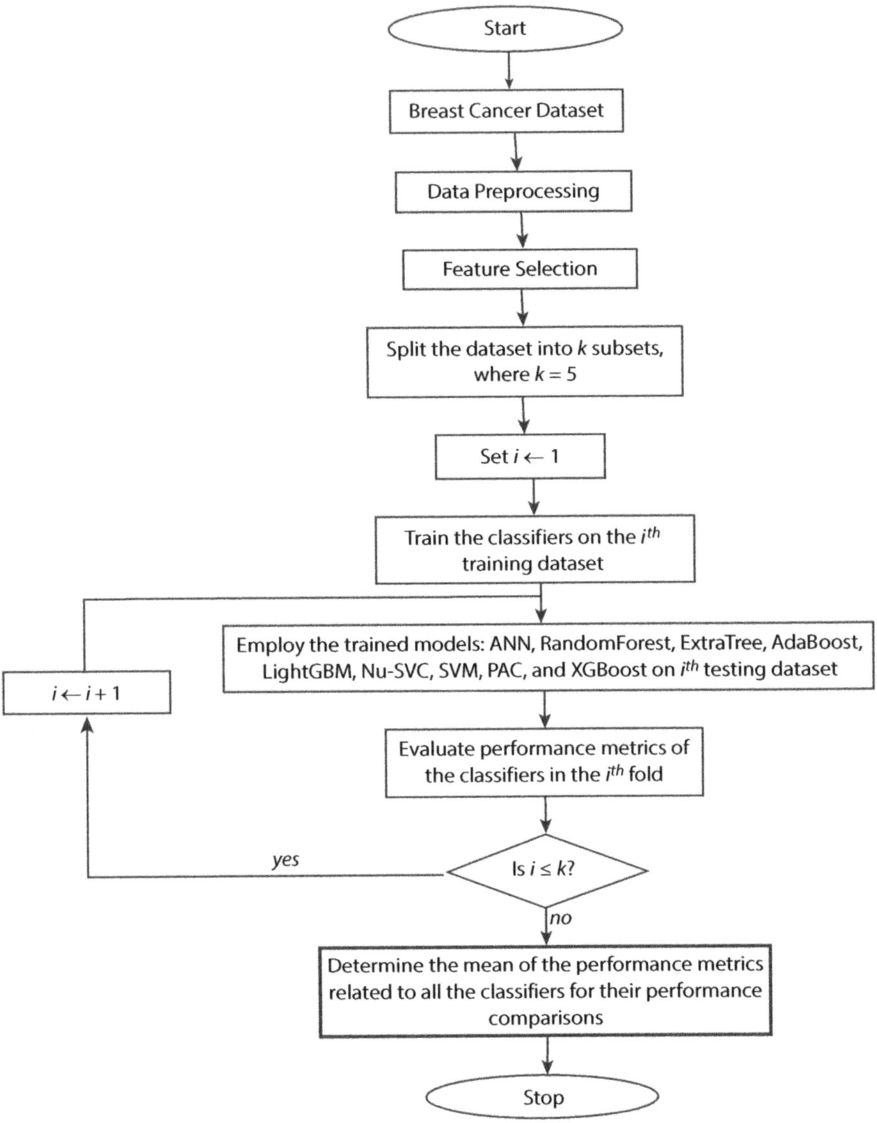

Figure 13.4 Workflow diagram for breast cancer classification using k-fold cross-validation and multiple machine learning models.

The workflow starts by setting $i = 1$, indicating the first fold. For each iteration, the classifiers are trained using the data from the ith training subset. These trained models, including artificial neural networks (ANNs), random forest, Extra Trees, AdaBoost, LightGBM, Nu-SVC, SVM, passive-aggressive classifier (PAC), and XGBoost, are then applied to predict outcomes on the ith testing subset. The performance metrics of these classifiers, such as accuracy, precision, recall, and F1-score, are evaluated for the current fold.

The process iterates by incrementing i until all k folds are processed. After completing the cross-validation, the mean of the performance metrics across all folds is calculated to ensure a fair comparison of the classifiers' performance. This approach facilitates the identification of the most effective model for classifying breast cancer data.

In this study, relevant features were extracted based on correlation analysis and are presented in Table 13.1. These features include statistical measures such as mean, standard error, and maximum values for various attributes, including radius, symmetry, concavity, and fractal dimension, which are crucial for the classification tasks.

Table 13.1 Extracted features and their descriptions based on correlation analysis for classification tasks.

Feature name	Description
Radius mean	Mean value of distances from the center to points on the perimeter
Symmetry mean	Mean value of symmetry of the cells
Fractal dimension mean	Mean value of the "coastline approximation" - 1.
Radius	Standard error of the radius
Smoothness	Standard error of local variation in radius lengths
Concavity	Standard error of the severity of concave portions of the contour
Concave points	Standard error of the number of concave portions of the contour
Symmetry	Standard error of symmetry of the cells
Fractal dimension	Standard error of fractal dimension
Symmetry worst	Maximum value of symmetry of the cells
Fractal dimension worst	Maximum value of fractal dimension

This study evaluated nine machine learning models, with hyperparameters adjusted as detailed in Table 13.2. Essential layouts comprise the artificial neural network (ANN) set to $hidden\_layer\_sizes = (50,50)$ and $max\_iter = 1000,$, the XGBoost Classifier utilizing n_estimators=100, and the random forest classifier also configured with $n\_estimators = 100$. Further scenarios, including AdaBoost, LightGBM, and NuSVC, underwent optimization with their specific hyperparameter configurations. Table 13.3 reports the performance measures of all the classifiers. Here, it can be observed that SVM becomes superior in terms of accuracy, recall, F1-score, and MCC. In contrast, Nu-SVC emerges as the better model in terms of precision. Subsequently, the graphical representation of the data reported in Table 13.3 is presented in Figure 13.5.

Table 13.2 Hyperparameters of the machine learning models.

Classifier	Hyperparameter values
Artificial neural network (ANN)	$hidden\_layer\_sizes = (50,50)$, $max\_iter = 1000$, $random\_state = 42$
XGBoost classifier	$n\_estimators = 100$, $learning\_rate = 0.3(default)$, $random\_state = 42$
Extra trees classifier	$n\_estimators = 100$, $random\_state = 42$
AdaBoost classifier	$n\_estimators = 50$, $learning\_rate = 1.0(default)$
LightGBM classifier	$num\_leaves = 31$, $learning\_rate = 0.1$, $n\_estimators = 100$
Random forest classifier	$n\_estimators = 100$, $random\_state = 42$
Support vector machine (SVM)	$kernel =' rbf$, $C = 1.0$, $gamma = 'auto'$
Passive-aggressive classifier (PAC)	$C = 1.0(default)$, $random\_state = 42$
Nu-support vector classifier (Nu-SVC)	$nu = 0.5$, $kernel = 'rbf'$

Table 13.3 Performance metrics of all the classifiers.

Sl. no.	Classifier	Accuracy	Precision	Recall	F1-score	MCC
1	ANN	0.955	0.945	0.938	0.940	0.906
2	XGBoost	0.939	0.920	0.919	0.919	0.871
3	Extra tree	0.939	0.936	0.922	0.916	0.870
4	Adaboost	0.932	0.914	0.904	0.908	0.855
5	LightGBM	0.928	0.958	0.881	0.909	0.863
6	Random forest	0.928	0.921	0.885	0.901	0.847
7	Nu-SVC	0.910	**0.987**	0.791	0.871	0.832
8	Passive-aggressive	0.762	0.863	0.684	0.693	0.526
9	SVM	**0.961**	0.971	**0.964**	**0.965**	**0.932**

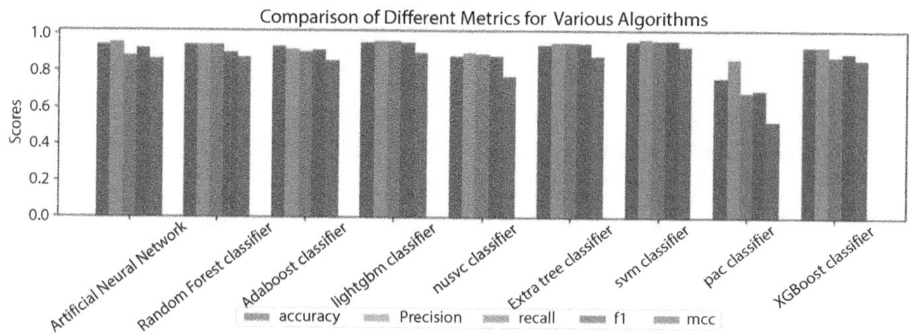

Figure 13.5 Graphical illustration of the performance metrics of the classifiers.

The distribution of the data-points of each feature (cf. Table 13.1) with respect to the target vector having malignant and benign are graphically illustrated in Figure 13.6.

The accuracy performance of various classifiers in 5-fold cross-validation is illustrated in Figure 13.7. Extra Trees, XGBoost, AdaBoost, and LightGBM demonstrate the most consistent performance, achieving average accuracy values exceeding 92% with minimal variation across the folds. Random forest consistently surpasses these leading classifiers across

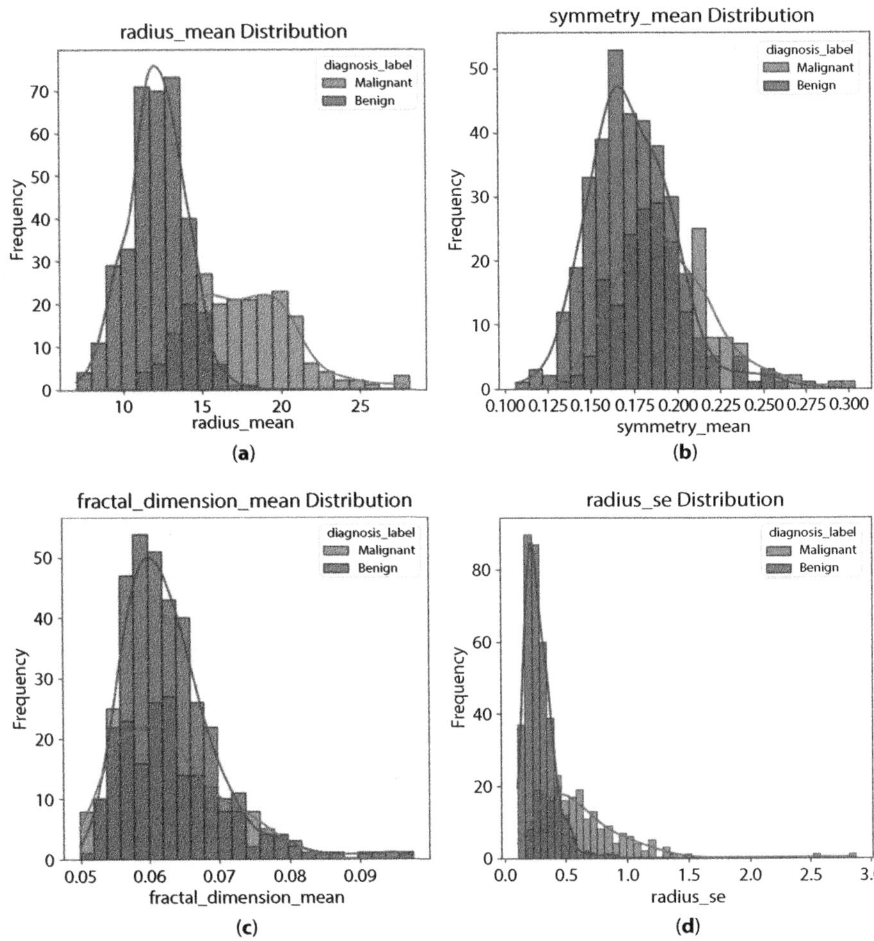

Figure 13.6 Distribution of each feature of the dataset with respect to malignant (M) and benign (B). *(Continued)*

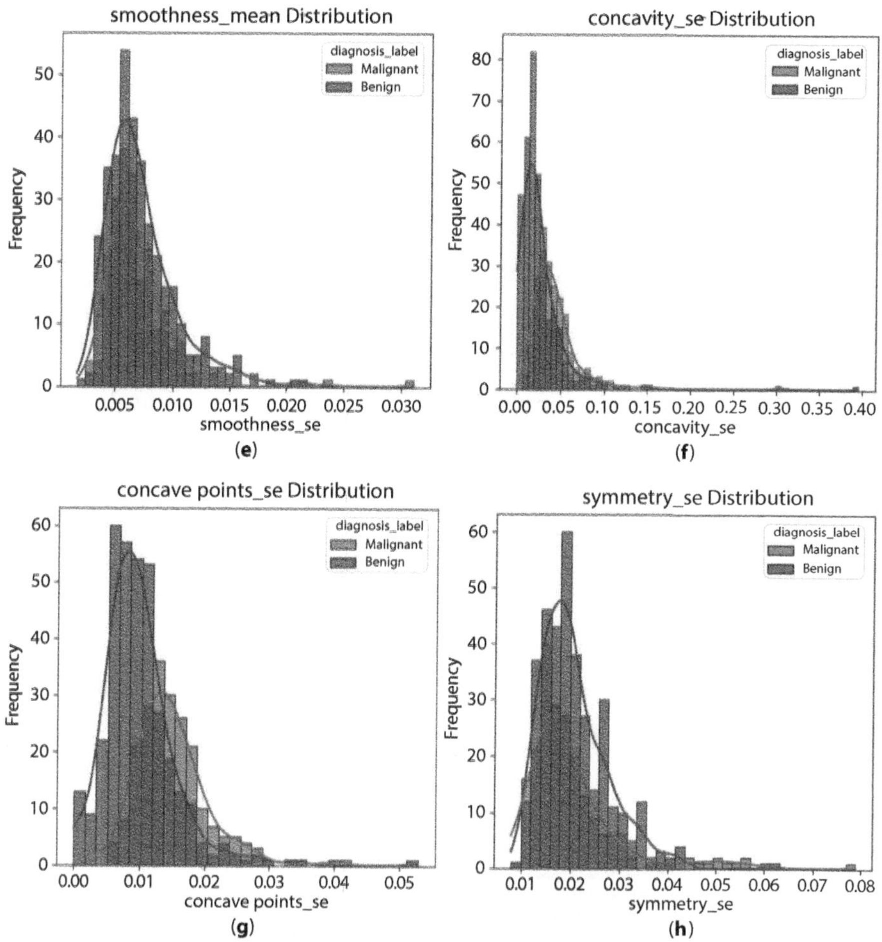

Figure 13.6 (Continued) Distribution of each feature of the dataset with respect to malignant (M) and benign (B). *(Continued)*

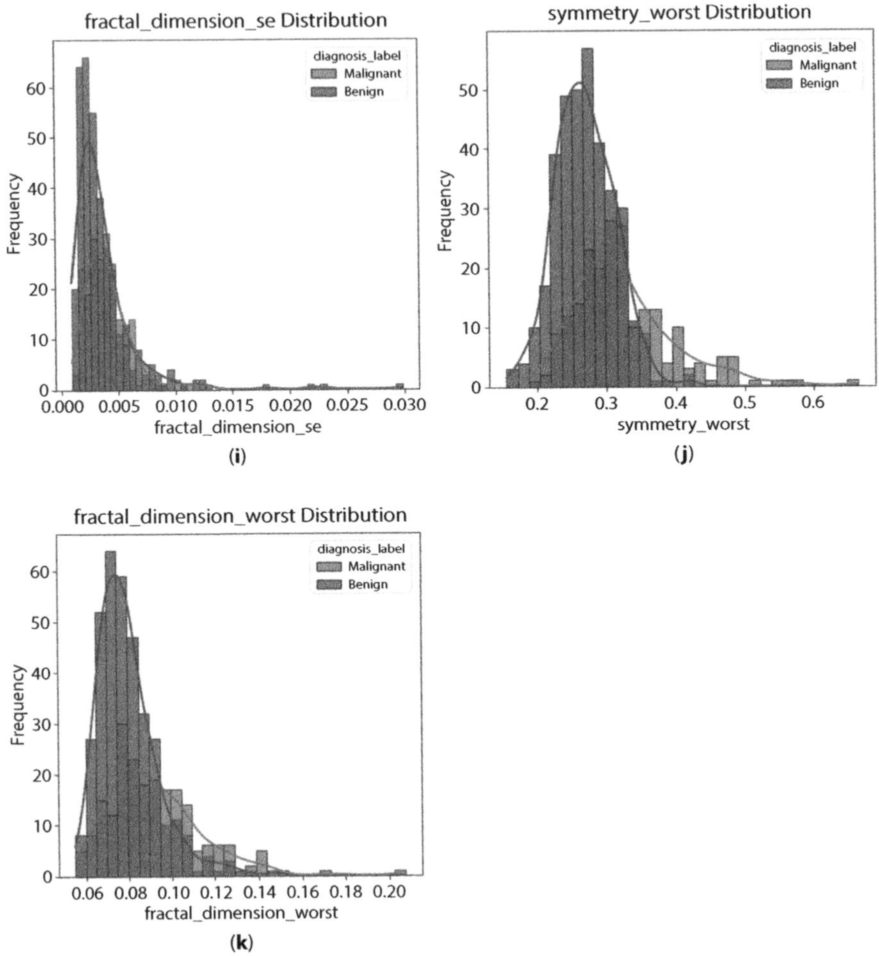

Figure 13.6 (Continued) Distribution of each feature of the dataset with respect to malignant (M) and benign (B).

all folds; however, its average accuracy is notably lower at 92.27%. SVM achieves an accuracy of 88.57%, slightly outperforming others, yet remains inferior to the leading classifiers; NuSVC exhibits a lower average accuracy of 85.41% and shows limited stability. PAC exhibits the lowest stability among classifiers, evidenced by a mean accuracy of 71.54% and significant variability across folds. Despite its significant unpredictability, the neural network (ANN) achieves an accuracy of 90.37%; however, it does not match the performance of the leading classifiers. Ensemble methods such as LightGBM, XGBoost, AdaBoost, and Extra Trees demonstrate superior

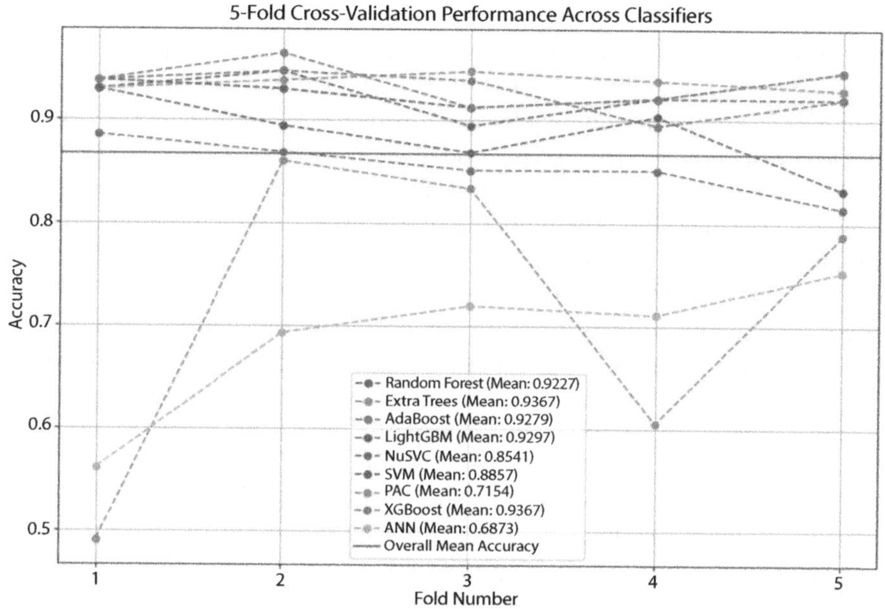

Figure 13.7 Fivefold cross-validation accuracy of classifiers - accuracy analysis.

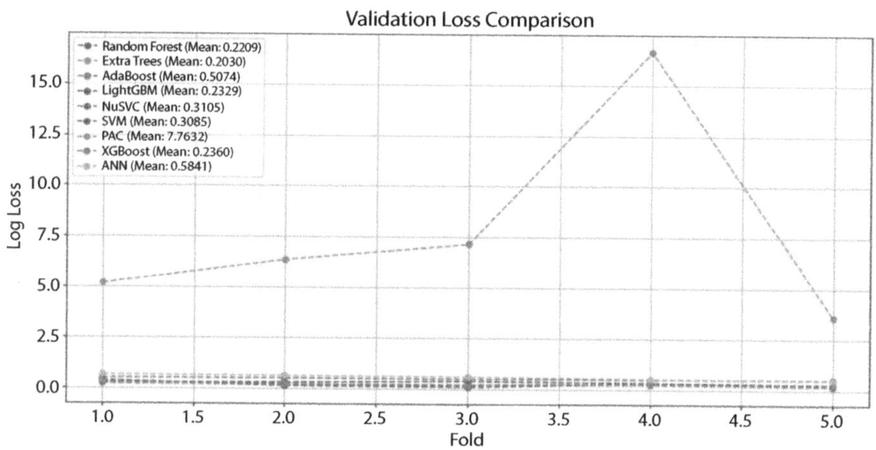

Figure 13.8 Fivefold cross-validation accuracy of classifiers -loss analysis.

stability and accuracy. Meanwhile, other classifiers such as PAC and SVM [7] exhibit low accuracy and are less useful for tasks with high precision.

Figure 13.8 presents a comparison of the validation log-loss for different models evaluated through 5-fold cross-validation. Among the models, Extra Trees exhibits the lowest mean log-loss at 0.2030, followed by random forest at 0.2209 and XGBoost at 0.2360, suggesting their effectiveness in reducing prediction errors. LightGBM demonstrates effective performance, achieving a mean log-loss of 0.2329. In comparison, NuSVC and SVM exhibit marginally elevated mean log-loss values of 0.3105 and 0.3085, respectively. The ANN model demonstrates a mean log-loss of 0.5841, whereas AdaBoost shows a lower log-loss of 0.5074. The passive-aggressive classifier (PAC) exhibits the poorest performance, evidenced by a notably high mean log-loss of 7.7632, rendering it the least effective model in this analysis. The results indicate that Extra Trees and random forest are the most effective models for reducing log-loss in this dataset.

The training and testing times for various classifiers are shown in Table 13.4. Analyzing the data, we observed that classifiers such as the passive-aggressive classifier (PAC), support vector machine (SVM), and Nu-support vector classifier (NuSVC) have significantly shorter training times compared to other methods. Specifically, PAC has the shortest training time (0.0089 s), followed by SVM (0.0547 s) and Nu-SVC (0.1102 s).

A voting classifier was developed to enhance prediction accuracy, utilizing both hard and soft voting techniques. The hard voting method entails each individual model casting a vote for a specific class, with the ultimate

Table 13.4 Training and testing times of different classifiers when implemented on the dataset.

Sl. no.	Classifier	Train time (s)	Test time (s)
1.	ANN	1.3039	0.1760
2.	XGBoost	0.6313	0.0210
3.	Extra trees	0.4858	0.0528
4.	AdaBoost	0.6417	0.0781
5.	LightGBM	0.2870	0.0142
6.	Random forest	0.4225	0.0313
7.	NuSVC	0.1102	0.0259
8.	PAC	**0.0089**	0.0048
9.	SVM	0.0547	0.0160

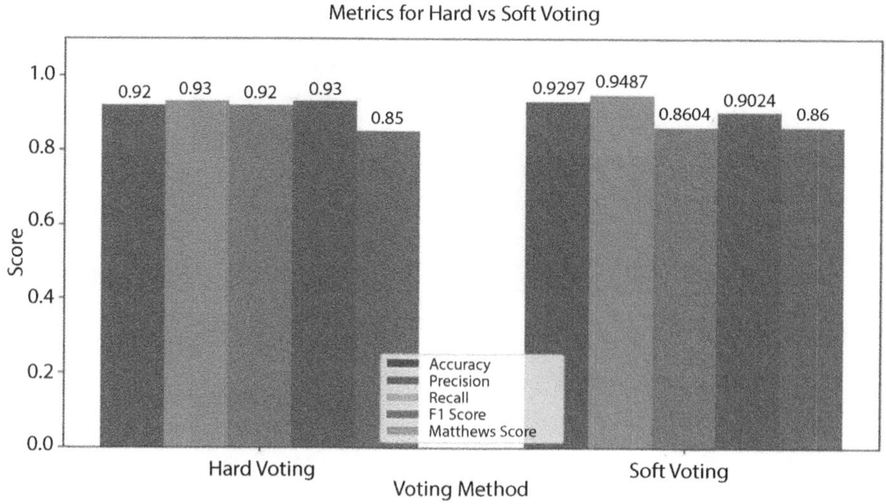

Figure 13.9 Performance of hard voting and soft voting of the voting classifier.

prediction determined by the majority vote. Soft voting differentiates itself from alternative approaches by combining the probability estimates from each model and calculating their average to generate a final prediction. This method has the potential to improve performance through the incorporation of the confidence levels associated with each classifier.

As illustrated in Figure 13.9, the soft voting strategy emerges as a more sophisticated ensemble technique. Its ability to synthesize probabilistic predictions from multiple models suggests a more refined predictive strategy compared to the traditional hard voting method. The incremental improvements in performance metrics, particularly the enhanced recall, underscore the potential of this approach in medical diagnostic contexts.

The research findings highlight the significant promise of ensemble methods, specifically soft voting, in improving the reliability and precision of predictive models for breast cancer diagnosis. By harmonizing the strengths of individual models through a probabilistic framework, this approach represents an advanced computational strategy for medical classification tasks.

Finally, in Figure 13.10, we represent the confusion matrices of all the classifiers, which graphically illustrate the fact that SVMs become superior in four out of five performance metrics, including accuracy, recall, F1-score, and MCC.

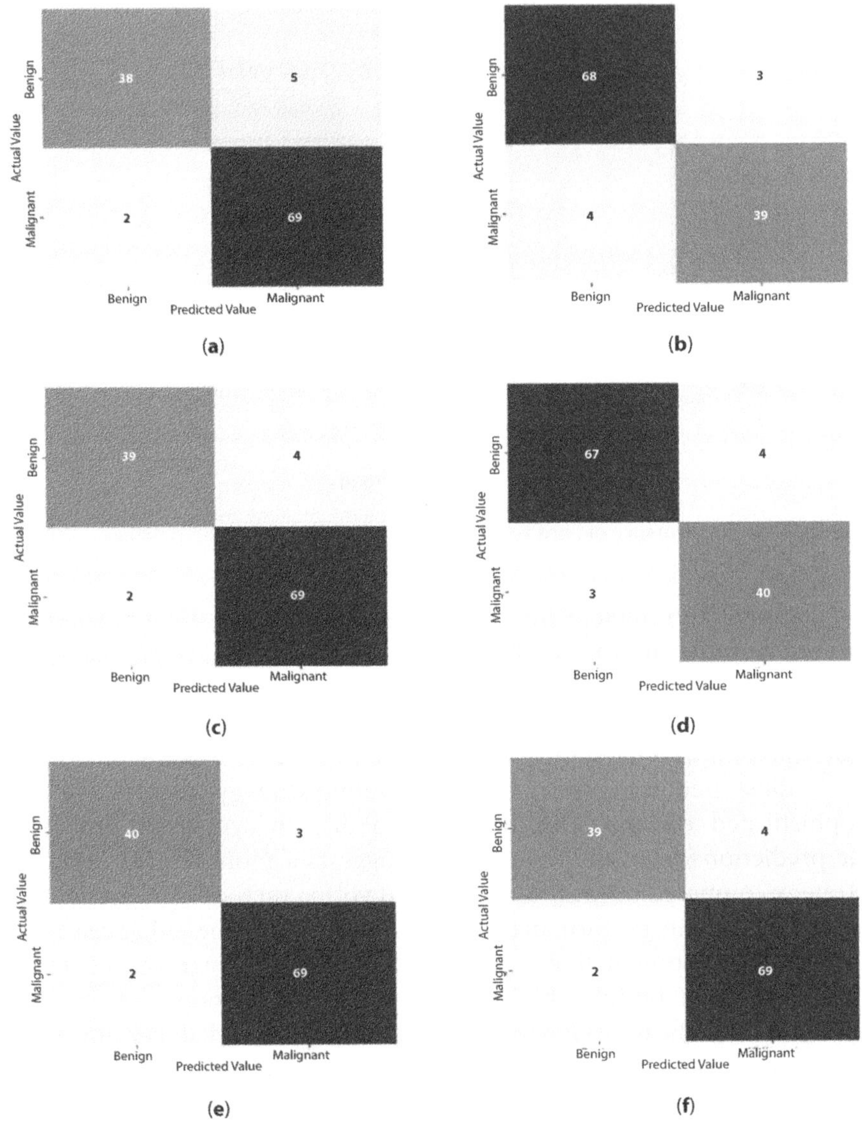

Figure 13.10 Confusion matrices of (a) ANN, (b) XGBoost, (c) Extra tree, (d) Adaboost, (e) LightGBM, (f) random forest, (g) Nu-SVC, (h) passive-aggressive, and (i) SVM. (*Continued*)

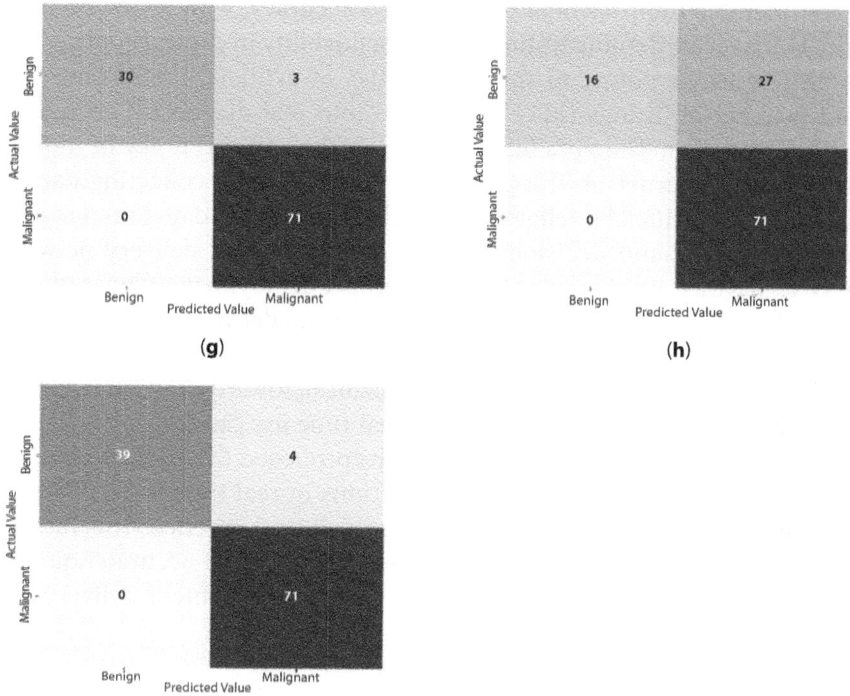

Figure 13.10 (Continued) Confusion matrices of (a) ANN, (b) XGBoost, (c) Extra tree, (d) Adaboost, (e) LightGBM, (f) random forest, (g) Nu-SVC, (h) passive-aggressive, and (i) SVM.

13.7 Conclusion

Using the Breast Cancer Wisconsin Diagnostic dataset, this research investigated the efficacy of a number of different machine learning (ML) classifiers for the identification of breast cancer. In this study, we proved that sophisticated classifiers, such as artificial neural networks (ANNs) and XGBoost, considerably improve diagnostic precision. This was accomplished by applying a comprehensive technique that included data preprocessing, feature selection, model training, hyperparameter tweaking, and cross-validation. Artificial neural networks (ANNs) achieved the greatest accuracy (95%), recall (93%), and Matthews correlation coefficient (91%), but XGBoost came in a close second with good performance metrics. Ensemble approaches, in particular soft voting classifiers, achieved a considerable improvement in prediction accuracy, with 92.9% accuracy and 94.8% recall, demonstrating their dependability in clinical applications.

This research points to the possibilities of artificial intelligence (AI) and Internet-of-Things (IoT) integration into real diagnostic procedures for breast cancer. The possibilities are more than the accuracy of individual models. Internet-of-Things devices are constantly collecting data on patients, and artificial intelligence models are using the data to refine diagnostic potential and decision-making power. Content delivery networks (CDNs) have a critical role to play in the proper deployment of those types of technologies, providing a mechanism to route and deliver the major streams of data generated by systems like artificial intelligence and the Internet of Things quickly and securely. And in the medical, towards low-latency delivery of diagnostic information, enabling real-time use cases.

Future work will focus on validating the proposed ML models on external datasets and developing implementations in real time to facilitate the seamless integration of AI-IoT solutions in clinical practice. This integration can enhance breast cancer diagnostics with more accurate, quicker, and customized healthcare solutions by employing content delivery networks (CDNs) for scalable data delivery.

AI disclosure statement: The author confirms that no artificial intelligence (AI)-generated content or AI-assisted writing tools were used in the preparation of this chapter (including summaries and the abstract). All text, analysis, and conclusions are the result of the author's original work.

References

1. Sadhukhan, S., Upadhyay, N., Chakraborty, P., Breast cancer diagnosis using image processing and machine learning, in: *Emerging Technology in Modelling and Graphics (Advances in Intelligent Systems and Computing)*, J.K. Mandal and D. Bhattacharya (Eds.), pp. 113–127, Springer, Singapore, 2020.
2. Bhushan, A., Gonsalves, A., Menon, J.U., Current state of breast cancer diagnosis, treatment, and theranostics. *Pharmaceutics*, 13, 5, 723, 2021.
3. Ahmad, J., Akram, S., Jaffar, A., Rashid, M., Bhatti, S.M., Breast cancer detection using deep learning: An investigation using the DDSM dataset and a customized AlexNet and support vector machine. *IEEE Access*, 11, 108386–108397, 2023.
4. Guizani, S., Guizani, N., Gharsallaoui, S., A Hybrid CNN-SVM Prediction Approach for Breast Cancer Ultrasound Imaging, in: *2023 International Wireless Communications and Mobile Computing (IWCMC)*, Marrakesh, Morocco, pp. 1574–1578, 2023.

5. Bilal, A., Imran, A., Baig, T.I., *et al.*, Breast cancer diagnosis using support vector machine optimized by improved quantum inspired grey wolf optimization. *Sci. Rep.*, 14, 10714, 2024. (Journal).
6. Sinha, D., Majumder, S., Bandyopadhyay, C., Sharma, H.K., Arithmetic Harris Hawks-Based Effective Battery Charging from Variable Sources and Energy Recovery Through Regenerative Braking in Electric Vehicles, Implying Fractional Order PID Controller. *Fractal and Fract.*, 9, 8, Article 525, 2025. https://doi.org/10.3390/fractalfract9080525.
7. Dai, B., Chen, R.-C., Zhu, S.-Z., Zhang, W.-W., Using random forest algorithm for breast cancer diagnosis, in: *2018 IS3C*, pp. 449–452, 2018.
8. Sharma, D., Kumar, R., Jain, A., Breast cancer prediction based on neural networks and extra tree classifier using feature ensemble learning. *Meas.: Sens.*, 24, 100560, 2022.
9. Schölkopf, B., Smola, A.J., Williamson, R.C., Bartlett, P.L., New support vector algorithms. *Neural Comput.*, 12, 5, 1207–1245, 2000.
10. Kadhim, R., Rania, K., Kamil, M., Comparison of machine learning models for breast cancer diagnosis. *IAES Int. J. Artif. Intell.*, 12, 415–421, 2023.
11. Ghavidel, A. and Pazos, P., Machine learning (ML) techniques to predict breast cancer in imbalanced datasets: A systematic review. *J. Cancer Surviv.*, 2023.
12. Hussain, S., Ali, M., Naseem, U., Nezhadmoghadam, F., Jatoi, M.A., Gulliver, T.A., Tamez-Peña, J.G., Breast cancer risk prediction using machine learning: A systematic review. *Front. Oncol.*, 14, 1343627, 2024.
13. Nassif, A.B., Abu Talib, M., Nasir, Q., Afadar, Y., Elgendy, O., Breast cancer detection using artificial intelligence techniques: A systematic literature review. *Artif. Intell. Med.*, 127, 102276, 2022.
14. Almarri, B., Gupta, G., Kumar, R., *et al.*, The BCPM method: Decoding breast cancer with machine learning. *BMC Med. Imaging*, 24, 248, 2024.
15. Bhattacharya, S., Sinha, D., Bandyopadhyay, C., Majumder, S., Biswas, A., On the Maintenance Oversight of the Healthcare Sector Based on Artificial Intelligence, in: *Decision Making Under Uncertainty Via Optimization, Modelling, and Analysis (Studies in Systems, Decision and Control)* Sahoo, L., Senapati, T., Pal, M., Yager, R.R. (Eds.), vol. 558, Springer, Singapore, 2025. https://doi.org/10.1007/978-981-96-0085-4_21.
16. Bhattacharya, S., Sinha, D., Bandyopadhyay, C., Secured Home Automation with Voice Recognition Using ML and IoT Devices, in: *Proceedings of the IEEE International Conference on Computer Vision and Machine Intelligence (CVMI)*, IEEE, pp. 1–6, Prayagraj, India, 2024. doi: 10.1109/CVMI61877.2024.10782875.
17. Bharadwaj, H.K., Sharma, N., Kumar, P., A Review on the Role of Machine Learning in Enabling IoT Based Healthcare Applications. *IEEE Access*, 9, 38859–38890, 2021. doi: 10.1109/ACCESS.2021.3059858.

14

Rethinking Cybersocialization in Posthumanist: An Evolution of University Curriculum Based on AIIoT Automation

Ahmad Al Yakin[1]*, Ali Said Al Matari[2], Abd. Ghofur[3], Muthmainnah Muthmainnah[1], Ahmed J. Obaid[4] and Souvik Ganguli[5]

[1]Training and Education, Polewali Mandar, Sulawesi Barat, Indonesia
[2]Center for Language and Foundations Studies, A'Sharqiyah University, Ibra, Oman
[3]Faculty of Computer Science and Mathematics, University of Kufa, Najaf, Iraq
[4]Education Department Pamekasan, State Islamic University of Madura, East of Java, Indonesia
[5]Department: Electrical and Instrumentation Engineering, Thapar Institute of Engineering and Technology, Patiala, India

Abstract

This study explores how AIIoT-enabled networked learning environments foster creativity and engagement in undergraduate students' content creation and monetization of artifacts. Cybersocialization and the commercialization of learning outcomes in Indonesian schools have received little attention. We fully understand its impact on students' motivation, academic achievement, and the monetization of educational artifacts on AIIoT-driven platforms and YouTube. This study conducted a survey of 23 students at Al Asyariah Mandar University using a quantitative method. We tasked participants with creating and sharing educational content on social media after their intervention by AIIoT. To evaluate the extent to which students participated in cybersocialization and its impact on academic performance, a questionnaire was used and analyzed using SPSS. This study found that students have a thorough understanding of how AI technologies like ChatGPT can improve learning outcomes (mean = 3.96), and they also have a thorough understanding of how to use these tools (mean = 4.48). The researchers found that commercializing educational content on sites such as YouTube

**Corresponding author*: ahmadalyakin76@gmail.com

Tanusree Chatterjee, Rajib Banerjee and Souvik Ganguli (eds.) Content Delivery Networks Using Artificial Intelligence and Internet of Things: Revolutionizing Content Delivery, (307–340) © 2026 Scrivener Publishing LLC

increases enthusiasm and creativity (mean = 4.35). As significant as the research is in exploring the interplay between technology, education, and commercialization, this study contributes to the expanding body of research on artificial intelligence and cybersocialization in the classroom.

Keywords: Artificial intelligence, Internet of Things, cybersocialization, curriculum, education

14.1 Introduction

In this digital era, teachers and students share learning artifacts, such as projects, instructional videos, and artwork, on social media platforms such as YouTube, Instagram, and TikTok [43]. One example of this trend is cybersocialization, which is problematic for students and teachers in Indonesia. The integration of AIIoT technology into the content production process is driving this change, part of a broader trend in educational distribution. This technology facilitates the creation, development, and monetization of educational content more easily. A new paradigm in student and educator engagement with technology and the dissemination of knowledge has emerged through cybersocialization [20]. This movement sees technology as essential to modern human contact and learning. A recent phenomenon has seen an increase in the use of online portfolios by students and teachers in Indonesia. Students who want to attract the attention of their classmates and potential audiences are more likely to make their learning projects and films publicly available, according to a new study. The visibility and monetization options it provides, educational content is gaining more importance on platforms like YouTube, Facebook, Instagram, and TikTok. Following the posthumanist trend, students can now upload their original work to a wide-reaching online community. In this view, the boundaries between human agency and technological interaction are increasingly blurred [4].

Learning outcome projects, such as instructional films, interactive classes, and multimedia presentations, are driving the increasing use of AI in Indonesian education. Content design platforms and video creation software, such as CapCut or Canva, are examples of AI technologies that educators and students can use to speed up the production process [38]. These technologies simplify the creation of high-quality educational materials for platforms like YouTube, eliminating the need for substantial technical knowledge. AI improve the accessibility and quality of student work by automating content production procedures, including video editing, subtitling, and dubbing.

The evolution of the curriculum has had an impact on socialization in the form of creating instructional videos or tutorials for sites like YouTube, which is the first step in making the transition from documentation to monetization. Students can use these platforms to track their progress in class and even make money from them through advertising, sponsorships, or premium memberships. Students, for example, can create dedicated channels for the purpose of presenting course content, showcasing class projects, or offering virtual tutorials; they can then monetize these channels depending on the number of subscribers and engagement they receive. Students are encouraged to see their learning outcomes as digital products that can be sold, not just as academic exercises.

As cybersocialization and conventional teaching methods are reimagined, the world of education in Indonesia is undergoing a dramatic upheaval. Digital and interactive lesson plans are born from the synergy between AI technology and human creativity, fostered by the integration of the AIIoT network into classroom teaching. Today, students are not just receiving information; they are actively participating in the creation, distribution, innovation, and commercialization of new knowledge that characterizes Education 6.0 [25] and the larger movement that encompasses this revolution; this movement emphasizes real-world application of knowledge and personalized learning to help students adapt to the technologically changing Society 5.0 [35].

The purpose of this study is to investigate students' ability to adapt to AIIoT used by educators and students to improve educational outcomes. Cybersocialization in the classroom is a reflection of the changes in how people interact with each other, where AIIoT technology plays an increasingly important role. Students demonstrate their learning in innovative and entrepreneurial ways using technology, while teachers leverage AI to coach students to design their learning artifacts by personalizing feedback and creating multimedia classes. Students gain digital literacy and marketable skills for a connected and globalized world through these adaptive educational procedures, highlighting the symbiosis of humans and technology at the heart of modern education. This study of cybersocialization in Indonesia, as examined by Al Yakin *et al.* [5], highlights the significant impact of AI on changing social practices around education and content production. A new territory has opened up at the intersection of education, technology, and commerce, with students increasingly uploading learning materials and selling their educational content through platforms such as YouTube. This trend challenges the limitations of established paradigms of teaching and learning.

14.2 Literature Review

14.2.1 Cybersocialization and AIIoT in the Curriculum

By advocating for the use of technology in education, the perspective of Serrano and Mårell [56] challenges conventional wisdom about human-centered learning. Based on this perspective, it illustrates digital platforms, artificial intelligence, the power of IoT, and human teachers working together to influence classroom dynamics. By examining how technology facilitates learning, creation, and social interaction in online environments, posthumanism reimagines the roles of educators and students in the context of a networked ecosystem. Below, we outline key ideas for understanding the view [52] of learning outcomes under this paradigm. Students and teachers have a symbiotic relationship with technology, from Saritepeci and Yildiz Durak [55] and Osado *et al.* [47].

According to Manoharan *et al.* [40], the modern educational network ecosystem that encourages collaboration and audience engagement in the classroom includes teachers, students, technology, and flexible digital platforms. According to Muthmainnah *et al.* [44], students are involved in creating and sharing multimedia projects through social media sites such as YouTube, Instagram, and Pinterest platforms. Students are not only connected to a larger community but also connected to where learning occurs through technological interactions and audiences around the world, and they also provide evidence of flexible learning using these methods. The term "cybersocialization" refers to the way people behave and engage with each other in virtual communities where technology acts as a mediator of learning. Al Yakin *et al.* [5] view cybersocialization as an inherent part of the educational environment. Both students and educators can use social media to share projects, learn from each other, and join a worldwide digital knowledge network. For instance, posting a video of an environmental science group project on YouTube and receiving likes, comments, and shares impacts students' social engagement and learning outcomes.

Michalcewicz *et al.* [42] stated that in a learning ecosystem, both human and nonhuman agents share the concept of agency; when students, teachers, and AI technologies collaborate, the result is learning outcomes. The technologies students use, such as data analytics, AI video editing, and design software, become co-creators with students, although teachers continue to provide input into content development. A perspective from Georgara *et al.* [24] views both creators and tools as equal partners in the pursuit of knowledge. This perspective highlights how networked ecologies transform project-based learning.

Today, students create content for a wider online audience, not just their instructors or classmates. Students participate in cybersocialization as they share their creations online and connect with viewers and commentators on sites like YouTube, TikTok, and educational social networks. They gain comfort with technology and develop critical thinking skills, considering the potential audience for their creations [54]. A group of high school students researching a historical figure, for example, might use AI to create a documentary and then share it on YouTube, where they can get feedback from people around the world. They incorporate this worldwide feedback loop into their learning process, which in turn influences the network's dynamic learning outcomes. Posthumanism also offers a fresh perspective on the monetization of learning outcomes. In addition to gaining new knowledge, students can make money when they share their learning projects (films, artwork, presentations) on sites like YouTube. Students gain financial literacy and marketable skills by integrating monetization into the learning process, which can take the form of advertising revenue, subscriptions, or sponsored posts [46, 57]. AI and digital tools enable students to explore the practical application of their ideas, thereby shaping their future careers and educational approaches.

The emerging digital era certainly requires a person's level of "digital literacy" that demonstrates their ability to use digital tools for research, evaluation, creation, and communication [6]. This includes media literacy, critical thinking, information literacy, and technical skills. Today, being able to navigate the digital landscape is not just a nice-to-have but a necessity. Technology permeates our daily lives, impacting everything from our professional and political lives to our educational experiences and even our health. With the rise of digital literacy, cybersocialization has become ubiquitous, impacting every aspect of our daily lives. A person's "cybersocialization" can be defined as their engagement in social contact, communication, and connection building using digital platforms such as social media, online forums, and the like.

The ability to communicate effectively, obtain information, and engage in global interactions through digital means, as well as engage in online social interactions safely and securely, has become a critical skill for success in the modern era. 'Digital literacy' refers to the ability to effectively use contemporary technologies and critically evaluate information obtained from these sources [22]. Competencies encompass a range of abilities, such as proficiently searching and evaluating online information, being aware of internet safety and security measures, effectively using digital communication tools for collaboration and communication, and creating and critically analyzing digital content [1]. Having a strong understanding of digital literacy in the modern era allows one to access and understand the vast amount of information available online, as well as participate in the digital economy [45].

Having a thorough understanding of digital literacy leads to better outcomes in the classroom and workplace. Enhanced employability is another benefit of digital literacy, as many modern jobs require proficiency with digital tools and techniques [58]. To keep up with the rapid evolution of the information and technology environment, it is essential to have a strong understanding of digital literacy, which is a prerequisite for continuing education. Cybersocialization includes not only digital literacy but also the following: learning and practicing good online etiquette, being aware of how one's online activities affect others, and developing ways to engage positively with others online [68]. Responsible cybersocialization is essential in combating cyberbullying, harassment, and other harmful online behaviors [36]. People who are well-connected socially and mentally tend to be more friendly, kind, and generous when interacting with others online, as shown in research [71]. Workplaces expect employees to use digital communication technologies professionally and appropriately, making cybersocialization important. Al Yakin *et al.* [5] found that as a result of engaging in cybersocialization, people were more likely to practice digital citizenship, which includes protecting their own privacy and security, considering the rights of others, and making positive contributions to online communities [27].

The emerging digital era certainly requires a person's level of "digital literacy" that demonstrates their ability to use digital tools for research, evaluation, creation, and communication. This includes media literacy, critical thinking, information literacy, and technical skills. Today, being able to navigate the digital landscape is not just a nice-to-have but a necessity. Technology permeates our daily lives, impacting everything from our professional and political lives to our educational experiences and even our health [19]. With the rise of digital literacy, cybersocialization has become ubiquitous, impacting every aspect of our daily lives. A person's "cybersocialization" can be defined as their engagement in social contact, communication, and connection building using digital platforms such as social media, online forums, and the like. In the modern digital age, cybersocialization has transformed human interaction, collaboration, and socialization, bringing with it both advantages and disadvantages.

Cybersocialization, or socializing through digital means, is a fast and revolutionary phenomenon that is changing the way people connect and interact. In addition to, or perhaps as a replacement for, more conventional means of social interaction—such as in-person meetings, online communities, messaging applications, and social media—have become increasingly commonplace in modern society [4, 70]. One of the first examples of digital socialization is the emergence of online bulletin boards and forums in the early 1990s. These platforms' users may have text-based conversations about topics relating to their interests. As internet use has grown, more complex forms of digital socialization have emerged, such as instant messaging, blogging, and social networking sites like Facebook and MySpace. Sites such as TikTok, Instagram, and Twitter attract billions of users every day, making social media a typical way for individuals to connect online [70].

14.2.2 Expanded Learning Outcomes with the AIIoT Network Ecosystem

Learning outcomes, in the view of Alharbi [7], go beyond conventionally measured cognitive capacities. Among these skills is the capacity to work within networks, worked ecosystems, created content, and made money off of it. In a globally networked world, these results also take into account the social and ethical consequences of making and sharing digital artifacts. Privacy, copyright, and cybersecurity are just a few of the digital citizenship problems that students learn about as they grow into responsible online content creators. Networked learning allows students to take their feedback loops outside of the traditional classroom setting. According to Magalhães *et al.* [39], data analytics (such as user engagement, video completion rates, audience demographics, etc.) can be used by AI-driven tools to instantly assess student material and provide tailored feedback. Teachers and students can use this AI-assisted feedback to fine-tune their projects using real-world interaction metrics, which improves both the content and the understanding and assessment of learning.

Online communication technologies, such as social media, instant messaging, and online forums, are crucial points of convergence for digital literacy and cybersocialization. Because these systems allow people to connect, communicate, and share information in virtual places, they are crucial to modern social interactions [28]. To participate in cybersocialization, people must have digital literacy skills, which include the ability to understand and effectively use these systems. The most important thing that everyone needs to know is how to develop and manage their social media presence, understand and respond to communications sent online, and determine the reliability of materials found online. Lack of proficiency in virtual literacy can negatively impact one's ability to socialize online, form meaningful relationships, and contribute to online organizations for leadership. Digital literacy and cybersocialization are concepts that touch on how people use and understand the internet to access and process information from AIIoT. It is essential to have excellent digital literacy skills to evaluate and confirm the accuracy of online statistics in this era of fake news and misinformation. Cyber socialization exposes people to a variety of information, ideas, and perspectives, which can influence their information processing, including education. Having digital literacy competencies, including media literacy and statistical literacy, allows people to make informed decisions in the digital age, evaluate the reliability of sources, and detect bias.

Cybersocialization and digital literacy have complex interactions in modern times, influencing people's practical communication styles, Internet of Things connections, and interactions with younger generations in online spaces [30]. The ability to navigate online communication systems, express oneself digitally, and compare online data are all essential components of cybersocialization, which requires individuals to have high levels of digital literacy. The intersection of digital literacy and cybersocialization will continue to define people's online behaviors and their impact on society at large as digital technology continues to change and influence how individuals participate in online environments.

An example would be a student using Instagram to share a science experiment video. In addition to the usual likes and comments, they might also gain insights from AI analytics that show them how well their work explains important scientific concepts. Ecosystems for techno-human learning from a [8] point of view, the classroom is best seen as a techno-human ecosystem in which technological elements significantly impact students' attitudes, actions, and knowledge. Learning is dynamic, interactive, and interconnected in adaptive environments where students' learning patterns are considered by AI tools. Redefining the teacher-student interaction, tools like AI-powered editing software or AI tutors become vital in the development of knowledge [62].

Posthumanism promotes thinking about how learning ecosystems may make technology use ethical and sustainable. The cybersocialization process has social, cultural, and environmental consequences that both students and educators should be aware of. This encompasses issues of digital equity, making sure that all students can participate by having access to the platforms and tools they need, and making sure that their data and content are used ethically in networked ecosystems. A study by Söğüt [59] on the view of learning outcomes in networked ecosystems advocates for a collaborative, techno-human approach in which technology is seen as more than simply an instrument but as an integral component of the educational journey. Cybersocialization is a key factor in defining contemporary educational experiences, and both students and teachers flourish in these online settings.

14.2.3 Revolutionized Artificial Intelligence and the Internet of Things in Education

Pattanaik *et al.* [49] stated that the integration of AI and IoT into education is driving a revolution in traditional teaching and learning methodologies, which in turn has given rise to the Education 4.0 and Education 6.0 frameworks. Personalized learning experiences driven by technological acceleration are the goals of these frameworks, which leverage AI and IoT to make education more accessible and of higher quality. In this study, we examined the basic ideas and theories of integrating AI and IoT into educational environments in the era of cybersocialization. The goal of artificial intelligence (AI) researchers is to create computer systems that can learn, reason, and solve problems in ways that humans can.

According to Štuikys and Burbait [64], AI/IoT is becoming increasingly important in the modern education sector, with examples such as intelligent tutoring systems and personalized learning platforms. AI can automate mundane administrative tasks, simplify procedures, enhance personalized learning plans, and unleash students' imaginations. AI-powered educational platforms, such as adaptive learning technologies and AIIoT intelligent tutoring systems, can provide meaningful learning experiences.

These programs examine information about students' learning habits, progress, and preferences to personalize lessons and feedback. Since each student receives individual feedback and evaluation, it enhances the quality of their educational experience [14]. For example, according to Teng [65], students can improve their language skills, creativity, and critical thinking with the help of AI-based writing aids and ChatGPT, which offer real-time feedback. Regarding the application of AI in the classroom, Howard Gardner's Theory of Multiple Intelligences is instructive. Students, according to this hypothesis, have multiple intelligences (e.g., linguistic, logical-mathematical, spatial), and AI tools can help accommodate these unique learning styles [18]. With the help of AI-powered platforms, teachers can create lessons that suit students' strengths by adapting the types of multimodal content (videos, simulations, interactive tests, etc.) they provide.

AI aligns with constructivist methods, which emphasize the need for active learning through interaction and problem-solving. With the help of AI-powered simulations and virtual environments, students can engage in constructivist learning practices such as experimenting and solving problems in real time. In addition to promoting social learning, AI can help students work collaboratively in groups by matching them with others who share similar interests and academic goals based on computational intelligence [63]. The system of networked devices that can collect and exchange data is called the Internet of Things (IoT). The Internet of Things (IoT) helps educators design smart learning environments that are data-driven, adaptive, and interactive. Educational procedures are becoming more efficient with the help of Internet of Things (IoT) devices such as smart boards, wearables, and sensors, which enhance classroom instruction and administrative tasks [50].

The Internet of Things (IoT) is paving the way for what is known as the "smart classroom," where students, instructors, and course materials can engage in real-time dialogue and engagement for the advancement of technology [37]. Teachers can see their students' engagement levels and progress in learning on the go with the use of smart gadgets, which also provide immediate feedback on understanding. Educators can devote more time to actually teaching when these technologies handle things like automating classroom lighting, temperature control, and attendance tracking [69]. When it comes to the Internet of Things (IoT) in the classroom, it is important to keep in mind the cognitive load theory [53]. Theoretically, the Internet of Things (IoT) enables the environment to manage this load by accelerating the learning process, which in turn helps students learn better by reducing excessive cognitive load. For example, by automating routine tasks and reducing distractions, smart classrooms free up students' attention to concentrate on the subject matter.

An education ecosystem that combines AI and IoT opens new possibilities for data-driven, personalized, and real-time learning experiences by facilitating a seamless flow of data between students, teachers, and devices. To provide insights that improve educational outcomes, AI algorithms examine data obtained from IoT devices. For example, wearable devices can monitor students' vital signs and activity levels, which AI may then utilize to propose individualized lesson plans and exercises [48]. Educators can monitor their students' progress and pinpoint where they need more help with the use of learning analytics powered by AI and the Internet of Things. Teachers can learn a lot about their students' interests and actions from the data acquired by Internet of Things (IoT) devices such as classroom sensors or learning management systems. AI uses this information to make predictions that help educators help kids who are having difficulties in class [26].

Society 5.0 and Education 6.0, as part of cybersocialization, focus on how AI and the Internet of Things (IoT) may work together to create a more equitable and human-centered society that uses technology to address both economic growth and social issues. Education 6.0 is a student-centered paradigm that utilizes technology to enable personalized learning, global connectivity, and lifetime learning opportunities. It is anticipated that education systems will include AI and IoT in this framework. Here, AI and the Internet of Things help level the playing field by making educational resources equally available to students from all socioeconomic levels.

To explain how individuals accept and utilize new technologies, Asad *et al.* [10] developed the Technology Acceptance Model (TAM), with value and usefulness being the most important factors in technology adoption, according to TAM. The use of AI and the Internet of Things in the classroom will depend on how much teachers and students perceive these technologies as tools to enhance effective lesson delivery and student achievement. As an illustration, when educators anticipate time savings and increased student engagement from AIIoT-enabled tools and access to learning resources, they are more likely to incorporate them into their lessons. Nevertheless, there are several obstacles that can hinder the widespread adoption and efficient use of AI and IoT in classrooms, including a lack of appropriate infrastructure, inadequate digital literacy, and worries about personal data protection. Consequently, it is imperative that researchers in the future investigate the challenges associated with using AI-IoT in various educational settings and propose strategies to overcome them.

Personalized, data-driven, and technology-centered learning is the hallmark of Education 7.0, which is being rapidly advanced thanks to the incorporation of AI and the Internet of Things into classrooms [13]. Learning efficiency, student creativity, and personalized feedback are all improved by AI technologies, while smart learning environments are created by IoT devices. Some important theories that can shed light on how to best use these technologies to enhance educational outcomes are Gardner's multiple intelligences, constructivist learning, and cognitive load theory [66].

In order to understand how teachers and students view and utilize these technologies, the TAM is crucial in cybersocialization. To solve problems with infrastructure, digital literacy, and data privacy, future studies should investigate the long-term effects of AI and the Internet of Things on students' ability to study.

Khan *et al.* [31] identified four functions of AI in the context of student learning as the focus of their review study. AI facilities serve to facilitate communication between humans and machines and encourage greater flexibility and engagement in virtual spaces, which is very much in line with how ChatGPT enhances education. Offering human-machine interaction: For example, with the help of AI chatbots, students can engage in educational discourse with computers. According to several studies [3], these chatbots help students hone their communication skills through ongoing dialogue. ChatGPT can provide students with responses that mimic human speech. Digital environments are becoming more adaptive and interactive, with features such as AI being used to collect data on student learning and make it easier for them to engage with the material. Data collected from students can be used by sophisticated digital learning environments to create learning tasks [41]. To engage students, ChatGPT can use a question-and-answer method, which involves asking students questions sequentially along an evolving inquiry path. To better meet students' needs and encourage them to have a positive learning experience, ChatGPT uses their answers and questions to explain their challenges [34]. Thus, these two functions are integral to the theoretical foundation of this inquiry.

The use of AI in the classroom is the second important area, and it serves three distinct functions [12]. Neither enhancing instructors' capacity to teach nor providing adaptive teaching strategies is a function that ChatGPT can fulfill; however, ChatGPT can assist teachers in their professional growth. Systematic research has found that AI technology is used to support teachers' ongoing professional development activities, which aids their professional development. By analyzing real-time classroom data, AI agents offer teachers feedback and ideas on their teaching methods, such as improving their questioning skills [61]. New teaching ideas, such as instructional design techniques and self-directed learning tasks/activities, can be found in ChatGPT, which can inspire teachers and aid their professional growth. Third, there is AI in assessment, which discusses how AI tools evaluate students' work. Unfortunately, ChatGPT may not be able to support assessment tasks. Nevertheless, ChatGPT can have an innovative impact in this area. This domain is thus retained in the conceptual framework [32].

Artificial intelligence in management is the last area. According to the evaluation, ChatGPT can support three main functions. Platform for improving management performance: Research reviews [21] found that AI has significantly improved the performance of management platforms by automating routine tasks, improving security, and completing labor-intensive work. Executive teams may find ChatGPT useful for (i) brainstorming potential ideas and proposals, and (ii) conducting data analysis and report generation. Providing personalized and easy-to-use services: A review study also found that AI-powered activity recommendation systems can help students find the most appropriate activities for them [67]. By presenting their situation, the management team can ask ChatGPT to suggest actions or decisions. This review study revealed that school management teams can benefit from AI technology by presenting information to support their decision-making process; they leveraged student data to estimate the likelihood of students skipping school and to guide students in choosing the right classes and extracurriculars. When making decisions or offering academic advice, ChatGPT can compile and analyze data from multiple sources, which is then shared with the management team [17].

The research examined components such as academic achievement, motivation and engagement, and 21st-century skills. The use of AI has a beneficial impact on these outcomes [15]. Students who are struggling academically or have low self-esteem report feeling more comfortable and confident after interacting with a robot or chatbot. According to research [33], this leads to significant improvements in students' academic performance. The evaluated research also demonstrates that AI-supported feedback learning systems enhance students' problem-solving abilities, communication skills, ability to learn independently, and creativity [9]. AI-powered chatbots boost students' self-esteem and ease their fears when learning. Furthermore, ChatGPT, when used properly, can improve students' learning outcomes. This is why the conceptual framework includes all four of these educational goals.

There are three parts to the second outcome, namely teacher learning: workplace efficiency, classroom competence, and perspectives on AI. Sok and Heng [60] found that online class administration, student attendance, course registration, and automated grading are all tasks that can minimize instructors' teaching load. The first outcome relates to these. The framework does not incorporate this, and neither does ChatGPT. Furthermore, ChatGPT supports the premise that encouraging instructors to reflect on their own practices and providing them with new ideas and perspectives can improve their teaching ability [16]. In addition, the evaluated studies showed that most teachers found using these new AI tools (such as ChatGPT) to be an intriguing and fun way to educate. Asad *et al.* [11] stated that they were interested in learning more about pedagogy and technology.

14.3 Methodology

In Indonesia, GenAI and AIIoT are still new to educators and institutions. In the education sector, most of the current discussions take place in higher education and focus on policy development and pedagogy. There is a serious lack of relevant studies on what the future of education, especially in schools, looks like. GenAI is an emerging and disruptive technology that presents opportunities for advancing education and strengthening innovative practices, while also posing challenges for schools, teachers, and students. Understanding the perspective of undergraduate students is crucial in understanding any innovative technology, such as AI or ChatGPT. In this study, ChatGPT is used for learning, teaching, assessment, and administration, particularly in the context of the phenomenon of students uploading their learning outcomes on YouTube. Therefore, the objective of this quantitative study is to explore how GenAI, particularly ChatGPT, transforms our educational experience from the viewpoint of undergraduate students at universities. Therefore, the main research question is how do students respond to the improvement of learning outcomes in the network ecosystem of cybersocialization? Does commercialization affect student learning from the documentation of learning outcomes and efforts to monetize these artifacts?

14.3.1 Participants

Subject selection for this study relied on purposeful sampling to gather diverse perspectives from undergraduate students involved in an experiment involving artificial intelligence and education, who varied widely in terms of their academic performance. A total of 23 undergraduate students participated in the study. They ranged in age from 18 to 19. All participants were required to attend a 2-hour sociology course and two workshops on ChatGPT prior to quantitative data collection. Instructors defined artificial intelligence (AI) technologies, including GenAI, discussed their data processing methods, and outlined the social and ethical issues raised by this technology.

During the learning process, participants discussed and demonstrated several forms of promotion using ChatGPT and designed their learning outcomes' content. Not only that, but they also voiced some of their concerns and discussed how they incorporated them into their lessons. The aim of this environment was to verify their familiarity with GenAI, ChatGPT, or any other type of AI, and their ability to utilize it for the creation of educational projects. After six meetings of learning, participants' opinions were collected in these two phases of investigation. The conceptual framework [29] proposes four important domains and two significant learning objectives, and as a start, participants filled out an open-ended survey to share their thoughts on these topics. Students were asked to modify the video project-based learning outcome system and upload it to their social media accounts. In addition, the survey collected their thoughts on the use of ChatGPT and other AI in the classroom and possible motivations for monetization. The survey should take no more than half an hour to complete. The researchers reviewed the survey responses and conducted quantitative analysis.

14.3.2 Data Analysis

In the data analysis section, after data collection was completed, the initial data acquisition exported responses from Google Forms and then imported them into Microsoft Excel. The quantitative survey data were carefully and thoroughly rechecked before being analyzed using the IBM SPSS version 26 application. By doing this step, it was ensured that the data set was ready for accurate statistical analysis and allowed us to closely observe the survey results and make them more reliable.

14.4 Findings and Discussion

The results of this study highlight the need to teach students how to use technology and communicate online effectively to meet the challenges of the modern labor market. Cyber socialization refers to the use of digital devices for meaningful interpersonal interactions, while digital literacy signifies the ability to collect, evaluate, and appropriately utilize digital data.

Incorporating cybersocialization and digital literacy into the curriculum can improve academic achievement, social and emotional development, and critical thinking skills and prepare students for the workforce. This study shows that students who are good at using technology have better grades; therefore, it is important to teach these skills to them, as discussed below.

This section provides a detailed explanation of the statistics. Table 14.1 shows details for 20 statements about what students know about generative AI (GenAI), how they use AIIoT (Artificial Intelligence and the Internet of Things) in the classroom, and how much they participate in cybersocializing. The data came from 23 people who filled out the survey, and the most important metrics were the minimum and maximum scores and the mean and standard deviation for each statement. First, knowledge of GenAI and its educational applications showed a modest level of familiarity. It appears that some students have a thorough understanding of GenAI, while others may not have much experience with it. The results of the second statement score showed that the majority of respondents believed that AI technology can significantly improve their learning outcomes, as indicated by the high score for this statement. This low standard deviation indicated that participants largely agreed. The third statement highlights how the use of AI technology, specifically ChatGPT, enhances learning approaches and assignments, resulting in a positive impact on students' academic work approaches. This statement is supported by a relatively high mean and low standard deviation.

The fourth question, "GenAI Trust," with an average score of 4.48 and a standard deviation of 0.67, demonstrates that students have a high level of confidence in using GenAI products. The standard deviation indicates that everyone in the group has given the same answer. Statement No. 5 revealed that students are motivated to enhance their academic performance through the upload of learning outcomes on sites like YouTube. The effect is not as big as that of the AI device, but the rather high average indicates that students are motivated to share their academic work on public platforms. The sixth statement about the commercialization of learning objectives boosts motivation and creativity, with a high average score from students, suggesting that they become more creative when they have the chance to earn money from their work. Responses vary to a moderate level, according to the standard deviation. In question no. 7, students generally stated that sharing and documenting learning outcomes has a positive effect on academic achievement (average = 2.70, standard deviation = 1.22).

Table 14.1 Descriptive statistics of cybersocialization and curriculum design by AIIoT.

Statements	N	Min	Max	Mean	Std. deviation
1. I am familiar with the concept of Generative AI (GenAI) and its application in education	23	1.00	5.00	3.2609	1.25109
2. I believe GenAI (e.g., ChatGPT) can improve my learning outcomes at university	23	2.00	5.00	3.9565	.92826
3. The use of AI tools such as ChatGPT improves the way I approach learning and assignments	23	2.00	4.00	3.5652	.58977
4. I feel confident using GenAI to help with my learning and assignments	23	3.00	5.00	4.4783	.66535
5. Uploading learning outcomes on platforms such as YouTube motivates me to perform better academically	23	2.00	4.00	3.5652	.58977
6. I believe commercialization of learning outcomes (e.g., monetizing YouTube videos) can increase learning motivation and creativity in designing learning artifact content	23	2.00	5.00	4.3478	.77511
7. The process of documenting and sharing learning outcomes on social media has a positive impact on my academic performance	23	1.00	4.00	2.6957	1.22232

(Continued)

Table 14.1 Descriptive statistics of cybersocialization and curriculum design by AIIoT. (*Continued*)

Statements	N	Min	Max	Mean	Std. deviation
8. I think there is a significant difference between learning outcomes when GenAI is integrated and when it is not integrated	23	1.00	5.00	3.3478	1.02730
9. The use of AI tools for educational purposes should be more widely adopted in schools and universities in Indonesia	23	2.00	5.00	3.0870	.79275
10. Monetization of educational content (such as learning projects) motivates cybersocialization (online interaction and collaboration) and has enhanced my ability to work with my peers on projects	23	1.00	5.00	3.3478	.98205
11. Commercialization of digital learning artifacts can result in better quality work by students	23	2.00	5.00	3.8696	.91970
12. I believe that the lack of relevant studies on GenAI in the Indonesian education sector hinders its implementation in schools	23	2.00	5.00	3.4348	.66237
13. I feel that the commercialization of learning outcomes can sometimes distract students from core educational goals	23	2.00	5.00	3.2609	.81002

(*Continued*)

Table 14.1 Descriptive statistics of cybersocialization and curriculum design by AIIoT. (*Continued*)

Statements	N	Min	Max	Mean	Std. deviation
14. Cybersocialization has helped me improve my digital literacy and understanding of online platforms	23	2.00	5.00	3.6522	.88465
15. I think Indonesian educators and institutions are not yet fully ready to integrate GenAI into teaching and learning	23	2.00	5.00	4.0435	1.02151
16. GenAI tools such as ChatGPT should be part of the standard curriculum in higher education	23	2.00	5.00	3.6522	.93462
17. I am motivated to create content (videos, presentations) that showcases my learning outcomes on public platforms	23	2.00	5.00	3.4783	.79026
18. The possibility of earning money from my learning outcomes increases my engagement with content education	23	3.00	5.00	3.7391	.68870
19. I believe the role of AI in education will continue to grow and have a significant impact on learning outcomes in the future in Indonesia	23	1.00	5.00	3.2174	1.16605

(*Continued*)

Table 14.1 Descriptive statistics of cybersocialization and curriculum design by AIIoT. (*Continued*)

Statements	N	Min	Max	Mean	Std. deviation
20. Uploading learning outcomes on platforms such as YouTube motivates me to improve academically	23	3.00	5.00	4.0000	.73855
Valid N (listwise)	23				

However, a lower average suggests that there is less consensus that sharing and documentation alone can improve academic success. The high standard deviation indicates a wide range of views.

Participants responded in statement no. 8 about how GenAI integration affects learning outcomes. Although their responses varied in intensity, all participants acknowledged that GenAI integration significantly improved learning findings. The subsequent statement reveals that students are increasingly utilizing artificial intelligence tools in the classroom, despite some participants' reluctance. However, the majority believes that the use of AI in Indonesian classrooms should increase. When asked about the importance of monetization in encouraging peer collaboration, the average response (average = 3.35, standard deviation = 0.98) revealed a slightly different perspective. Statement no. 11 (average = 3.87, standard deviation = 0.92) indicates that students believe that selling digital learning artifacts will enhance the quality of their work.

The responses are quite consistent with each other. The lack of relevant studies in Indonesia hinders the use of GenAI (average = 3.43, development standard = 0.66). Clearly, everyone agrees that the lack of research on GenAI poses a significant obstacle to its application in Indonesian schools. However, data from statement no. 13 revealed the risk of commercialization harming educational objectives (average = 3.26, development Standard = 0.81). Although opinions differed among participants, some were concerned that monetization could divert educational goals. The average digital literacy score ranged from 3.65 to 0.88, indicating some variation in views. Most participants agreed that AIIoT in the era of cybersocialization helps improve digital literacy.

The question of GenAI integration is still in its infancy among Indonesian educators (mean = 4.04, standard of development = 1.02), as shown in

Table 14.1. With such a high mean, it is clear that most agree that teachers face significant barriers when trying to properly incorporate GenAI into their curriculum. This statement suggests that public awareness serves as a moderate motivator for content creation. With a mean of 3.74 and a standard deviation of 0.69, monetization does increase engagement with educational content. There is a fair amount of consistency in the responses from participants regarding the idea that the prospect of financial gain increases engagement. Statement 19 addresses the potential impact of AI on education in the future. This lower mean reflects the fact that there is much disagreement about how AI will ultimately impact classroom instruction. The research on question twenty reveals a high mean score, indicating that students' academic achievement improves when they upload content to platforms like YouTube. It appears that most people believe that uploading academic projects to sites like YouTube can improve their grades.

The data results shown in Table 14.2 present reliability data from a 20-item survey on AIIoT curriculum design and cybersocialization. The primary metric, Cronbach's alpha, demonstrated excellent internal consistency at 0.900. Using Cronbach's alpha, which measures the degree of association between a group of items, can help determine the reliability of a questionnaire. Surveys with reliability coefficients above 0.9 can measure students' knowledge of AI, their comfort level with its use, and their views on the impact of monetization and cybersocialization on their education.

A high Cronbach's alpha of 0.900, indicating consistent responses across the 20 items, suggests that participants shared similar perspectives on cybersocialization, artificial intelligence technology, curriculum design, and monetization of learning outcomes. This reliability provides confidence in the validity of the dimensions being investigated, including students' perceptions of the role of AI in the classroom and the incentive effects of cybersocialization-based monetization. Such high reliability ensures that survey results in educational research are credible and accurately represent participants' perspectives on these topics. Table 14.3 shows the mean score of cybersocialization effects to the university curriculum by AIIoT development in education. Students totally agreed and support the correlation between curriculum and AIIoT.

Table 14.2 Reliability statistics on cybersocialization and curriculum design with AIIoT.

Cronbach's alpha	No. of items
0.900	20

Table 14.3 The mean score.

	N	Mean	Std. deviation	Std. error mean
Cybersocialization	23	72.0000	10.47942	2.18511

Table 14.4 One-sample test.

	Test value = 0					
					95% confidence interval of the difference	
	t	df	Sig. (2-tailed)	Mean difference	Lower	Upper
Cybersocialization and curriculum design by AIIoT	32.950	22	0.000	72.00000	67.4684	76.5316

Table 14.4 displays the sample size of these 23 studies and presents the results of the t-test, which examines the relationship between cybersocialization and curriculum design by AIIoT (artificial intelligence and the Internet of Things). With a t-value of 32.950 and a p-value (Sig. 2-tailed) of 0.000, the results are significant at the 0.05 level of statistical analysis. This indicates that there is a highly significant difference of 72.00 between the group mean scores and the test score of 0. The 95% confidence interval, ranging from 67.47 to 76.53, likely encompasses the actual mean difference, further confirming the observed strong effect.

Based on the data in Table 14.4 present cybersocialization score based on responses from 23 participants, with a mean of 72.00, a standard deviation of 10.48, and a standard error of 2.19. These data indicate that, on average, participants rated their engagement with cybersocialization relatively high, with a consistent spread of responses as reflected by the standard deviation.

14.5 Discussion

This study enhances the integration of online socialization into the classroom, fostering the development of critical thinking skills and the capacity for critical reflection. This is crucial because it enables students to assess, investigate, and form conclusions from diverse online sources, a skill that cybersocialization and virtual literacy promote. By finding solutions to challenges on the global web, finding reliable sources, and analyzing them rigorously, students can develop their critical thinking skills by investigating claims of truth and validity. Internet-based literacy programs provide students with the opportunity to develop critical thinking skills applicable to a variety of educational and professional contexts, in line with Reddy and Lakshmi's [51] study. Enhanced cybersocialization is important for students' future careers, as employers today place a premium on applicants with digital literacy skills.

As reported in Table 14.1, digital literacy will be a critical asset in the future workforce. Learners may be better prepared for jobs that can help them achieve professional success by integrating cybersocialization and digital literacy into the learning environment. Students who are competent in the use of digital resources for skills in digital communication, collaboration, and problem-solving in the workplace learn new technologies and enhance their cybersocialization skills. Furthermore, through online socialization, account creation, networking with other students and professionals, and the acquisition of digital citizenship skills, undergraduate students can build trusting relationships that are essential for thriving in the technology era.

This level of dependability has far-reaching consequences. The first takeaway is that the impact of AI and cybersocialization on students' motivation and interest in learning is well-known. As a second benefit, it verifies that there is a consistent scale to the questions about ChatGPT and other AI tools and their impact on curriculum development, further establishing the relationship between technology and contemporary pedagogy (Table 14.1 statement number 16). It ensures that the data used for subsequent analyses, including correlational studies, are valid. Networked learning theory, self-determination theory, and the technology acceptance model (TAM) are among the theoretical frameworks that lend credence to the validity of the survey. Students' consistent comments about the value of AI tools align with TAM's emphasis on how user perceptions of usefulness and ease of use drive technology adoption. Students are increasingly engaging in cybersocialization, where they use technology to share and monetize learning outcomes, and networked learning theory reflects this trend (statement number 10).

According to this study, students' drive for monetization lends credence to the premise that giving them agency over the creation and sharing of digital information increases their engagement with instructional content. The strong internal consistency of the Cronbach's alpha score of 0.900 confirms the survey's reliability in measuring students' views on cybersocialization and AIIoT curriculum design (Table 14.2). This reliability lends credence to the study's originality by aligning with modern educational theories and highlighting the revolutionary impact of AI and the Internet of Things in the classroom. This study aligns with how people accept and use technology, as described by Adouani and Khenissi [2]. There are two primary considerations: perceived usefulness of the system (how much a person thinks it will improve their performance) and perceived ease of use (how much a person thinks it will require little or no effort). The high mean scores in statements 2 (3.96) and 4 (4.48) support the TAM model's explanation of how people adopt new technologies. They show that students thought GenAI could help them learn and were comfortable with it. Active learning practices are central to constructivist learning theory; according to this approach, students construct their own knowledge and understanding of the world as they go through life.

Statements 3 (3.57) and 5 (3.57) demonstrate the power of AI-enhanced activities and public work sharing in empowering students to take ownership of their learning using ChatGPT and other AI-based tools. High mean scores for statements such as 6 (4.35) and 18 (3.74) indicate that students are more motivated to engage in academic work when they have the opportunity to earn money and gain public notoriety through sites such as YouTube. While money provides tangible benefits, it also fosters intrinsic motivation by giving individuals more freedom to express themselves creatively.

Using AIIoT to strengthen the connection between students, educators, and subject matter is a central tenet of networked learning theory. Several statements, such as 10 (3.35) and 14 (3.65), lend credence to the idea of cybersocialization, where people learn to communicate and collaborate online. This is in line with the posthumanist view, discussed in this paper, where humans and technology work together to create new knowledge. There is congruence between the results of this study and the digital literacy framework, which emphasizes the importance of digital competence in navigating today's technological environment. According to study participants (Statement 14, Mean = 3.65), cybersocialization has improved students' digital literacy, demonstrating the critical role of technology in cultivating skills needed for today's economy. This study stands out as the first to investigate how GenAI, cybersocialization, and the commercialization of education impact the dynamic educational environment in Indonesia.

Students improved their academic performance and explored monetization opportunities by combining AI tools with public platforms, such as YouTube. Supporting the shift to Education 6.0 and posthumanist curricula, the results demonstrate how technology is transforming educational outcomes while adding new dimensions to motivation, creativity, and digital literacy. All of this makes this study stand out as an original addition to the literature on digital literacy. The study stands out from other research on artificial intelligence and education in the Indonesian context due to these factors.

Based on the findings presented in Table 14.1, which reference constructivist learning theory and networked learning theory, it is evident that students learn best when they collaborate and share what they have learned through the use of digital tools (such as AI and the Internet of Things). The high mean differences and statistically significant t-values in this study indicate that students actively participated in cybersocialization, lending credence to the idea that digital technology creates a more dynamic classroom where students can work together to achieve better outcomes. The data also support constructivist learning theory, which asserts that students learn most effectively when they actively construct knowledge through real-world interactions. The goal of the AIIoT-enhanced curriculum design was to combine theoretical and practical knowledge through student interaction with AI technology and sharing digital learning content.

Based on these findings, AI-powered learning has the potential to improve educational outcomes in the future. Students in the study believed that AI helped their learning process (mean = 4.48). Another finding highlights how AI-powered platforms, such as YouTube, allow students to monetize their educational content, which increases their motivation and creativity. This study is in line with Self-Determination Theory (SDT), which states that the greatest motivator for learners is a sense of control and mastery over their own learning independently. Because AIIoT goes beyond normal classroom requirements and promotes ownership by allowing students to grow and monetize their work massively, it encourages students to produce high-quality learning material evaluation artifact content (mean = 4.35). Motivating students to actively participate in their learning exposes them to simple academic principles and practical applications. Cybersocialization, where students work together to share what they have learned in an online environment, is believed to improve students' digital literacy, as found in this study.

The results of this study are in line with the main focus of Ruiz-Rojas *et al.* [54] on learning through digital relationships and effective collaborative networks. Cybersocialization activities supported by AI and the

Internet of Things (AIIoT) can help students prepare for careers in these technology-driven industries. However, this study highlights several significant obstacles that hinder the full implementation of AIIoT in Indonesian schools. Significant barriers that limit access to digital platforms and AI technologies in underserved and rural areas are the lack of infrastructure, technology, and digital literacy (SDT) competencies. Current curriculum reforms are needed to incorporate project-based learning (PBL) and AIIoT literacy, according to the study. By incorporating AIIoT literacy into the curriculum, students will gain the skills to utilize AIIoT tools while understanding their real-world uses and the ethical considerations that come with them.

AIIoT-based learning is consistent with constructivist pedagogy because it promotes problem-solving in context, which is the essence of PBL. Students' interest, creativity, and ability to apply what they have learned will increase when these project-based assignments incorporate AIIoT technologies in a stimulating manner. Commercialization of student learning must uphold several ethical considerations, one of which is the right of students to receive an honest education. There is a need to incorporate AIIoT and TAM principles into the curriculum in schools and universities to ensure the ethical and appropriate use of AI, especially for assessing student work. Finally, a major focus should be on ensuring that AIIoT resources are available to children from all socioeconomic backgrounds. Financial assistance or subsidies for students from low-income families should be a policy priority to bridge the digital achievement gap—cybersocialization.

The most effective way to make AI and IoT devices more available in schools is through serious public-private collaboration. To ensure that no student falls behind in the digital revolution of education, public-private collaboration is essential. As a result of integrating AI and IoT into its education system, Indonesia has a golden opportunity to foster digital literacy, student engagement, and individualized learning. However, there is still a significant journey ahead for AIIoT to reach its full potential in the country. These include re-evaluating the curriculum, improving physical facilities, and training AIIoT teachers. By implementing these problem-solving strategies, Indonesian education can grow into a more vibrant, inventive, and inclusive environment for all students equally. With the right policies and financial support, Indonesia can realize the importance of AIIoT and its potential to transform students' readiness for a digital and connected world, as this study clearly shows.

14.6 Conclusions

The study found that the integration of AI and IoT in Indonesian universities enhances cybersocialization, which in turn impacts academic achievement. By monetizing the outcomes of course-based projects on sites like YouTube, there is great potential to integrate student artifacts with new technologies into the curriculum to revolutionize education, motivate students, and maximize creativity. Students were generally pleased with artificial intelligence (AI) tools like ChatGPT because they felt they enhanced their learning experience and encouraged more active participation in the digital environment. Students can leverage the transformation of digital platforms like YouTube to monetize their learning outcomes and gain valuable insights into the practical implications of their academic creative work. This study demonstrates how AIIoT tools like ChatGPT and YouTube can enhance educational outcomes and motivate students to think creatively, critically, and innovatively, adding to the ongoing discussion on the role of AI in curriculum reform. The study emphasizes the growth of media and technology literacy with the encouragement of students' digital creativity as they learn to use social media and AI to create, share, and monetize content and facilitate the use of AIIoT for educational purposes.

AI disclosure statement: The author confirms that no artificial intelligence (AI)-generated content or AI-assisted writing tools were used in the preparation of this chapter (including summaries and the abstract). All text, analysis, and conclusions are the result of the author's original work.

Bibliography

1. Abdullah, M.A., Digital maturity of the Egyptian universities: Goal-oriented project planning model. *Stud. Higher Educ.*, 49, 8, 1463–1485, 2024, https://doi.org/10.1080/03075079.2023.2268633.
2. Adouani, Y. and Khenissi, M.A., Investigating computer science students' intentions towards the use of an online educational platform using an extended technology acceptance model (e-TAM): An empirical study at a public university in Tunisia. *Educ. Inf. Technol.*, 1–25, 2024, https://doi.org/10.1007/s10639-023-12437-6.
3. Akpan, I.J., Kobara, Y.M., Owolabi, J., Akpan, A.A., Offodile, O.F., Conversational and generative artificial intelligence and human–chatbot interaction in education and research. *Int. Tran. Oper. Res.*, 2024, https://doi.org/10.1111/itor.13522.

4. Al Yakin, A.M., Rasyid, A.R., Massyat, M., Muthmainnah, M., Aquino, A.B., Noche, T.H., Cardoso, L., The importance of collaboration between human intelligence and GenAI in digitalization of education, in: *Impact and Potential of Machine Learning in the Metaverse*, pp. 129–160, IGI Global, USA, 2024, https://doi.org/10.4018/979-8-3693-5762-0.ch006.
5. Al Yakin, A., Muthmainnah, Ganguli, S., Cardoso, L., Asrifan, A., Cybersocialization through smart digital classroom management (SDCM) as a pedagogical innovation of "Merdeka Belajar Kampus Merdeka (MBKM)" curriculum, in: *Digital Learning-Based Education: Transcending Physical Barriers*, pp. 39–61, Springer Nature Singapore, Singapore, 2023, https://doi.org/10.1007/978-981-19-8967-4_3.
6. Alarcón-López, C., Krütli, P., Gillet, D., Assessing ChatGPT's influence on critical thinking in sustainability-oriented activities, in: *2024 IEEE Global Engineering Education Conference (EDUCON)*, 2024, May, IEEE, pp. 1–10, https://doi.org/10.1109/EDUCON60312.2024.10578790.
7. Alharbi, F., Integrating internet of things in electrical engineering education. *Int. J. Electr. Eng. Educ.*, 61, 2, 258–275, 2024, https://doi.org/10.1177/002072092090342.
8. Aljapurkar, A.V. and Ingawale, S.D., Revolutionizing the techno-human space in human resource practices in Industry 4.0 to usage in Society 5.0, in: *Digital Transformation: Industry 4.0 to Society 5.0*, pp. 221–257, Springer Nature Singapore, Singapore, 2024, https://doi.org/10.1007/978-981-99-8118-2_10.
9. Altinay, Z., Altinay, F., Sharma, R.C., Dagli, G., Shadiev, R., Yikici, B., Altinay, M., Capacity building for student teachers in learning, teaching artificial intelligence for quality of education. *Societies*, 14, 8, 148, 2024, https://doi.org/10.3390/soc14080148.
10. Asad, M.M., Naz, A., Shaikh, A., Alrizq, M., Akram, M., Alghamdi, A., Investigating the impact of IoT-based smart laboratories on students' academic performance in higher education. *Universal Access Inf. Soc.*, 23, 3, 1135–1149, 2024, https://doi.org/10.1007/s10209-022-00944-1.
11. Asad, M.M., Shahzad, S., Shah, S.H.A., Sherwani, F., Almusharraf, N.M., ChatGPT as artificial intelligence-based generative multimedia for English writing pedagogy: Challenges and opportunities from an educator's perspective. *Int. J. Inf. Learn. Technol.*, 2024, https://doi.org/10.1108/IJILT-02-2024-0021.
12. Ayanwale, M.A., Adelana, O.P., Molefi, R.R., Adeeko, O., Ishola, A.M., Examining artificial intelligence literacy among pre-service teachers for future classrooms. *Comput. Educ. Open*, 6, 100179, 2024, https://doi.org/10.1016/j.caeo.2024.100179.
13. Ayeni, O.O., Al Hamad, N.M., Chisom, O.N., Osawaru, B., Adewusi, O.E., AI in education: A review of personalized learning and educational technology. *GSC Adv. Res. Rev.*, 18, 2, 261–271, 2024, https://doi.org/10.30574/gscarr.2024.18.2.0062.

14. Baig, A., Cressler, J.D., Minsky, M., The future of AI in education: Personalized learning and intelligent tutoring systems. *AlgoVista J. AI Comput. Sci.*, 1, 2, 2024, https://algovista.org/index.php/AVJCS/article/view/12.
15. Cantas, C., Soyer, C., Batur, O., Examination of undergraduate students' artificial intelligence anxiety, multidimensional 21st-century skills, and lifelong learning levels in terms of various variables. *Turk. Online J. Educ. Technol.-TOJET*, 23, 3, 29–53, 2024, https://eric.ed.gov/?id=EJ1434130.
16. Casey, D., ChatGPT in public policy teaching and assessment: An examination of opportunities and challenges. *Aust. J. Public Adm.*, 2024, https://doi.org/10.1111/1467-8500.12647.
17. Casheekar, A., Lahiri, A., Rath, K., Prabhakar, K.S., Srinivasan, K., A contemporary review on chatbots, AI-powered virtual conversational agents, ChatGPT: Applications, open challenges, and future research directions. *Comput. Sci. Rev.*, 52, 100632, 2024, https://doi.org/10.1016/j.cosrev.2024.100632.
18. Chandrasekera, T., Hosseini, Z., Perera, U., Bazhaw Hyscher, A., Generative artificial intelligence tools for diverse learning styles in design education. *Int. J. Archit. Comput.*, 2024, https://doi.org/10.1177/14780771241287345.
19. Da Silva, S.P., *Pedagogical resilience, from class to digital room: A comparative study of technology adoption in Brazilian higher education before and during the pandemic*, Editora Dialética, Brazil, 2024, https://books.google.co.id/books?id=N_wQEQAAQBAJ.
20. Dilnoza, I., Scientific-pedagogical principles of cyber socialization of future teachers in the conditions of digital education. *Eur. J. Innovation Nonformal Educ.*, 4, 3, 294–296, 2024, http://www.inovatus.es/index.php/ejine/article/view/2731.
21. Elbanna, S. and Armstrong, L., Exploring the integration of ChatGPT in education: Adapting for the future. *Manage. Sustain. An Arab Rev.*, 3, 1, 16–29, 2024, https://doi.org/10.1108/MSAR-03-2023-0016.
22. Ferilli, G.B., Palmieri, E., Miani, S., Stefanelli, V., The impact of FinTech innovation on digital financial literacy in Europe: Insights from the banking industry. *Res. Int. Bus. Finance*, 69, 102218, 2024, https://doi.org/10.1016/j.ribaf.2024.102218.
23. Fernández-Batanero, J.M., Montenegro-Rueda, M., Fernández-Cerero, J., López Meneses, E., Adoption of the Internet of Things in higher education: Opportunities and challenges. *Interact. Technol. Smart Educ.*, 21, 2, 292–307, 2024, https://doi.org/10.1108/ITSE-01-2023-0025.
24. Georgara, A., Santolini, M., Kokshagina, O., Haux, C.J.J., Jacobs, D., Biwott, G., Rodriguez-Aguilar, J.A., Optimising team dynamics: The role of AI in enhancing challenge-based learning participation experience and outcomes, 2024, HAL Archives Ouvertes, https://hal.science/hal-04691209/.
25. Goel, P.K., Education research and Industry 6.0, in: *Sustainable innovation for Industry 6.0*, pp. 92–108, IGI Global, USA, 2024, https://doi.org/10.4018/979-8-3693-3140-8.ch006.

26. Goel, P.K., Singhal, A., Bhadoria, S.S., Saraswat, B.K., Patel, A., AI and machine learning in smart education: Enhancing learning experiences through intelligent technologies, in: *Infrastructure possibilities and human-centered approaches with Industry 5.0*, pp. 36–55, IGI Global, USA, 2024, https://doi.org/10.4018/979-8-3693-0782-3.ch003.
27. Håkansson Lindqvist, M., Mozelius, P., Jaldemark, J., Cleveland Innes, M., Higher education transformation towards lifelong learning in a digital era: A scoping literature review. *Int. J. Lifelong Educ.*, 43, 1, 24–38, 2024, https://doi.org/10.1080/02601370.2023.2279047.
28. Haque, M.A., Ahmad, S., Hossain, M.A., Kumar, K., Faizanuddin, M., Islam, F., Nazeer, J., Internet of Things-enabled e-learning system for academic achievement among university students. *E-Learning Digital Media*, 2024, https://doi.org/10.1177/20427530241280078.
29. Järvelä, S. and Hadwin, A., Triggers for self-regulated learning: A conceptual framework for advancing multimodal research about SRL. *Learn. Individual Differ.*, 115, 102526, 2024, https://doi.org/10.1016/j.lindif.2024.102526.
30. Kandil, O., Rosillo, R., Abd El Aziz, R., De La Fuente, D., Investigating the impact of the Internet of Things on higher education: A systematic literature review. *J. Appl. Res. Higher Educ.*, 2024, https://doi.org/10.1108/JARHE-05-2023-0223.
31. Khan, A.S., Alnmer, S., Khan, S.A., Maximizing learning potential: Integrating multiple intelligences theory in EFL teaching and learning. *Educ. Adm. Theory Pract.*, 30, 2, 852–858, 2024, https://doi.org/10.53555/kuey.v30i2.2509.
32. Khan, M., Chaudhry, M.N., Ahsan, M., Ahmad, R., ChatGPT and the future of impact assessment. *Environ. Sci. Policy*, 157, 103779, 2024, https://doi.org/10.1016/j.envsci.2024.103779.
33. Kim, A. and Su, Y., How implementing an AI chatbot impacts Korean as a foreign language learners' willingness to communicate in Korean. *System*, 122, 103256, 2024, https://doi.org/10.1016/j.system.2024.103256.
34. Kohler, K., You only need to change your direction: A look at the potential impact of ChatGPT on education. *Technol. Lang. Teach. Learn.*, 6, 1, 1–18, 2024, https://doi.org/10.29140/tltl.v6n1.1103.
35. Kukreja, M., Arumugam, S.K., Tyagi, A.K., Sivakumar, M., Future perspectives of ChatGPT towards Industry 4.0, Society 5.0, and modern education, in: *Impacts of generative AI on creativity in higher education*, pp. 101–140, IGI Global, USA, 2024, *https://doi.org/10.4018/979-8-3693-2418-9.ch005.*
36. Kulkarni, M., Durve, S., Jia, B., Cyberbully and online harassment: Issues associated with digital wellbeing, arXiv Preprint arXiv:2404.18989, 2024, https://doi.org/10.48550/arXiv.2404.18989.
37. Kurni, M. and KG, S., IoT for education, in: *The Internet of Educational Things: Enhancing Students' Engagement and Learning Performance*, pp. 1–22, Springer Nature Switzerland, Switzerland, 2024, https://doi.org/10.1007/978-3-031-67387-0_1.

38. Lavric, A., Beguni, C., Zadobrischi, E., Căilean, A.M., Avătămăniței, S.A., A comprehensive survey on emerging assistive technologies for visually impaired persons: Lighting the path with visible light communications and artificial intelligence innovations. *Sensors*, 24, 15, 4834, 2024, https://doi.org/10.3390/s24154834.
39. Magalhães, A., Alves, J.M., Andrade, A., The Internet of Things in education: Technology, pedagogy, and learning. *Rev. Port. Investig. Educ.*, 27, 1–25, 2024, https://doi.org/10.34632/investigacaoeducacional.2024.16057.
40. Manoharan, G., Razak, A., Rajchandar, K., Nithya, G., Durai, S., Ashtikar, S.P., Digital learning for professional development in varied fields of service sectors: Embracing technological advancements, in: *Embracing Technological Advancements for Lifelong Learning*, pp. 111–137, IGI Global, USA, 2024, https://doi.org/10.4018/979-8-3693-1410-4.ch006.
41. Matthew, U.O., Oyekunle, D.O., Akpan, E.E., Oladipupo, M.A., Chukwuebuka, E.S., Adekunle, T.S., Onumaku, V.C., Generative artificial intelligence (AI) on sustainable development goal 4 for tertiary education: Conversational AI with user-centric ChatGPT-4, in: *Impacts of Generative AI on Creativity in Higher Education*, pp. 259–288, IGI Global, USA, 2024, https://doi.org/10.4018/979-8-3693-2418-9.ch010.
42. Michalcewicz-Kaniowska, M.M., Komarnicka, A., Toader, C.S., Prospects of using remote teaching in school education: Research findings. *Zesz. Nauk. Organ. Zarz. Politech. Śl.*, 2024, https://doi.org/10.29119/1641-3466.2024.193.6.
43. Muthmainnah, Ganguli, S., Al Yakin, A., An effective investigation on YIPe-learning based for twenty-first century class, in: *Digital Learning Based Education: Transcending Physical Barriers*, pp. 21–38, Springer Nature Singapore, Singapore, 2023, https://doi.org/10.1007/978-981-19-8967-4_2.
44. Muthmainnah, M., Darmawati, B., Khang, A., Al Yakin, A., Obaid, A.J., Elngar, A.A., Shaping artificial intelligence-perceived hybrid learning environment at university toward the global talent development strategy, in: *AI-Oriented Competency Framework for Talent Management in the Digital Economy*, pp. 219–243, CRC Press, London, 2023, https://doi.org/10.1201/9781003440901-14.
45. Norrie, C.S., Deckers, S.R., Radstaake, M., van Balkom, H., A narrative review of the sociotechnical landscape and potential of computer-assisted dynamic assessment for children with communication support needs. *Multimodal Technol. Interact.*, 8, 5, 38, 2024, https://doi.org/10.3390/mti8050038.
46. Ofulue, J. and Benyoucef, M., Data monetization: Insights from a technology-enabled literature review and research agenda. *Manage. Rev. Q.*, 74, 2, 521–565, 2024, https://doi.org/10.1007/s11301-024-00278-x.
47. Osado, A.G., Merlo, L.S.B., Fernández, C.V., Promoting learning and soft skills through student-created videos in a science course, in: *EDULEARN24 Proceedings*, IATED, pp. 2573–2581, 2024, https://doi.org/10.21125/edulearn.2024.12345.

48. Ouyang, F. and Zhang, L., AI-driven learning analytics applications and tools in computer-supported collaborative learning: A systematic review. *Educ. Res. Rev.*, 44, 100616, 2024, https://doi.org/10.1016/j.edurev.2024.100616.
49. Pattanaik, S., Mohammed, M., Sood, V., Artificial intelligence and machine learning in Industry 6.0, in: *Industry 6.0*, pp. 37–54, CRC Press, London, 2024, https://doi.org/10.1201/9781003298765-3.
50. Rajamohan, K., Rangasamy, S., Manoj, C.N., Mary, L., Sabu, R., Jose, S.S., Internet of Things (IoT) as a game changer in the education sector, in: *Advanced IoT Technologies and Applications in the Industry 4.0 Digital Economy*, pp. 251–274, CRC Press, London, 2024, https://doi.org/10.1201/9781003298765-12.
51. Reddy, K.V. and Lakshmi, G.S., Introduction to critical thinking skills, in: *Critical Thinking for Professional and Language Education: A Machine-Generated Literature Overview*, pp. 1–19, Springer Nature Switzerland, Switzerland, 2024, https://doi.org/10.1007/978-3-030-12345-6_1.
52. Resendiz-Calderón, C.D., Farfan-Cabrera, L.I., Cazares-Ramírez, I.O., Nájera-García, P., Okoye, K., Assessing benefits of computer-based video training and tools on learning outcomes and motivation in mechanical engineering education: Digitalized intervention and approach. *Front. Educ.*, 9, 1292405, 2024, https://doi.org/10.3389/feduc.2024.1292405.
53. Rock, L.Y., Tajudeen, F.P., Chung, Y.W., Usage and impact of the internet-of-things-based smart home technology: A quality-of-life perspective. *Universal Access Inf. Soc.*, 23, 1, 345–364, 2024, https://doi.org/10.1007/s10209-024-00967-y.
54. Ruiz-Rojas, L.I., Salvador-Ullauri, L., Acosta-Vargas, P., Collaborative working and critical thinking: Adoption of generative artificial intelligence tools in higher education. *Sustainability*, 16, 13, 5367, 2024, https://doi.org/10.3390/su16135367.
55. Saritepeci, M. and Yildiz Durak, H., Effectiveness of artificial intelligence integration in design-based learning on design thinking mindset, creative and reflective thinking skills: An experimental study. *Educ. Inf. Technol.*, 1–35, 2024, https://doi.org/10.1007/s10639-024-11045-z.
56. Serrano-Ausejo, E. and Mårell-Olsson, E., Opportunities and challenges of using immersive technologies to support students' spatial ability and 21st-century skills in K-12 education. *Educ. Inf. Technol.*, 29, 5, 5571–5597, 2024, https://doi.org/10.1007/s10639-024-11111-9.
57. Shukla, S., Bisht, K., Tiwari, K., Bashir, S., Data monetization, in: *Data Economy in the Digital Age*, pp. 37–62, Springer Nature Singapore, Singapore, 2023, https://doi.org/10.1007/978-981-997676-8_3.
58. Sinitsyna, E., Connected digitally and dedicated loyally: The features and impact of internal online events on employee outcomes. *Manage. Res. Rev.*, 2024, https://doi.org/10.1108/MRR-12-2023-0598.

59. Söğüt, S., Generative artificial intelligence in EFL writing: A pedagogical stance of pre-service teachers and teacher trainers. *Focus ELT J.*, 6, 1, 58–73, 2024, https://doi.org/10.14748/felt.v6n1.12345.
60. Sok, S. and Heng, K., Opportunities, challenges, and strategies for using ChatGPT in higher education: A literature review. *J. Digital Educ. Technol.*, 4, 1, ep2401, 2024, https://doi.org/10.29333/jdet.v4n1.ep2401.
61. Song, Y., Weisberg, L.R., Zhang, S., Tian, X., Boyer, K.E., Israel, M., A framework for inclusive AI learning design for diverse learners. *Comput. Educ.: Artif. Intell.*, 100212, 2024, https://doi.org/10.1016/j.caeai.2024.100212.
62. Sperling, K., Stenberg, C.J., McGrath, C., Åkerfeldt, A., Heintz, F., Stenliden, L., In search of artificial intelligence (AI) literacy in teacher education: A scoping review. *Comput. Educ. Open*, 100169, 2024, https://doi.org/10.1016/j.caeo.2024.100169.
63. Sree, M.S., Reddy, C.K.K., Hanafiah, M.M., Lippert, K., Smart education systems using the Internet of Behavior-based computational intelligence, in: *Internet of Behavior-Based Computational Intelligence for Smart Education Systems*, pp. 213–236, IGI Global, USA, 2025, DOI: 10.4018/979-8-3693-8151-9.ch008.
64. Štuikys, V. and Burbait, R., *Evolution of STEM-driven computer science education*, Switzerland: Springer, Cham, 2024, DOI: https://doi.org/10.1007/978-3-031-48235-9.
65. Teng, M.F., ChatGPT is the companion, not enemies": EFL learners' perceptions and experiences in using ChatGPT for feedback in writing. *Comput. Educ.: Artif. Intell.*, 7100270, 2024, https://doi.org/10.1016/j.caeai.2024.100270.
66. Walela, A., Multiple intelligence in the teaching and learning process: A study of Howard Gardner's thought, challenges and opportunities. *Int. J. Educ. Lang. Lit. Arts Cult. Soc. Humanit.*, 2, 4, 133–155, 2024, DOI: https://doi.org/10.59024/ijellacush.v2i4.1006.
67. Wei, Q., Yao, Z., Wei, B., Jin, Z., Xu, X., Evaluation of ChatGPT-generated medical responses: A systematic review and meta-analysis. *J. Biomed. Inf.*, 104620, 2024, https://doi.org/10.1016/j.jbi.2024.104620.
68. Zawish, M., Dharejo, F.A., Khowaja, S.A., Raza, S., Davy, S., Dev, K., Bellavista, P., AI and 6G into the metaverse: Fundamentals ,challenges ,and future research trends. *IEEE Open J. Commun. Soc.*, 5, 730–778, 2024, https://doi.org/10 .1109/OJCOMS .2024 .1234567.
69. Zhang, X., Ding, Y., Huang, X., Li, W., Long, L., Ding, S., Smart classrooms: How sensors and AI are shaping educational paradigms. *Sensors*, 24, 17, 2024, Basel, Switzerland, https://doi.org/10 .3390/s24173891.
70. Jai, B.R. and Shih, M.F., Technology: Limited or Infinite? *Emerging Media*, 2, 1, 55–69, 2024.
71. Sundararajan, L., Yeh, K.H., Ting, R. S.K., Terpstra-Tong, J., Bond, M.H., Culture as an ontological universe: Toward a broader model of culture in the globalizing era. *Theory Psychol.*, 2025, 09593543251374680.

15

Future Trends and Innovation Education 7.0 by Adopting AIIoT Networks

Besse Darmawati[1*], Ade Mulyanah[1], Adri Adri[1], Ratnawati Ratnawati[1], Resti Nurfaidah[1], Muthmainnah Muthmainnah[2] and Souvik Ganguli[3]

[1]*Organisasi Riset Arkeologi, Bahasa, dan Sastra, Badan Riset dan Inovasi Nasional, Jakarta, Indonesia*

[2]*Teacher Training and Education Faculty, Polewali Mandar, Sulawesi Barat, Indonesia*

[3]*Electrical and Instrumentation Engineering Institute, Thapar Institute of Engineering and Technology, Patiala, India*

Abstract

The purpose of this study is to evaluate the impact of AIIoT networks in smarter, more connected classrooms by enhancing personalized, adaptive, and real-time learning environments. This study used a pre-experimental design to assess the conceptual understanding and AI competency of seven literature instructors from the postgraduate students at *Yayasan Pendidikan Ujung Pandang (YPUP)* Makassar for the academic year 2023. Surveys and interviews offered qualitative insights, whereas pre- and posttests yielded quantitative data. With posttest scores significantly higher than pretest scores, the data suggest that instructors' AI expertise has improved dramatically. Teachers' capacity to tailor lessons to each student's needs, streamline administrative processes, and encourage active learning using AIIoT platforms has improved significantly. However, the study uncovered challenges such as data security, resource allocation, and the need for teacher training. By demonstrating the utility of AIIoT networks in Education 7.0, this study adds to the existing body of knowledge by highlighting the importance of AI in automating classroom operations and the value of personalized learning. There is great hope that AI and IoT networks will transform education by facilitating the development of personalized, adaptive, and interactive learning environments for individual students.

**Corresponding author*: bess001@brin.go.id

Tanusree Chatterjee, Rajib Banerjee and Souvik Ganguli (eds.) Content Delivery Networks Using Artificial Intelligence and Internet of Things: Revolutionizing Content Delivery, (341–374) © 2026 Scrivener Publishing LLC

***Keywords*:** AIIoT, Education 7.0, personalized learning, adaptive learning, teacher training, data security, AI in education

15.1 Introduction

New opportunities have arisen in many fields because of the convergence of IoT and AI technologies [44]; one of these fields is education. This chapter delves into the dynamic landscape resulting from the integration of IoT and AI in education, examining the challenges encountered and the potential opportunities that lie ahead. Recent years have seen significant shifts in the education business due to advancements in technology [3]. The interoperability of smart devices in these online educational centers enhances traditional classroom settings. The key components of this change are Internet-of-Things (IoT) devices, which encompass sensors, wearables, and other networked devices that can collect and transmit data in real time [20].

AI processing and evaluation of the data stream support data-driven insights and informed decision-making. The exponential development of digital technology over the last several decades has influenced almost every facet of human life, but perhaps none more so than the field of education [56]. In order to better serve their students and themselves, today's teachers can take advantage of the plethora of digital materials and technologies at their fingertips. There has always been a lot of curiosity about the potential effects of technology in the classroom. This interest led to the establishment of the field of Artificial Intelligence in Education (AIED) over 30 years ago. AIED aims to provide new tools that improve teaching and learning by combining advanced AI technology with pedagogical practices [32].

The Internet of Things (IoT) and artificial intelligence (AI) have the potential to revolutionize the way students and educators learn. More than ever, we can realistically pursue goals like personalization, interaction, and real-time flexibility to enhance engagement and improve learning outcomes. However, we must overcome the challenges this integration has presented. Concerns about personal information, data security, technological obstacles, and the need to train educators are all aspects that might affect the outcome of a deployment [69]. Concerns about data security and privacy are the focus of this chapter as it analyzes the challenges associated with integrating AI and the Internet of Things into educational settings. "Computing systems that are able to engage in human-like processes such as learning, adapting, synthesizing, self-correction, and use of data for complex processing tasks" [59] is an integrative definition among several others in the literature. Findings from AIED studies have the potential to improve educational resources, adaptive instruction techniques, and individualized learning. Notable studies that thoroughly review the various roles, benefits, challenges, and overall impact of AI in educational settings include [2, 5, 42, 73].

These studies critically examine and classify AI's integration into education, offering valuable perspectives on its influence and potential to shape teaching and learning practices. A subfield of artificial intelligence known as generative AI (GenAI) uses machine learning algorithms to create media that sound, look, and read like a person. In response to sophisticated and multimodal stimuli, these algorithms assess and duplicate the structure and properties of certain categories of information to create new content that replicates the original [64]. Many other sectors and industries could benefit from GenAI's many uses, including healthcare, manufacturing, the arts, and more.

Numerous studies [41, 55, 74] have demonstrated that the integration of GenAI into educational settings is increasing, indicating its potential to enhance education in various ways. The intervention of GenAI in education seems to be deficient in classifying these tools according to their prospective usage; thus, this study fills that void. The study focuses on GenAI tools that could function as assistants for instructors. Although these tools are easily accessible and user-friendly, it's crucial to integrate them into authentic teaching practices, documented through a lesson plan, to guarantee their effectiveness. Creating a lesson plan, detailing the day's lessons, the content to cover, the class structure, and any required materials, is an essential part of any teacher's preparation for the profession [61]. Teachers' decisions about class structure significantly influence the integration of technology in the classroom, establishing a close connection between these two factors and the use of technology in education [19]. Lesson plans can help both current and future educators incorporate technology into their teaching by showcasing educators' goals to do so [37].

This study aims to emphasize the use of GenAI tools in our lesson planning. In contrast to teacher-centered techniques, they promote better comprehension of the subject matter through student engagement in knowledge generation [47]. This research primarily employs the student-centered learning approach because of its significant benefits for students, including improved motivation and learning outcomes. However, its effectiveness heavily relies on the role and competencies of educators, posing challenges for its implementation in real-world settings [14]. This study provides practical information on how to effectively deploy GenAI tools in real-world educational scenarios by providing instances of their use in multiple contexts.

15.2 Literature Review

15.2.1 AIIoT Network in Education: A New Center Study Design

The use of AI and the Internet of Things in classrooms is still young, but its breakthroughs stem from earlier technological advances. The late 20th century saw the introduction of computers and the internet into classrooms, which sparked the idea of utilizing technology to improve education [22]. Their uses in the classroom expanded in tandem with the development of these technologies. Initially, teachers used computers for simple data entry and word processing, but later, they incorporated more advanced educational applications and online learning platforms [30].

As more and more gadgets became internet-enabled in the early 2000s, the foundations of the Internet of Things (IoT) in the classroom started to take form. The concept of a "smart classroom" arose, in which different gadgets could exchange data and cooperate to improve the quality of instruction and student engagement. For instance, smartboards replaced traditional chalkboards, enabling teachers to display multimedia content and interact with digital tools during lessons [24]. Sensors and other Internet-of-Things devices ensured optimal learning environments by monitoring ambient parameters like lighting and temperature. Recently, the Internet of Things (IoT) has garnered attention for its potential to monitor students' health and activity levels using wearable devices, thereby enhancing PE programs and improving students' overall health.

Machine learning and data analytics have been the backbone of AI educational evolution. The 1960s acknowledged the potential educational benefits of AI, but the new millennium saw the emergence of real implementations. Adaptive learning systems were the first AI applications that could adjust the course material's level of difficulty based on the student's progress. The goal of these systems was to meet the unique needs of each student by creating individualized learning plans. Another development is the rise of intelligent tutoring systems (ITS), which may mimic human instructors to provide personalized instruction. The rapid advancement of AI, new and improved tools for grading, student assessment, and administrative chores, has made these available in the field of education [33].

There is a plethora of literature on the pros and cons of using AI and the Internet of Things in the classroom, as this survey of previous research shows. There is a mountain of evidence showing that these tools improve student engagement and performance in the classroom [17]. Dai *et al.* [15] found that the use of Internet-of-Things (IoT) devices in the classroom significantly increased student engagement and enthusiasm. Biswal *et al.* [11] found that adaptive learning systems powered by AI not only enhanced student performance but also provided personalized learning routes tailored to individual needs. Johnson *et al.* found that AI-driven analytics might help find students who were about to drop out of school so they could get help when they needed it, which ultimately led to higher retention rates [75].

The literature does, however, highlight several obstacles and restrictions. Since many AI and IoT systems necessitate the gathering and processing of massive volumes of personally identifiable information, data privacy and security are important issues. Studies like the one by Simsek [52] highlight the significance of strong security measures and the hazards of data breaches. The hefty price tag of these technologies is another obstacle; many schools, especially those in underdeveloped nations, just cannot afford to implement them. Educators and administrators may be hesitant to embrace new technology and adjust long-established pedagogical practices due to resistance to change.

There are still a lot of unanswered questions regarding the use of AI and the Internet of Things in the classroom, despite the wealth of literature on the topic. Unfortunately, there has been a dearth of research examining the long-term effects of these technologies on academic performance [70]. Questions regarding the long-term advantages and possible downsides remain unsolved because most of the current research focuses on short-term impacts. Another gap is the lack of studies exploring the moral implications of AI in the classroom. While some research does address fairness and bias, further in-depth analyses of how AI algorithms could exacerbate or alleviate educational disparities are required. There is also a dearth of studies examining how various educational settings, such as those in rural areas or those that serve underserved populations, might benefit from the Internet of Things and artificial intelligence. To fully grasp the advantages and disadvantages of IoT and AI in the classroom, it is essential to fill in these gaps.

The term "Internet of Things" (IoT) refers to a system of networked computing devices that can share and receive data for various purposes. The Internet of Things (IoT) encompasses a wide range of devices and systems that can enhance learning, streamline administration, and provide teachers with more insights into their students' actions and grades. Examples of Internet-of-Things (IoT) technology in the classroom include smartboards, sensors, wearable tech, smart lighting, and linked learning platforms [27]. These innovations pave the way for the development of "smart classrooms," in which the virtual and real worlds come together to offer a dynamic and engaging learning environment. When it comes to the Internet of Things (IoT), creating "smart classrooms" is among the most game-changing uses. By providing a more engaging and interactive alternative to static chalkboards, smartboards have mostly supplanted their more antiquated counterparts in the classroom [28]. Students and teachers can work together in real time on these boards, which can also show multimedia and connect to educational applications. Sensors can automatically adjust the temperature, lighting, and air quality in the classroom to create the optimal learning environment. Not only does this make schools more energy efficient and cost-effective, but it also makes students more comfortable and helps them concentrate better.

Interactive learning environments, enabled by the Internet of Things, are transforming students' engagement with educational information. To accommodate various learning styles, IoT devices may generate real-time feedback and design immersive learning environments. For instance, students can experience more engaging and memorable learning by donning AR or VR headsets, which can transport them to historical landmarks, scientific laboratories, or even space [63]. Wearable devices like fitness trackers or smartwatches can monitor students' health indicators and physical activity levels, facilitating the integration of PE with academics. Teachers can use the information gathered from students' wearable fitness trackers to create individualized wellness programs that foster students' physical and intellectual development.

Several case studies show how universities and schools have used the Internet of Things (IoT) to excellent effect. The implementation of Internet-of-Things (IoT) technology is a prime illustration of this. Connected classrooms, automated attendance systems, and energy-efficient smart lighting were all parts of the institute's smart campus solution [21]. The Internet of Things (IoT) allows the school's lighting system to adapt to different levels of natural light and occupancy, which further improves energy efficiency. Tablets and smartboards are examples of interactive learning tools that allow teachers to provide students with immediate feedback and create individualized lessons.

These examples show how the Internet of Things (IoT) may revolutionize classrooms. Educational institutions can improve the quality of their learning environments in terms of engagement, efficiency, and security by leveraging networked devices and real-time data. Nevertheless, meticulous preparation, financial backing, and continuous upkeep are necessary for the effective deployment of IoT in educational settings. In order for all students to reap the benefits of these innovations, educational institutions must tackle issues like data privacy and security, as well as the digital divide, and implement the appropriate policies and methods. The Internet of Things (IoT) can significantly enhance educational outcomes and prepare students for a technologically driven future.

15.2.2 GenAI Tools in EFL Education 7.0

The usage of artificial intelligence (AI) has been steadily increasing across several industries, including education, as a result of globalization [67]. The use of AI in language classrooms is one promising area. Artificial intelligence (AI) holds the potential to transform language education by customizing learning experiences and providing instant feedback [35]. The use of AI has the potential to revolutionize language learning by making it more interactive and captivating for students. Students can engage in real-life discussions with virtual language tutors, receive instantaneous comments on their grammar and pronunciation, and gain access to individualized study materials based on their specific learning requirements through AI-powered language learning platforms [31].

Artificial intelligence may also evaluate student performance data to find weak spots and personalize learning activities to meet individual language objectives [71]. In addition, by incorporating real-world language usage into the classroom, AI can make natural language processing technologies more accessible [7]. To help students in improving their language skills and gaining a better understanding of other cultures, they should study real language materials, like news articles, social media posts, and online debates [25].

AI is influencing the field of English language instruction, and artificial intelligence (AI) has the potential to drastically alter the pedagogical ambiance of English language learning classrooms. The ability to tailor one's English class experience is one of the most striking ways in which AI is revolutionizing education. Learning platforms driven by AI can adjust to the unique requirements of each student, tailoring their comments and suggestions to help them progress at their own speed. Because students are more invested and driven to study when they receive individualized guidance, this personalized approach has the potential to foster a more encouraging and happier classroom environment [9].

In addition, by offering real-time translation and pronunciation help, AI may significantly improve the ambiance in the English lesson. This can be particularly helpful for students who are not native speakers, since it allows them to confidently and productively interact with their classmates. Because it helps to bridge language barriers and promotes better understanding among students, AI could support pronunciation and translation, creating an inclusive setting. Further, by enabling dynamic and engaging learning activities, AI can help alter the ambiance of the English class. An example of this would be the incorporation of gamified features, such as educational development trends, into AI-powered language learning platforms. The introduction of digital tools into the classroom in the form of games, quizzes, and interactive exercises has the potential to engage students and bring the material to life. This can make class more fun and energetic for everyone because students are more engaged and more likely to speak up and collaborate [29].

One of the most noteworthy developments in this industry is the creation of AI-powered platforms that provide language learners with genuine, stimulating dialogue. To begin with, platforms driven by AI have the potential to give language learners an immersive and lifelike conversational experience. Students can practice their English speaking and listening abilities in a safe, controlled setting with the help of these platforms, which use NLP and ML algorithms to mimic real-life conversations. Because students may converse with AI-generated characters and receive immediate comments on their language skills, this real stimulation conversation platform has the potential to make learning more engaging and participatory [46].

On top of that, AI can tailor conversation simulations to each student's unique requirements and objectives for learning. AI-driven dialog simulators can adjust to address individual students' language challenges by evaluating their current proficiency level, learning pace, and areas for improvement. When students are able to pinpoint their specific language learning needs, they are more likely to progress toward their goals, which in turn increases the efficacy of language learning overall. Furthermore, AI-driven systems can provide a plethora of conversational examples and themes, all designed to mimic the kinds of real-life circumstances that students could face while interacting with native English speakers. Students can practice their communication skills in a variety of circumstances, from casual chats to formal business meetings, through these platforms. Due to the many realistic contexts in which learners can practice and improve their language skills, this could be particularly helpful for students preparing for real-world conversation [60]. In addition, AI can help teachers monitor their students' progress and pinpoint where they need improvement by providing insightful analytics and data on how well they do in conversation simulations. Teachers can learn more about their students' language skills, fluency, and communication skills by seeing how they engage with the AI-generated characters. Educators can use this data-driven strategy to make informed decisions about education and give students individualized support to boost their English proficiency [39].

AI has revolutionized the creation of qualitative study materials that cater to all four language levels: reading, speaking, listening, and writing. AI-driven platforms possess the capacity to generate exceptional personalized resources tailored to each student's unique needs and goals. The use of AI allows for the creation of individualized study guides that target certain language difficulties. By identifying specific areas that need development, students can make significant progress in their language skills through this tailored method, which can enhance the effectiveness of language learning. AI is also capable of producing intriguing and varied educational resources that address many different language-related subjects and situations. Students can practice their communication skills in real-world contexts using these resources, which cover a wide range of topics from casual chats to business encounters [51].

While students work with the course materials, AI-powered platforms can offer immediate comments and assistance. Students can get real-time feedback on their language skills, fluency, and communication tactics *via* these systems that analyze their interactions with AI-generated content. They can identify areas for improvement and make adjustments. This real-time support can enhance students' language skills and enrich their learning experience [4]. Example apps include "AWord," "Skyeng," "Anki," "LinguaLeo," and "Memrise," all of which can evaluate a person's speech and the way they pronounce specific words [34]. These vocabulary applications track the student's mistakes and suggest reviewing certain words and phrases in the next class. Many online conferencing apps, like Zoom and Google Meet, "grab" the presenter mid-discussion and place them front and center on the screen.

15.2.3 Smart Learning Curriculum with AIIoT Network

The rise of AI and the expansion of the Internet of Things (IoT) have been two of the most game-changing innovations in recent history, altering whole markets and completely changing the way we live our lives. When applied to the realm of education, these technologies can revolutionize the way teachers learn and teach, ushering in a new era of efficient and tailored instruction [26]. When the fields of artificial intelligence (AI) and the Internet of Things (IoT) began to intersect, they coined the term "Artificial Intelligence of Things" (AIIOT). AIIOT harnesses the combined capabilities of smart decision-making and the extensive network of linked sensors and devices. The Internet of Things (IoT) has significantly impacted learning and teaching, particularly in higher education. Educators should rethink and redesign educational systems using Internet-of-Things (IoT) technology to make them more responsive to rapid change, versatile, adaptable, and flexible. The Internet of Things has had a significant influence in many domains, but one of the most important is technology-enhanced learning (TEL). Smart learning environments and online learning have been made better by computer-assisted learning systems that are Internet-of-Things (IoT)–enabled. These systems let teachers remotely check and monitor students' academic progress and behavior and use smart devices and Internet-of-Things apps to make learning more interactive and personalized. Maker education, which focuses on STEM (science, technology, engineering, and mathematics) subjects and encourages students to solve problems creatively and through hands-on learning, has also benefited greatly from the Internet of Things (IoT).

Conversely, AI has the potential to radically alter the nature and function of education, which in turn would impact how humans learn and teach. Artificial intelligence (AI) uses new techniques and methods to understand data, while the Internet of Things (IoT) changes the physical landscape and the reach of data-generating devices. Artificial intelligence's (AI) capacity to process massive volumes of data, evaluate trends, and make "intelligent" decisions presents a tremendous opportunity to optimize teaching methods and improve the learning experience. The rapid development of AI has opened a world of possibilities for reshaping the transmission, acquisition, and application of knowledge. Through adaptable, customized learning experiences, AI has the potential to revolutionize both education and training. Instead of taking a cookie-cutter approach, which is common in traditional education, AI can help tailor learning programs to each student's unique background, interests, and strengths. Educational platforms can now use artificial intelligence (AI) algorithms and machine learning techniques to personalize the curriculum, tempo, and delivery of lessons to meet the needs of individual students. This allows students to learn at their own speed and improves their overall learning outcomes [47].

In addition, tools driven by AI can facilitate the production of content, the development of interactive learning materials, and the provision of immediate feedback. Also, AI can help with smart teaching systems that provide students with tailored advice and assistance all the way through their educational careers. The development of artificial intelligence has the potential to revolutionize the educational system. The combination of artificial intelligence (AI) and Internet-of-Things (IoT) technology allows for the creation of "intelligent" classrooms that can detect and respond to each student's unique needs and learning preferences. Smart classrooms powered by the Internet of Things (IoT) can gather real-time information on students' involvement, focus, and understanding from a variety of IoT devices, including sensors, wearables, and interactive whiteboards. By using this data, educators may better personalize their lessons to meet the needs of their students and increase their chances of success in the classroom. Educational institutions can improve the learning experience by collecting and analyzing real-time data through the integration of AI algorithms with IoT devices. This allows for the provision of tailored recommendations and feedback. These applications are raising some privacy and ethical concerns that should not be ignored [40].

AI is experiencing exponential growth and improvement. AI is the engine that propels the smart industry to establish itself in many different areas, but it also speeds up the progress from the "IoT" to "AI" [52]. The web's status as a national strategic data industry is critical to the growth of national fiscal policy. With advancements in AIIOT, the trend of combining AI and IoT is becoming increasingly difficult to ignore. Every nation is striving to integrate the field of artificial intelligence with the Internet of Things (IoT), aiming to create opportunities, inspire the community to actively promote "artificial intelligence" (AI) within the startup culture, and optimize and integrate AI to its fullest potential applications, such as the distribution of public funds on the web that connects everything.

Cutting-edge technological innovations, fresh concepts, and new developments in various sectors propel the progress of innovative technology and the economy, proposing more ambitious and modern standards for the education of engineers' skills and innovative abilities in the workplace. The curriculum for the Internet of Things (IoT) should foster the development of innovative and creative skills in engineers. These individuals should be high-quality and adaptable, possess excellent practical and creative skills, and excel in global competition. This is particularly important as the demands of the technological revolution continue to drive change in the manufacturing sector. The Internet of Things (IoT) and related technologies play a crucial role in fostering innovation and achieving artificial intelligence, as evidenced by their extensive use in various industry sectors such as contemporary medicine, food production, energy, manufacturing, transportation, and shipment [23]. It demands an unprecedented level of competence from workers who can utilize the Internet of Things (IoT) in various fields [48]. Experts in this field play a crucial role in implementing the innovation-driven development approach, providing solid technical assistance and skilled personnel security for the interconnected intelligent ecosystem.

According to Abaee *et al.* [1], there are primarily two schools of thought when it comes to the human-AI interaction in the field of AIED: the replacement school and the augmentation school. An earlier study in the

AIED looked at creating a rule-based personal intelligent teacher, individualized help for each student, and AI systems that could provide step-by-step tutorials [6] for the organization of material. While this method may seem promising in theory, it would effectively replace educators by automating classroom processes and offering personalized feedback and recommendations to students. As an example, current studies address the delicate balance between automating certain tasks and allowing teachers to maintain some degree of autonomy in the classroom to ensure success and avoid losing the flexibility to establish their own learning objectives.

However, inadequate automation may force teachers to perform tasks they would prefer not to perform and limit their ability to personalize their lessons [43]. A teacher's ability to strike this equilibrium may also be highly task-specific. Being aware of the need to build AI systems that can directly assist students and maximize teachers' complementary strengths, a new viewpoint known as the augmentation perspective has emerged. This viewpoint emphasizes the role of AI in making teaching processes easier by improving teachers' cognition through faster data collection, analysis, and the translation of that data into meaningful insights and actions. The AI systems in education and IT collaborate with human teachers, observing, learning, restricting, and supervising each other for optimal performance.

According to Tariq *et al.* [58], when humans and AI work together, they are able to coordinate complicated educational activities and accomplish tasks that would be impossible for each of them to do on their own. For example, Vonitsanos *et al.* [62] investigated the optimal design of intelligent tutoring systems (ITSs) to allow instructors to coordinate student-to-teacher and group-to-group activities in class sessions. Dhakshan *et al.* [16] developed a mixed-reality tool for AI-enhanced classrooms that alerts instructors in real time to potential issues that AI systems might struggle with and to keep in mind the augmentation perspective, which holds that artificial intelligence and human cognition are deeply interdependent in the context of education, and consider, build, and study the AI system as a hybrid human-AI system, incorporating both teachers and AI, as it can be challenging to disentangle AI-integrated instruction from human cognition alone. However, how to optimally integrate AI and human educators remains an open subject.

AIED research has paid less attention to instruction [16]. Based on the Wawak *et al.* [65] student-AI collaboration (SAC) model, which includes the following four dimensions for understanding TAC in the classroom, this study aims to examine how to design learning and teaching to support teacher-AI collaboration (TAC). According to Kumar *et al.* [36], curriculum design is the process of creating a course outline that considers a variety of factors, including but not limited to reflecting educational philosophies and beliefs, aligning course material with learning objectives, incorporating pedagogy and methodology into the syllabus, and developing assessments to measure student progress.

In addition to AIED's primary goal of assisting each students, recent studies have focused on using AIED to enhance students' higher-order thinking abilities [68], including creativity [10], computational thinking (CT) [66], and problem-solving skills. Through various learning activities, AIED broadly categorizes into (1) organizing AI, (2) using AI, and (3) building AI. This approach allows students to acquire subject-matter knowledge and skills according to their learning status, preferences, or personal characteristics. Conceptually, students should recognize and understand AI by considering the following: computers use sensors to perceive the world; agents maintain representations of the world and reason from them; machines learn from data; intelligent agents require a variety of knowledge to interact naturally with humans; and AI applications can have both positive and negative impacts on society. Furthermore, AIED aims to empower students to actively participate in the development and application of AI and to function as problem solvers and collaborators in project-based learning. This approach is particularly beneficial for younger students who may lack a solid understanding of abstract concepts.

There are three distinct kinds of interactions between teachers and AI approaches: cognitive, socio-emotional, and artifact-mediated. Socio-emotional interaction includes how people in a group show and shape their emotions, as well as the social and emotional climate in which they work together [8]. This is different from cognitive interaction, which involves trying to understand through reasoning, thinking, and co-elaborating knowledge. In order to facilitate cognitive engagement between AI and educators, it is essential that educators possess AI literacy. This competency allows educators to define problems, determine when and how to address them using AI, and assess the results that have a significant impact on AI [79, 80]. Along these lines, the ethical considerations of educators when choosing, developing, and using AI in the classroom are foundational, and they influence both the degree to which students rely on and feel empowered by AI in their relationships and decision-making [38]. Conversely, it is critical to have an appropriate interface that allows for two-way communication between educators and AI.

Rather than focusing on the physical area, like a classroom, educators typically talk about the social, psychological, or conceptual context as the learning and teaching environment [18]. To examine the learning space, the institution, and the culture as three distinct aspects of the environment that impact the structuring and implementation of Technology-Assisted Communication (TAC) [12]. The first is the rapid development of smart classroom systems based on IoT technology, which enable the integration of AI into real classroom education and its collaboration with teachers. Using IoT and AI power, a school gathers data in real-time from various school devices such as digital textbooks, laptops, and mobiles. It then analyzes this data to assist teachers in understanding student behaviors, predicting their learning, and creating instructional logic in real-time, thereby providing personalized adaptive assistance.

15.3 Research Methodology

15.3.1 Participants

This study employed a pre-experimental research approach to conduct a special training course for postgraduate students at *Yayasan Pendidikan Ujung Pandang (YPUP)* Makassar for the academic year 2023 with the aim of enhancing their practical skills. The study involved 7 literature teachers with extensive experience in teaching, studying, and researching literature. The sample was randomly selected from teachers who were enrolled in postgraduate studies in English education. Therefore, this study was able to avoid using the usual control group and instead concentrate on how the training course affected the participants. According to data collected from respondents with an average of 5–10 years of teaching experience, the teachers in this study were comparable to their peers at similar universities in terms of their educational background, computer skills, and interest in artificial intelligence. The study's results require careful interpretation, as the participants, not randomly selected, collaborated on all learning tasks. Each group received mobile devices and computers equipped with the latest AIIoT platform to support their initiatives.

15.3.2 Instrument

- *Conceptual Understanding of AI*

The study developed a 20-item test for conceptual understanding of AI, with a maximum possible score of 100. There is a scale from 0 (little understanding) to 5 (very good understanding) for each item. The purpose of the test is to determine how well teachers understand machine learning, neural networks, and AI applications, among other basic AI ideas in teaching.

For example, one item assesses students' knowledge of machine learning by asking them to distinguish between supervised and unsupervised learning using a given scenario. One task in neural networks asks teachers to determine the purpose of each layer of a basic feed-forward network. Regarding AI, one question tests teachers' capacity to assess its consequences in practical situations.

- *Observational AI Competence*

This study conducted the assessment of teachers' knowledge, skills, and perspectives related to AI through a competency checklist. The curriculum of this short educational session served as the inspiration for the development of this checklist, which includes a variety of competency-based questions. For instance, a knowledge assessment question prompts teachers to differentiate between traditional teaching methods and machine learning-assisted learning. Finally, one question asks teachers to rate their opinions about the social consequences of AI. On this scale, 0 means very little competence, and 5 means very much. Careful calibration ensured that each item accurately reflected the range and depth of AI competencies taught in the session and evaluated each lecturer's performance using predetermined criteria to provide a consensus score for their competence.

- *Questionnaire*

To this end, the survey was administered to teachers regarding their impressions of the AIIOT platform. The survey consisted of 20 questions with a Likert scale rating ranging from 1 (strongly disagree) to 5 (strongly agree). This survey has an acceptable level of internal consistency, as indicated by Cronbach's alpha, which is 0.864.

15.3.3 Procedure

To determine how well the AIIOT platform improves teachers' AI skills, this study used a pre-experimental research approach based on a single-group pre-posttest research design. Only conceptual understanding tests could benefit from this layout. Initially, teachers would spend about half an hour on a pretest before the training to gauge their AI understanding. This study used the pretest as a baseline assessment to identify participants' initial status prior to the implementation of the AIIOT learning activity, thereby determining its exact impact. Then, they evaluated students' opinions about the platform and their AI competencies using a posttest design only. Rather than suggesting the potential impact of the AIIOT platform, these metrics sought to indicate teachers' course learning and post-course perspectives.

Using an AI competency checklist, instructors monitored each group's progress in real time during the seminar. Before meeting to discuss their evaluations, each educator independently evaluated the participants using established criteria. As a result of this debate, they were able to resolve any differences in their initial evaluations and reach a group consensus score. Each teacher was tasked with operating the AIIOT activity through a team competition, dividing participants into two teams, which relied on cooperative activities. So, the goal of this lesson was to teach students the theory behind AIIOT, help them hone their practical skills, and encourage the right mindset when it comes to learning new things and practicing what they've learned.

Teachers spent about half an hour on the post-lesson test and fifteen minutes on the AI perception survey and interviews. Although there was no control group in this design, they were able to account for confounding variables by comparing the results before and after the intervention. Assuming there was no large gap between the tests, the researchers could more reliably attribute any changes seen in the posttest to the AIIOT learning activity rather than random chance. This study was thus able to assess the efficacy of the AIIOT learning activity by comparing the pre- and posttest results. A significant shift in posttest scores would indicate that the AIIOT learning exercise was effective.

15.4 Findings and Discussion

This study analyzes seven literature teachers from the postgraduate program at *Yayasan Pendidikan Ujung Pandang (YPUP)* in Makassar for the academic year 2023. The goal is to determine how the AIIOT platform has affected their AI literacy and teaching capacity. The following sections present the data and analysis.

Table 15.1 displays the pretest results, with an average score of 60.71 and a standard deviation of 6.07. A pretest score between fifty and sixty-five shows that there is adequate background knowledge of artificial intelligence. The posttest results showed an SD of 5.77 and an average of 87.29. Participants' post-training understanding of AI was significantly higher, with posttest values ranging from 80.00 to 96.00. The findings of the post-intervention tests demonstrate a statistically significant improvement

Table 15.1 Pretest and posttest.

	N	Minimum	Maximum	Mean	Std. deviation
Pretest	7	50.00	65.00	60.7143	6.07493
Posttest	7	80.00	96.00	87.2857	5.76525
Valid N (listwise)	7				

in AI-related knowledge and competency. The AIIOT platform significantly enhances teachers' comprehension of AI, with an average increase of 26.57 points compared to the pretest. These findings are in line with previous studies in the field of education that have demonstrated the positive effects of AI tools on professional growth and personalized learning [76]. The statistical benefits shown here provide further evidence that, with proper implementation, AIIOT systems can revolutionize pedagogy.

Table 15.2 presents the Analysis of Variance (ANOVA) evaluation of the variation in pretest scores among the 7 individuals in this study. The analysis of variance shows that only group differences explain pretest score variability. All participants had slightly different total scores, but there was no within-group variance, indicating that they did very well within their own groups. These data show that the integration of artificial intelligence (AI) and the Internet of Things (IoT) in education, known as AIIoT in Education 7.0, creates an intelligent, interactive, and personalized learning environment.

By utilizing the AIIoT platform, it is possible to collect real-time data from Internet-of-Things (IoT) devices such as smart boards, sensors, and wearables, and then apply AI to evaluate this data. Personalized learning

Table 15.2 ANOVA pretest.

	Sum of squares	df	Mean square	F	Sig.
Between Groups	221.429	6	36.905	.	.
Within groups	0.000	0	.		
Total	221.429	6			

experiences enhance student engagement and learning outcomes. By tracking each student's progress and adjusting their level of difficulty and feedback, AI can tailor learning materials. Adaptive learning systems have demonstrated improvements in academic performance, aligning with this finding [57]. AI in Education 7.0 presents the possibility of highly personalized lessons. To assess students' understanding, pace, and requirements, AIIoT platforms can monitor their real-time interactions with learning materials. Studies have shown that students are more engaged and retain more information when they receive this level of customization.

To validate significance, a more thorough analysis with complete statistical data (F-values and p-values) is required, but incomplete ANOVA findings suggest that there were some differences in pretest performance between subjects.

Table 15.3 explains that the t-value of 26.442 indicates that there is a statistically significant difference between 0 and the pretest mean score (60.71). This difference is highly statistically significant, as seen by the very low p-value ($p < 0.001$). It seems that the participants had a strong understanding of AI before the training. There is 95% certainty that the actual mean difference (pretest score) is between 55.10 and 66.33 based on the confidence interval. Additional evidence supporting the relevance of the results is the absence of overlap with 0. The posttest mean score (87.29) is substantially different from 0, as shown by the t-value of 40.057, which is even higher than the pretest t-value. Participants' knowledge and skills in AI showed significant improvements after the intervention, as shown by the highly significant difference with a p-value of less than 0.001. Participants' post-training knowledge was far from zero, as indicated by the confidence interval for the posttest (81.95 to 92.62), confirming the strong positive impact of the AIIOT platform. The AIIOT platform significantly enhanced participants' knowledge of AI, as evidenced by the highly significant

Table 15.3 One-sample test.

	Test value = 0					
					95% confidence interval of the difference	
	t	**df**	**Sig. (two-tailed)**	**Mean difference**	**Lower**	**Upper**
Pretest	26.442	6	.000	60.71429	55.0959	66.3327
Posttest	40.057	6	.000	87.28571	81.9537	92.6177

t-values for both the pretest and posttest and the substantial increase in mean scores from 60.71 to 87.29. This transition greatly enhanced educators' ability to incorporate AI into their teaching techniques, demonstrating the revolutionary potential of AIIOT technology in education.

A data-driven style of education optimizes learning and teaching using AI and IoT technology. With the ongoing development of AIIOT technologies, immersive technologies, such as AR and VR, will likely play a significant role in future classrooms. These technologies will enable students to engage with subjects through more hands-on and interactive exploration. Smart classrooms where students' real-time feedback optimizes the learning environment for optimal focus and comfort could be a reality thanks to these improvements. Ambient factors, including lighting, temperature, and seating configurations, change automatically.

The findings of the one-sample t-test show that there were substantial changes between the instructors' scores on the pretest and posttest, indicating that the AIIOT platform greatly enhanced their understanding of AI. This shows that AIIOT can help educators improve their skills, which is crucial to Education 7.0 and the widespread use of these technologies in the classroom.

Table 15.4 displays the reliability statistics of the AIIOT (artificial intelligence and Internet of Things) platform in Education 7.0. The platform is concerned with a 20-item questionnaire that assesses the views and abilities of educators regarding the incorporation of AI in the classroom. The reliability of the data is 0.864, which indicates a high level of reliability and indicates the close connection and consistent measurement of the same underlying notion among the items in the questionnaire. A Cronbach's alpha number of 0.8 indicates strong dependability, whereas a value above 0.7 is generally deemed satisfactory.

This reliability shows that we accurately conducted assessments of teachers' views and competencies related to AIIOT platforms within the framework of AIIOT in Education 7.0. Researchers can then use the survey results with confidence to draw conclusions regarding the efficacy of AIIOT in the classroom. The AIIOT architecture in Education 7.0 combines the Internet of Things (IoT) and artificial intelligence (AI) to create

Table 15.4 Reliability statistics of AIIOT.

Cronbach's Alpha	N of items
0.864	20

a personalized and dynamic learning environment. Teachers repeatedly commended the platform's capacity to boost pedagogical approaches through AI-driven insights and real-time data from IoT devices, including sensors, wearables, and smartboards, due to the high reliability of the data.

15.4.1 Findings

Three respondents provided the results for this research interview, with question 1 highlighting the potential of AIIoT in classroom 70. They responded that Education 7.0 will greatly benefit from the revolutionary possibilities presented by AIIoT and mostly humanized AI, as discussed in the first question. Educators can design individual learning spaces that change in response to each student's strengths and weaknesses by leveraging data collected in real-time through connected devices such as sensors, wearables, and smart boards. According to the respondents, they can make necessary adjustments to their learning paths because of the rapid feedback they receive through this personalized method. By automating tedious processes such as assessments and lesson modifications, teachers can not only spend more time with students but also engage them on a deeper level. The adaptive nature of AIIoT-enhanced classrooms boosts teacher performance and motivation, making learning more interactive and engaging.

Question 2 inquired about the challenges involved in implementing AIIoT. The second question started a conversation about the main challenges that schools face when trying to use AIIoT networks. As AI and IoT systems rely on the collection and analysis of large amounts of student data, data privacy and security are important concerns. Concerns about the handling of personal information are increasing, and the education sector is becoming more vulnerable to data breaches as a result. Furthermore, inadequate infrastructure poses a significant obstacle, particularly in impoverished schools or rural regions that lack the financial means to purchase the cutting-edge technology essential for AIIoT. Adopting these methods at scale may be challenging for these institutions due to resource allocation issues. A lack of teacher training also contributes to educators' inability to successfully integrate AI-driven tools into their classrooms. Furthermore, ethical concerns, including the potential for AI bias, necessitate careful policymaking and safeguards to prevent the exacerbation of current educational disparities. To ensure that all educational settings have equitable access to the benefits of AIIoT, it is important to address these barriers.

The next question is, "How does the Internet of Things affect teachers' AI competence?" The discussion of the third question highlighted the positive influence of AIIoT on teachers' AI competence. After using the AIIoT platform, instructors' understanding of AI principles and how to apply them in educational settings significantly improved, indicating that they (teachers) were able to automate assessments, engage students more effectively, and personalize learning with the use of AI insights. The platform proved its power to revolutionize teaching methods by making them more data-driven and adaptable to the unique needs of each student, while also helping instructors improve their AI literacy. The ability to integrate AI into everyday teaching lays the foundation for future AI-enhanced education models, enabling teachers to optimize their teaching tactics through technology and thereby contributing to the advancement of Education 7.0.

15.4.2 Discussion

The Internet of Things (IoT) paves the way for adaptive learning by letting teachers collect and evaluate data from students' interactions as they happen. Using these tools, educators can personalize lessons for each student, meeting their unique learning requirements [72]. Credible assessments of educators' AIIOT platform experiences provide credence to the idea that AIIOT has the potential to greatly improve educational practices. Teachers view AIIOT platforms as useful tools to enhance teaching efficiency and student engagement, as evidenced by the consistent replies across all 20 categories. The focal areas of the questionnaire probably mirrored the features of AIIOT in Education 7.0, which include personalized learning paths, automated assessment, and real-time feedback.

All these platforms can help teachers identify students who are struggling and provide them with targeted interventions to improve their performance and retention, similar to the Johnson *et al.* [75] study. Implementing AIIOT in the classroom is difficult, but the benefits are great, and it carefully addresses ethical considerations and data privacy when collecting student data. Concerns exist that AI algorithms might exhibit bias and unintentionally exacerbate existing educational disparities [45]. Schools and educators require adequate infrastructure and training to successfully utilize AIIOT platforms. In underfunded schools or rural areas, these technologies may be unaffordable due to the digital divide. Therefore, it is imperative to address resource gaps before implementing such systems on a large scale. Education 7.0 signifies a shift toward highly interactive education.

The study found that educators can better manage classroom dynamics, personalize learning, and improve student outcomes with the help of AI-driven tools in AI-enhanced learning environments. This aligns with the notable rise in posttest scores [76]. By combining the Internet of Things (IoT) with artificial intelligence (AI), AIIOT in Education 7.0 creates an adaptive learning environment and real-time data. Here, AIIOT platforms can collect information from a variety of IoT devices (including smartboards, sensors, and wearables) to instantly modify lesson plans to suit the needs of individual students. With the emphasis on student engagement and customized learning paths in Education 7.0, this level of personalized instruction becomes critical. Teachers can better manage heterogeneous classes with the help of adaptive AI systems that provide personalized learning recommendations and feedback to each student. Educators can reduce some of the administrative burden by automating assessments and attendance monitoring using AIIOT platforms. Teachers will have more time for high-level tasks such as student guidance and curriculum development as their confidence in using AI-based technologies increases.

Teachers seem to be aware of the platform's usefulness, since they scored highly in areas like being confident when utilizing AI tools, knowing how AI affects student results, and being able to differentiate between traditional and AI-assisted learning. The very high reliability score suggests that the results of the study should be applicable in other classrooms. Given the positive effects of AIIOT platforms on teacher engagement and student learning outcomes in this study, we can reasonably assume that other diverse educational settings with more widespread adoption may achieve comparable outcomes. In contexts where student demands are very varied, educators may find AIIOT appealing because of its capacity to improve instructional flexibility and streamline administrative tasks, such as grading, attendance, and student progress monitoring. There are obstacles to deploying AIIOT in Education 7.0, notwithstanding the positive outcomes. To guarantee that educators can make effective use of these tools, infrastructure and training are crucial. IoT devices collect massive volumes of student data, raising ethical considerations related to data privacy and security. As the use of AIIOT systems grows in popularity, we must also carefully consider concerns about student autonomy, bias in AI algorithms, and fair access to the technology across different educational contexts.

The reliability results demonstrate the high trustworthiness of the 20-item questionnaire in evaluating the AIIOT platform in Education 7.0, as evidenced by its high Cronbach's alpha score of 0.864. The instrument effectively captured teachers' opinions on the benefits of AIIOT, including improved pedagogical approaches, enhanced student engagement, and support for adaptive learning. It indicated consistent responses. To ensure the success of this novel approach to learning, it is vital to equip instructors with the necessary tools and address ethical concerns as AIIOT continues to revolutionize educational processes.

In the findings, there is the issue of how AIIOT learning differs from more traditional classroom methods. Score between 4.00 and 5.00, the mean is 4.57, and the standard deviation is 0.53. Educators' self-assurance in differentiating between conventional and AIIOT-based pedagogical practices was impressive. The larger literature [76] highlights that AI-enhanced teaching methods enable personalized learning and deeper engagement. Most teachers seem to have no trouble grasping these distinctions, since the modest standard deviation suggests that participants' answers are consistent with one another. Second, we need to evaluate the advantages and disadvantages of integrating AI. The score is 3.00 to 5.00, and the range is as follows: mean = 4.00 and SD = 0.58. The mean score is still positive, but there seems to be more variety in teachers' capacity to critically appraise the merits and downsides of AI in education, as indicated by the broader range (3.00 to 5.00) and higher standard deviation. Here is where things really get interesting; educators must weigh the benefits of AI in terms of personalization against obstacles like data privacy and ethical considerations [77].

Statement number 3 pertains to the assurance in discovering AI-powered resources to enhance education, with a standard deviation of 0.53 and a mean of 4.57. In these statements, educators were quite certain that they could find AI resources to supplement their pedagogical practices. This assurance shows that teachers are willing to embrace new technology to enhance their teaching methods, which is essential for the smooth incorporation of AI into classrooms. Next, a statement about recognizing the boundaries of existing educational AI, with the data of SD = 0.49 and mean = 4.29, shows educators generally agree within reasonable margins of error on the constraints imposed by AI in the classroom. Responsible use of artificial intelligence requires an awareness of its limitations, such as the fact that it cannot substitute human empathy and that algorithms may be biased [54].

Statement number 5 asked about being able to demonstrate the effects of AI on the results of student learning. The described standard deviation is 0.53, and the mean is 4.57. The participants are confident in their ability to explain how AI can impact student outcomes. AI-driven systems that provide tailored feedback and adaptive learning routes can enhance students' interest and performance [5, 49]. Teachers demonstrate a high level of comprehension, as seen by their consistent responses. Then, statement number 6 asked about belief in one's capacity to analyze AI's impact on society; the statistical data show that the mean is 4.14, and the standard deviation is 0.38. More training or conversation is needed regarding the broader societal implications of AI, such as its impact on job markets, privacy, and educational equity. Teachers were generally confident, but they were a little less so in this area. According to Popenici and Kerr [78], there is a significant amount of discussion in educational literature about the ethical considerations of AI.

In statement number 7, using artificial intelligence to assess students, the standard deviation is 0.53, and the mean is 4.57. The data described that educators were optimistic about their capacity to use AI in assessing students. There has been a recent uptick in the adoption of AI-powered systems for real-time student performance evaluations, with the promise of actionable information for educators in line with Chigbu *et al.* [13]. Then, there is the statement about a curriculum for AI competence, which had a standard deviation of 0.58. Educators felt they had a good grasp of the AI curriculum, but they could do better. Some teachers might need further help to understand the skills required for AI-driven instruction, according to the variant.

Question number 9, which asked about the function of the teacher in an AI-enabled lesson, showed that the standard deviation is 0.53, and the mean is 4.43. In an AI-driven classroom, teachers feel more at ease describing themselves as facilitators who guide students through customized learning paths, rather than as content deliverers. In settings where artificial intelligence handles most of the regular evaluation and feedback, this role is crucial. Then, question number 10 asked about elucidating AI in personalized learning settings, and it showed that the standard deviation is 0.53, and the mean is 4.57. In line with recent educational trends centering on individualized learning, participants showcased a solid grasp of how AI may be utilized to construct customized learning environments.

The evaluation of AIIOT platforms, specifically question numbers 11 to 20, described that educators were very pleased with the AIIOT platform's capacity to facilitate personalized learning, user-friendliness, and time-saving features, as indicated by their consistently high scores (ranging from 4.14 to 4.71). With a mean score of 4.71 and a standard deviation of 0.49, confidence in enabling AI-based learning with AIIOT platforms tops the list. Teachers found the platform valuable, but they need to improve its ability to meet individual student requirements. The lowest grade, which is still favorable, was for personalizing student learning experiences (mean = 3.71, SD = 0.76). Both students and educators can benefit from AIIoT technology. Artificial intelligence (AI) systems can greatly assist instructors by providing assessments, assessing student performance, and offering insights into student learning. According to Roopaei and Roopaei [50], this frees up teachers to spend more time engaging with students and coming up with innovative ways to teach.

The data collectively demonstrate that the AIIOT platform enhanced educators' understanding and familiarity with AI in the classroom. While the platform did a good job overall in terms of AI literacy development, there is still room for improvement in areas like tailoring student experiences and assessing the social ramifications of AI. With the ability to empower instructors to develop more tailored and efficient learning environments, this rigorous investigation underlines the promise of AIIOT networks to improve education.

15.5 Conclusion

Artificial intelligence and the Internet of Things are emerging fields of study, and this article shows how they can transform the way schools work by creating intelligent, personalized learning spaces. Innovation in Education 7.0 is a combination of real-time data collection and adaptive learning, which enables personalized experiences for students and improves their engagement and learning outcomes. The integration of AI and IoT enables this. This study primarily adds to the body of knowledge by providing concrete evidence that AIIoT is useful for enhancing educators' AI skills and fostering engaging and collaborative learning environments. A major leap forward in educational technology, the study shows that AIIoT can automate testing and feedback, reduce administrative tasks, and provide students with personalized learning pathways. The report highlights the necessity of developing policies related to data security and privacy, as well as extensive teacher training programs aimed at equipping educators with the necessary resources to integrate AIIoT into their teaching methods while simultaneously supporting initiatives to close the digital gap by ensuring equal access to technology for all students, especially those living in remote regions or with limited financial means.

AI disclosure statement: The author confirms that no artificial intelligence (AI)-generated content or AI-assisted writing tools were used in the preparation of this chapter (including summaries and the abstract). All text, analysis, and conclusions are the result of the author's original work.

References

1. Abaee, M., Saeedi, M., Taherdoost, H., Examining the Role of IoT and Cloud Computing in Achieving Sustainable Development Goals, in: *Reshaping Environmental Science Through Machine Learning and IoT*, pp. 74–97, IGI Global, USA, 2024, DOI: http://dx.doi.org/10.4018/979-8-3693-2351-9.ch005.
2. Abulibdeh, A., Zaidan, E., Abulibdeh, R., Navigating the confluence of artificial intelligence and education for sustainable development in the era of industry 4.0: Challenges, opportunities, and ethical dimensions. *J. Cleaner Prod.*, 140527, 2024, DOI: http://dx.doi.org/10.1016/j.jclepro.2023.140527.
3. Aithal, P.S., Prabhu, S., Aithal, S., Future of Higher Education through Technology Prediction and Forecasting. *Poornaprajna Int. J. Manag. Educ. Soc. Sci. (PIJMESS)*, 1, 1, 01–50, 2024, https://dx.doi.org/10.2139/ssrn.4901474.
4. Akpan, I.J., Kobara, Y.M., Owolabi, J., Akpan, A.A., Offodile, O.F., Conversational and generative artificial intelligence and human–chatbot interaction in education and research. *Int. Tran. Oper. Res.*, 2024, https://onlinelibrary.wiley.com/doi/abs/10.1111/itor.13522.
5. Al Yakin, A., Muthmainnah, M., Al-Matari, A.S., de Barros Cardoso, L.M.O., Yunus, N.H., Hajar, S., Harianti, H., Transferability Cybersocialization as a Digital Civility Through Artificial Intelligence-Based Design Thinking Mindset to Modern Pedagogy, in: *International Conference on Advances in Information Communication Technology & Computing*, 2024, April, Springer Nature Singapore, Singapore, pp. 237–250, http://dx.doi.org/10.1007/978-981-97-6106-7_13.
6. Allioui, H., Allioui, A., Mourdi, Y., Navigating transformation: unveiling the synergy of IoT, multimedia trends, and AI for sustainable financial growth in African context. *Multimedia Tools Appl.*, 1–45, 2024, https://doi.org/10.1007/s11042-024-19038-6.
7. AlTwijri, L. and Alghizzi, T.M., Investigating the integration of artificial intelligence in English as foreign language classes for enhancing learners' affective factors: A systematic review. *Heliyon*, 10, 10, 2024, https://doi.org/10.1016/j.heliyon.2024.e31053.
8. Arts, E., De Castro, B.O., Luteijn, E., Elsendoorn, B., Vissers, C.T., Interactive virtual reality training to improve socio-emotional functioning in adolescents with developmental language disorders: A feasibility study. *Clin. Child Psychol. Psychiatry.*, 29, 3, 1100–1120, 2024, doi:10.1177/13591045231220694.
9. Ayanwale, M.A., Adelana, O.P., Molefi, R.R., Adeeko, O., Ishola, A.M., Examining artificial intelligence literacy among pre-service teachers for future classrooms. *Comput. Educ. Open*, 6, 100179, 20242024, https://doi.org/10.1016/j.caeo.2024.100179.

10. Aytekin, A. and Topçu, M.S., Improving 6th Grade Students' Creative Problem Solving Skills Through Plugged and Unplugged Computational Thinking Approaches. *J. Sci. Educ. Technol.*, 33, 1–25, 2024, https://doi.org/10.1007/s10956-024-10130-y.
11. Biswal, A.K., Avtaran, D., Sharma, V., Grover, V., Mishra, S., Alkhayyat, A., Transformative Metamorphosis in Context to IoT in Education 4.0. *EAI Endorsed Trans. Internet Things*, 10, 2024, https://doi.org/10.4108/eetiot.4636.
12. Bui, T., English as a Foreign Language Teachers' Perspectives on the Use of Information and Communication Technology in Early Language Education: Vietnamese Context, Master's thesis, Itä-Suomen yliopisto, University of Eastern Finland, Philosophical Faculty, School of Applied Educational Science and Teacher Education, 2024, https://erepo.uef.fi/server/api/core/bitstreams/d8bf0db5-3449-4757-a6c6-2993b89d1e92/content.
13. Chigbu, B.I., Umejesi, I., Makapela, S.L., Adaptive Intelligence Revolutionizing Learning and Sustainability in Higher Education: Enhancing Learning and Sustainability With AI, in: *Implementing Interactive Learning Strategies in Higher Education*, pp. 151–176, IGI Global, Brazil, 2024, DOI: 10.4018/979-8-3693-3559-8.ch006.
14. da Silva, P.A., de Oliveira Filho, F.L.C., Theodorovski, R., Maravalhas, A.L.G., Hansel, T.F., Monteiro, R.R., Viudes, M.M., Integration of artificial intelligence and GPT API applied to education: challenges and potentialities of digital teaching. *Caderno Pedagógico*, 21, 2, e2708–e2708, 2024, https://doi.org/10.54033/cadpedv21n2-036.
15. Dai, Z., Yang, Y., Chen, Z., Wang, L., Zhao, L., Zhu, X., Xiong, J., The role of project-based learning with activity theory in teaching effectiveness: Evidence from the internet of things course. *Educ. Inf. Technol.*, 1–33, 2024, https://doi.org/10.1007/s10639-024-12965-9.
16. Dhakshan, S., Balamurugan, G., Mohan, J.S., Tyagi, A.K., Role of Emerging Technologies in Education 4.0: Challenges and Future Research Directions, in: *Architecture and Technological Advancements of Education 4.0*, pp. 131–154, 2024, DOI: 10.4018/978-1-6684-9285-7.ch006.
17. Fang, F. and Jiang, X., The analysis of artificial intelligence digital technology in art education under the internet of things. *IEEE Access*, 12, 22928–22937, 2024, doi: 10.1109/ACCESS.2024.3363655.
18. García-Monge, A., Bores-García, D., González-Calvo, G., Socio-emotional aspects articulated with content in an online teachers' community of practice: 'agreeable dialogue zones. *Teach. Dev.*, 28, 2, 215–243, 2023, https://doi.org/10.1080/13664530.2023.2286982.
19. Graefen, B. and Fazal, N., From Chat bots to Virtual Tutors: An Overview of Chat GPT's Role in the Future of Education. *Arch. Pharm. Pract.*, 15, 2, 43–52, 2024, https://doi.org/10.51847/TOuppjEDSX.

20. Gupta, A.K., Aggarwal, V., Sharma, V., Naved, M., Education 4.0 and Web 3.0 technologies application for enhancement of distance learning management systems in the post–COVID-19 era, in: *The Role of Sustainability and Artificial Intelligence in Education Improvement*, pp. 66–86, Chapman and Hall/CRC, London, 2024. https://www.taylorfrancis.com/chapters/edit/10.1201/9781003425779-4/education-4-0-web-3-0-technologies-application-enhancement-distance-learning-management-systems-post%E2%80%93covid-19-era-aditya-kumar-gupta-vivek-aggarwal-vinita-sharma-mohd-naved.
21. Hallal, L., Rhinelander, J., Venkat, R., Recent trends of authentication methods in extended reality: A survey. *Appl. Syst. Innovation*, 7, 3, 45, 2024, https://doi.org/10.3390/asi7030045.
22. Haque, M.A., Ahmad, S., Hossain, M.A., Kumar, K., Faizanuddin, M., Islam, F., Nazeer, J., Internet of things enabled e-learning system for academic achievement among university students. *E-Learning Digital Media*, 2024, Advance online publication, https://doi.org/10.1177/20427530241280007.
23. Hemamalini, V., Mishra, A.K., Tyagi, A.K., Kakulapati, V., Artificial intelligence–blockchain-enabled–internet of things-based cloud applications for next-generation society, in: *Automated Secure Computing for Next-Generation Systems*, pp. 65–82, 2024, http://dx.doi.org/10.1002/9781394213948.ch4.
24. Hossain, R., Adopting Industry 4.0: A strategic solution for transforming Smart Bangladesh: Prospective connections, opportunities, and challenges. *Pak. J. Life Social Sci.*, 22, 1, 3304–3323, 2024, https://doi.org/10.57239/PJLSS-2024-22.1.00240.
25. Huang, F., Wang, Y., Zhang, H., Modelling generative AI acceptance, perceived teachers' enthusiasm and self-efficacy to English as a foreign language learners' well-being in the digital era. *Eur. J. Education. Adv. Online Publ.*, 2024, https://doi.org/10.1111/ejed.12770.
26. Ironsi, C.S., Exploring the potential of generative AI in English language teaching, in: *Facilitating Global Collaboration and Knowledge Sharing in Higher Education with Generative AI*, pp. 162–185, IGI Global, USA, 2024, DOI: 10.4018/979-8-3693-0487-7.ch007.
27. Ismaeel, H., Salih, A.A., Ahmed, O.M., Yazdeen, A.A., Abdullah, R.M., Sami, T.M.G., A review of research methodologies for analyzing the sustainability benefits and challenges of AI, IoT, and enterprise systems integration. *J. Inf. Technol. Inf.*, 3, 2, 2024, Domain hilang.
28. Iyer, S.S., Singh, A.K., Subramanian, R., Reyes Jr., F.E., Khan, F., Tantry, A., Krishnan, A.S., The usefulness of big data and IoT/AI at Dubai University. *Kurdish Stud.*, 12, 2, 6198–6220, 2024, http://kurdishstudies.net/menu-script/index.php/KS/article/view/2828.
29. Javaid, Z.K., A systematic review on cognitive and motivational impact on English language learning through artificial intelligence. *Int. J. Literature Ling. Transl. Stud.*, 4, 1, 2024, https://www.ojs.wumardan.edu.pk/index.php/ijllts/article/view/286.

30. Joshi, K., Kumar, R., Bharany, S., Saini, D.K.J.B., Kumar, R., Ibrahim, A.O., Medani, M., Exploring the connectivity between Education 4.0 and Classroom 4.0: Technologies, student perspectives, and engagement in the digital era. *IEEE Access*, 2024, Advance online publication, https://doi.org/10.1109/ACCESS.2024.1234567.
31. Khazanchi, R. and Khazanchi, P., Generative AI to improve special education teacher preparation for inclusive classrooms, in: *Exploring New Horizons: Generative Artificial Intelligence and Teacher Education*, p. 159, 2024, https://www.learntechlib.org/primary/p/223928/ebook_223928.pdf#page=165.
32. Kim, J., Leading teachers' perspective on teacher-AI collaboration in education. *Educ. Inf. Technol.*, 29, 7, 8693–8724, 2024, https://doi.org/10.1007/s10639-024-11023-7.
33. Klimenko, A.Y., Technological change and its effect on education. *J. Educ. Psychol. Res. (JEPR)*, 6, 1, 1–18, 2024, https://jeprjournal.com/index.php/jepr/article/view/123.
34. Kolegova, I.A. and Levina, I.A., Using artificial intelligence as a digital tool in foreign language teaching. *Вестник Южно-Уральского государственного университета. Серия Образование. Педагогические науки*, 16, 1, 102–110, 2024, https://cyberleninka.ru/article/n/using-artificial-intelligence-as-a-digital-tool-in-foreign-language-teaching.
35. Kot, S.O. and Nykyporets, S.S., Utilization of artificial intelligence in enhancing English language proficiency in tertiary education, in: *Science and Education in the Third Millennium: Information Technology, Education, Law, Psychology, Social Sphere, Management*, pp. 250–274, 2024, Chap. 10, https://ir.lib.vntu.edu.ua/handle/123456789/42590.
36. Kumar, A., Rani, M., Sisodia, D.R., Perwej, Y., Kakde, A.C., Rakhimjonovna, F.M., Transforming education through IoT and AI opportunities and challenges. *Educ. Adm. Theory Pract.*, 30, 5, 11610–11622, 2024, https://doi.org/10.53555/kuey.v30i5.4982.
37. Largo, M.A., Sederia, T.M., Villadores, F.C., Andrada, M., Smart Chat, Bright Minds: Does Chat GPT Propel Students to Academic Heights? *J. Educ. Learn. Advancements*, 1, 1, 8–21, 2024, Retrieved from https://jelamagste.org/journalofed/index.php/jela/article/view/14.
38. Li, X., Li, Y., Wang, R., Li, J., Exploring fluctuations in collaborative engagement: How do cognitive and socio-emotional interaction intertwine in online collaborative learning? *Educ. Technol. Res. Dev.*, 1–25, 2024, https://doi.org/10.1007/s11423-024-10012-3.
39. Loor, M.A.M., Solorzano, D.M.A., Katherine, A., Moreira, V., Integration of artificial intelligence in english teaching. *J. Cleaner Prod.*, 289, 125834, 2024.
40. Matthew, U.O., Oyekunle, D.O., Akpan, E.E., Oladipupo, M.A., Chukwuebuka, E.S., Adekunle, T.S., Onumaku, V.C., Generative artificial intelligence (AI) on sustainable development goal 4 for tertiary education: Conversational AI with user-centric ChatGPT-4, in: *Impacts of Generative AI on Creativity in*

Higher Education, pp. 259–288, IGI Global, USA, 2024, DOI: 10.4018/979-8-3693-2418-9.ch010.
41. Mouta, A., Torrecilla-Sánchez, E.M., Pinto-Llorente, A.M., Design of a future scenarios toolkit for an ethical implementation of artificial intelligence in education. *Educ. Inf. Technol.*, 29, 9, 10473–10498, 2024, https://doi.org/10.1007/s10639-024-11045-1.
42. Murdan, A.P. and Halkhoree, R., Integration of artificial intelligence for educational excellence and innovation in higher education institutions, in: *2024 1st International Conference on Smart Energy Systems and Artificial Intelligence (SESAI)*, 2024, June, IEEE, pp. 1–6, Tidak terdeteksi.
43. Mustafa, A.A., Abdulqader, D.M., Ahmed, O.M., Ismael, H.R., Hasan, S., Ahmed, L.H., Based on principles of clouding and web technology: A review of using AI, IoT, and secure enterprise systems for energy efficiency focusing on smart buildings, sustainable future. *J. Inf. Technol. Inf.*, 3, 2, 2024, Tidak terdeteksi.
44. Muthmainnah, M., Yakin, A.A., Obaid, A.J., Ziboon, A.R.T., Shaping user behavior in EFL education as an evaluation of artificial intelligence, computational thinking, and computational participations frameworks in digital humanities, in: *AIP Conference Proceedings*, 2024, September, vol. 3207, No. 1, AIP Publishing, https://doi.org/10.1063/5.0234077.
45. Ogunleye, B., Zakariyyah, K.I., Ajao, O., Olayinka, O., Sharma, H., Higher education assessment practice in the era of generative AI tools, arXiv preprint arXiv:2404.01036, 2024, https://doi.org/10.48550/arXiv.2404.01036.
46. Palamar, S.P., Brovko, K., Semerikov, S.O., Enhancing foreign language learning in Ukraine: Immersive technologies as catalysts for cognitive interest and achievement, in: *Selected Papers of the X International Scientific Conference "Information Technology and Implementation" (IT&I-2023). Conference Proceedings. Kyiv, Ukraine, November 20–21, 2023*, No. 3624, pp. 69–81, 2024, CEUR Workshop Proceedings, https://lib.iitta.gov.ua/id/eprint/741724/.
47. Peters, M. A., Jackson, L., Papastephanou, M., Jandrić, P., Lazaroiu, G., Evers, C. W., ... & Fuller, S., AI and the future of humanity: ChatGPT-4, philosophy and education–Critical responses. *Educational Philosophy and Theory*, 56, 9, 828–862, 2024.
48. Rath, K.C., Khang, A., Roy, D., The role of Internet of Things (IoT) technology in Industry 4.0 economy, in: *Advanced IoT technologies and applications in the industry 4.0 digital economy*, CRC Press, pp. 1–28, 2024, https://www-taylorfrancis-com.ezproxy.ugm.ac.id/chapters/edit/10.1201/9781003434269-1/role-internet-things-iot-technology-industry-4-0-economy-kali-charan-rath-alex-khang-debanik-roy.
49. Rekha, K., Gopal, K., Satheeskumar, D., Anand, U.A., Doss, D.S.S., Elayaperumal, S., Ai-Powered Personalized Learning System Design: Student Engagement and Performance Tracking System, in: *2024 4th International Conference on Advance Computing and Innovative Technologies*

in Engineering (ICACITE), 2024, May, IEEE, pp. 1125–1130, https://doi.org/10.1109/ICACITE60783.2024.10617155.

50. Roopaei, M. and Roopaei, R., Gamifying AI Education for Young Minds: The TransAI Adventure in Learning, in: *2024 IEEE World AI IoT Congress (AIIoT)*, 2024, May, IEEE, pp. 129–134, https://doi.org/10.1109/AIIoT61789.2024.10578991.
51. Shafiee Rad, H., Revolutionizing L2 speaking proficiency, willingness to communicate, and perceptions through artificial intelligence: a case of Speeko application. *Innovation Lang. Learn. Teach.*, 18, 4, 364–379, 2024, https://doi-org.ezproxy.ugm.ac.id/10.1080/17501229.2024.2309539.
52. Shwedeh, F., Salloum, S.A., Aburayya, A., Fatin, B., Elbadawi, M.A., Al Ghurabli, Z., Al Dabbagh, T., AI Adoption and Educational Sustainability in Higher Education in the UAE, in: *Artificial Intelligence in Education: The Power and Dangers of ChatGPT in the Classroom*, Springer Nature Switzerland, Cham, pp. 201–229, 2024, https://doi.org/10.1007/978-3-031-52280-2_14.
53. Simsek, A., How Will Education Look Like in the Future? Contemporary Educational Technology. *Contemp. Educ. Technol.*, 16, 3, 2024, https://eric.ed.gov/?id=EJ1437435.
54. Sindiramutty, S.R., Prabagaran, K.R.V., Jhanjhi, N.Z., Murugesan, R.K., Malik, N.A., Hussain, M., Ethics and transparency in secure web model generation, in: *Reshaping cybersecurity with generative AI techniques*, T. Tariq (Ed.), pp. 411–464, IGI Global, USA, 2025, https://doi.org/10.4018/978-1-7998-1234-5.ch022.
55. Sok, S. and Heng, K., Opportunities, challenges, and strategies for using ChatGPT in higher education: A literature review. *J. Digital Educ. Technol.*, 4, 1, ep2401, 2024, https://doi.org/10.29333/jdet.2024.v4n1.2401.
56. Szalkowski, G.A. and Johansen, C., Defining and measuring the effects of digital technologies on social sustainability: A systematic literature review. *Sustainable Dev.*, 32, 3, 1678–1699, 2024, https://doi.org/10.1002/sd.2267.
57. Tariq, M.U. (Ed.), *AI-driven learning and engagement in higher education*, IGI Global, USA, 2024, https://doi.org/10.4018/978-1-7998-6789-0.
58. Tariq, R., Casillas-Muñoz, F., Hassan, S., Ramírez-Montoya, M., Synergy of Internet of Things and education: Cyber-physical systems contributing towards remote laboratories, improved learning, and school management. *J. Social Stud. Educ. Res.*, 15, 2, 305–352, 2024, https://doi.org/10.17499/jsser.v15i2.1234.
59. Tiwari, S., The rise of intelligent machines: An introduction to artificial intelligence, in: *Artificial intelligence and machine learning in drug design and development*, pp. 1–22, IGI Global, USA, 2024, DOI: 10.4018/979-8-3693-2333-5.ch003.
60. Urbaite, G., The role of technology in modern language education. *EuroGlobal J. Ling. Lang. Educ.*, 1, 1, 3–10, 2024, https://doi.org/10.69760/w00r1v81.

61. Van Wyk, M.M., Is ChatGPT an opportunity or a threat? Preventive strategies employed by academics related to a GenAI-based LLM at a faculty of education. *J. Appl. Learn. Teach.*, 7, 1, 2024, https://doi.org/10.37074/jalt.v7n1.a3.
62. Vonitsanos, G., Moustaka, I., Doukakis, S., Mylonas, P., Transforming Education in the Digital Age: Exploring the Dimensions of Education 4.0, in: *2024 IEEE Global Engineering Education Conference (EDUCON)*, 2024, May, IEEE, pp. 01–10, https://doi.org/10.1109/EDUCON60312.2024.10578747.
63. Waliu, A.O., Muojekwu, E.E., Matthew, U.O., Kazaure, J.S., Oseni, V.E., Ononiwu, C.C., Haruna, K., Clinical Simulation in Nursing Education: Immersive Educational Tech for Nurses and Midwives, in: *Creating Immersive Learning Experiences Through Virtual Reality (VR)*, pp. 43–76, IGI Global, USA, 2025, DOI: 10.4018/979-8-3693-6407-9.ch003.
64. Wang, Y., Wang, L., Siau, K.L., Human-Centered Interaction in Virtual Worlds: A New Era of Generative Artificial Intelligence and Metaverse. *Int. J. Human–Computer Interact.*, 1–43, 2024, https://doi-org.ezproxy.ugm.ac.id/10.1080/10447318.2024.2316376.
65. Wawak, S., Teixeira Domingues, J.P., Sampaio, P., Quality 4.0 in higher education: reinventing academic-industry-government collaboration during disruptive times. *TQM J.*, 36, 6, 1569–1590, 2024, https://doi.org/10.1108/TQM-07-2023-0219.
66. Weng, X., Ye, H., Dai, Y., Ng, O., Integrating Artificial Intelligence and Computational Thinking in Educational Contexts: A Systematic Review of Instructional Design and Student Learning Outcomes. *J. Educ. Comput. Res.*, 62, 6, 1640–1670, 2024, https://doi.org/10.1177/07356331241248686.
67. Williamson, B., The social life of AI in education. *Int. J. Artif. Intell. Educ.*, 34, 1, 97–104, 2024, https://doi.org/10.1007/s40593-023-00342-5.
68. Wu, T.T., Silitonga, L.M., Murti, A.T., Enhancing English writing and higher-order thinking skills through computational thinking. *Comput. Educ.*, 213, 105012, 2024, https://doi.org/10.1016/j.compedu.2024.105012.
69. Yadav, S., Reimagining education with advanced technologies: Transformative pedagogical shifts driven by artificial intelligence, in: *Impacts of Generative AI on the Future of Research and Education*, pp. 1–26, IGI Global, USA, 2025, DOI: 10.4018/979-8-3693-0884-4.ch001.
70. Yalli, J.S., Hasan, M.H., Badawi, A., Internet of Things (IoT): Origin, embedded technologies, smart applications and its growth in the last decade. *IEEE Access*, 2024, https://doi.org/10.1109/ACCESS.2024.3418995.
71. Yeh, H.C., Revolutionizing language learning. *Educ. Technol. Soc.*, 27, 3, 335–353, 2024, https://www.airitilibrary.com/Common/Click_DOI?DOI=10.30191%2fETS.202407_27(3).TP01.
72. Yigci, D., Eryilmaz, M., Yetisen, A.K., Tasoglu, S., Ozcan, A., Large language model-based chatbots in higher education. *Adv. Intell. Syst.*, 2400429, 2024, https://doi.org/10.1002/aisy.202400429.

73. Yu, J.H., Chauhan, D., Iqbal, R.A., Yeoh, E., Mapping academic perspectives on AI in education: Trends, challenges, and sentiments in educational research (2018–2024). *Educ. Technol. Res. Dev.*, 1–29, 2024, https://doi.org/10.1007/s11423-024-10425-2.
74. Zhang, P. and Tur, G., A systematic review of ChatGPT use in K-12 education. *Eur. J. Educ.*, 59, 2, e12599, 2024, https://doi.org/10.1111/ejed.12599.
75. Johnson, W.L., Data-driven development and evaluation of Enskill English. *Int. J. Artif. Intell. Educ.*, 29, 3, 425–457, 2019.
76. Luckin, R. and Holmes, W., Intelligence unleashed: An argument for AI in education, 2016.
77. Smith, M. and Neupane, S., Artificial intelligence and human development: Toward a research agenda. *International Development Research Centre*, 2018.
78. Popenici, S.A. and Kerr, S., Exploring the impact of artificial intelligence on teaching and learning in higher education. *Res. Pract. Technol. Enhanc. Learn.*, 12, 1, 22, 2017.
79. Long, D. and Magerko, B., What is AI literacy? Competencies and design considerations, in: *Proceedings of the 2020 CHI Conference on Human Factors in Computing Systems*, pp. 1–16, 2020, April.
80. Vazhayil, A., Shetty, R., Bhavani, R.R., Akshay, N., Focusing on teacher education to introduce AI in schools: Perspectives and illustrative findings, in: *2019 IEEE Tenth International Conference on Technology for Education (T4E)*, IEEE, pp. 71–77, 2019, December.

Index

Also of Interest

Check out these other related titles from Scrivener Publishing

Distributed Algorithms and Protocols for the Metaverse: A Comprehensive Guide, Edited by Abhishek Kumar, Sashi Tarun, Anita Sardana, Abhineet Anand, ISBN: 9781394392605. Master the construction of robust distributed systems with this technical roadmap, providing the large-scale algorithms and synchronization protocols needed to power a seamless, low-latency metaverse.

Secure Blockchain and Quantum Technologies for 6G-Enabled IoT, Edited by T. Perarsi, M. Karthiga, S. Sountharrajan, E. Suganya, and Balamurugan Balusamy, ISBN: 9781394310678. Future-proof your hyper-connected infrastructure against the looming quantum threat bwith this guide to mastering blockchain-driven security frameworks and quantum-resistant architectures essential for safeguarding data integrity in the 6G-powered era.

Attacks and Defenses in Explainable Artificial Intelligence, Edited by Aml Dattatraya Vibhute, Rajesh Kumar Dhanaraj, Malathy Sathyamoorthy, and Paramasivam A. ISBN: 9781394305582. Bridge the critical gap between AI transparency and security with this essential guide to the systematic defense frameworks and ethical strategies needed to protect explainable AI (XAI) systems from sophisticated adversarial attacks.

Generative Artificial Intelligence for Next-Generation Security Paradigms, Edited by Santosh Kumar Srivastava, Durgesh Srivastava, Manoj Kumar Mahto, Ben Othman Soufiane, and Praveen Kantha, ISBN: 9781394305643. Fortify your digital defenses with this essential book, which provides a roadmap for moving beyond the limitations of traditional encryption by leveraging generative AI algorithms to proactively anticipate, detect, and mitigate the next generation of cyber threats in real-time.

Security and Privacy in 6G Communication Technology, Edited by Parita Jain, Puneet Kumar Aggarwal, Mandeep Singh, Sushil Kumar Singh, and Amit Singhal, ISBN: 9781394311002. Future-proof your knowledge and expertise in telecommunications with this essential guide, which provides a comprehensive analysis of the critical security and privacy challenges in the transition to 6G communication.

Development of 6G Networks and Technology, Edited by Suman Lata Tripathi, Mufti Mahmood, C. Narmadha, and S. Albert Alexander, ISBN: 9781394230655. *Development of 6G Networks and Technology* provides an in-depth exploration of the potential impact of 6G networks on various industries, including healthcare, agriculture, transport, and national security, making it an essential resource for researchers, scholars, and students working in the field of wireless networks and high-speed data processing systems.

Quantum Machine Learning for 6G Networks, Edited by Pallavi Sapkale, Shilpa Mehta and S. Balamurugan, ISBN: 9781394238088.

Virtual Reality and Augmented Reality with 6G Communication, Edited by B. Sundaravadivazhagan, N. Gnanasankaran, C. Pethuru Raj, and A. Saleem Raja, ISBN: 9781394336050. Stay ahead of the technological curve with this essential book, which provides a comprehensive guide to the transformative convergence of Virtual Reality (VR), Augmented Reality (AR), and 6G communication.

INTEGRATED DEVICES FOR ARTIFICIAL INTELLIGENCE AND VLSI: VLSI Design, Simulation and Applications, Edited by Balwinder Raj, Suman Lata Tripathi, Tarun Chaudhary, K. Srinivasa Rao, and Mandeep Singh, ISBN: 9781394204359. With its in-depth exploration of the close connection between microelectronics, AI, and VLSI technology, this book offers valuable insights into the cutting-edge techniques and tools used in VLSI design automation, making it an essential resource for anyone seeking to stay ahead in the rapidly evolving field of VLSI design.

DECENTRALIZED SYSTEMS AND DISTRIBUTED COMPUTING, Edited by Sandhya Avasthi, Suman Lata Tripathi, Namrata Dhanda, and Satya Bhushan Verma, ISBN: 978139420436. This book provides a comprehensive exploration of next-generation internet, distributed systems, and distributed computing, offering valuable insights into their impact on society and the future of technology.

ELECTRIC VEHICLE DESIGN: Design, Simulation and Applications, Edited by Krishan Arora, Suman Lata Tripathi, and Himashu Sharma, ISBN: 9781394204373. This book will serve as a definitive guide to conceptual and practical knowledge about the design of hybrid electrical vehicles (HEV), battery electrical vehicles (BEV), fuel cell electrical vehicles (FCEV), plug-in hybrid electrical vehicles (PHEV), and efficient EV charging techniques with advanced tools and methodologies for students, engineers, and academics alike.

INDUSTRIAL CONTROL SYSTEMS, Edited by Vipin Chandra Pal, Suman Lata Tripathi, and Souvik Ganguli, ISBN: 9781119829256. This volume serves as a comprehensive guide in the journey of industrial control systems with a multidisciplinary approach to the key engineering problems in the 21st century.

MODERN AUTOMOTIVE ELECTRICAL SYSTEMS: Theory and Applications, Edited by Pedram Asef, Sanjeevikumar Padmanaban, and Andrew Lapthorn, ISBN: 1119801047. Presenting the concepts and advances of modern automotive electrical systems, this volume, written and edited by a global team of experts, also goes into the practical applications for the engineer, student, and other industry professionals.

NANODEVICES FOR INTEGRATED CIRCUIT DESIGN, Edited by Suman Lata Tripathi, Abhishek Kumar, K. Srinivasa Rao, and Prasantha R. Mudimela, ISBN: 9781394185788. Written and edited by a team of experts in the field, this important new volume broadly covers the design of nano-devices and their integrated applications in digital and analog integrated circuits (IC) design.

EXPLAINABLE MACHINE LEARNING MODELS AND ARCHITECTURES: Real-Time System Implementation, Edited by Suman Lata Tripathi and Mufti Mahmud, ISBN: 9781394185849. This cutting-edge new volume covers the hardware architecture implementation, the software implementation approach, and the efficient hardware of machine learning applications.

MACHINE LEARNING TECHNIQUES FOR VLSI CHIP DESIGN, Edited by Abhishek Kumar, Suman Lata Tripathi, and K. Srinivasa Rao, ISBN: 9781119910398. This cutting-edge new volume covers the hardware architecture implementation, the software implementation approach, and the efficient hardware of machine learning applications with FPGA or CMOS circuits, and many other aspects and applications of machine learning techniques for VLSI chip design.

INTELLIGENT GREEN TECHNOLOGIES FOR SMART CITIES, Edited by Suman Lata Tripathi, Souvik Ganguli, Abhishek Kumar, and Tengiz Magradze, ISBN: 9781119816065. Presenting the concepts and fundamentals of smart cities and developing "green" technologies, this volume, written and edited by a global team of experts, also goes into the practical applications that can be utilized across multiple disciplines and industries, for both the engineer and the student.

DESIGN AND DEVELOPMENT OF EFFICIENT ENERGY SYSTEMS, edited by Suman Lata Tripathi, Dushyant Kumar Singh, Sanjeevikumar Padmanaban, and P. Raja, ISBN: 9781119761631. Covering the concepts and fundamentals of efficient energy systems, this volume, written and edited by a global team of experts, also goes into the practical applications that can be utilized across multiple industries, for both the engineer and the student.

Electrical and Electronic Devices, Circuits, and Materials: Technical Challenges and Solutions, edited by Suman Lata Tripathi, Parvej Ahmad Alvi, and Umashankar Subramaniam, ISBN: 9781119750369. Covering every aspect of the design and improvement needed for solid-state electronic devices and circuit and their reliability issues, this new volume also includes overall system design for all kinds of analog and digital applications and developments in power systems.

Printed and bound by CPI Group (UK) Ltd, Croydon, CR0 4YY
28/07/2026
14931302-0002